GLIMPSES OF AGRICULTURE

GLIMPSES OF AGRICULTURE

Editors

P. Kannan Ph.D.

Assistant Professor (SS&AC), Dryland Agricultural Research Station
Tamil Nadu Agricultural University, Chettinad, Tamil Nadu

M. Govindaraj Ph.D.

Scientist (Small millets), ICRISAT, Hyderabad

P. Arunachalam Ph.D.

Assistant Professor (PB&G), Agricultural College Research Institute
Madurai, Tamil Nadu Agricultural University, Tamil Nadu

M. Rajesh Ph.D.

Assistant Professor (Plant Pathology) Centre of Excellence in Millets
Tamil Nadu Agricultural University, Tamil Nadu

L. Allwin Ph.D.

Assistant Professor (Agricultural Entomology)
Agricultural College and Research Institute, Killikulam
Tamil Nadu Agricultural University, Tamil Nadu

G. Mariappan Ph.D.

(Agricultural Officer) Department of Agriculture, Vellore, Tamil Nadu

A Paperback Division of

NEW INDIA PUBLISHING AGENCY

101, Vikas Surya Plaza, CU Block, LSC Market
Pitam Pura, New Delhi 110 034, India
Phone: + 91 (11)27 34 17 17 Fax: + 91(11) 27 34 16 16
Email: info@nipabooks.com
Web: www.nipabooks.com

Feedback at feedbacks@nipabooks.com

ISBN: 978-93-87973-30-5

Composed, Designed and Printed in India

Preface

Agriculture encompasses large number of disciplines. Thus one has to read the number of text books to prepare ourself to come out successful in competitive examination. This book covers all the disciplines of agriculture mainly soil and crop management, crop improvement, crop protection, agricultural extension and economics and agricultural statistics. Further allied subject of horticulture, animal husbandry and ago-forestry also narrated in this book. In this revised edition cutting edge information were included. The contributors of this book meticulously updated and gave the concised information in different disciplines of agriculture for the benefit of the readers.

Here, we would like to express our heartfelt thanks to our friends and colleague Zadda Kavitha, N.Kumaran, K. Manikandan, R. Dhanabal, V.Radha Krishnan, Dinesh and Alagu for their support and inspiration rendered for preparing this book.

Finally we wish to thank all the contributors, who responded to our request and provided unique information in a simple and easy language. We are thankful to the M/S New India Publishing Agency, New Delhi for designing and printing this book in perfect manner.

We hope this book will be handy to the aspirants preparing for the competitive examination. The suggestions from the readers are welcome to improve this book in the future editions.

New Delhi
29 May 2018

EDITORS

Contents

List of Contributors

1. Alex Albert, V., Ph.D., Assistant Professor (SST), Agricultural Engineering College of Research Institute, Tamil Nadu Agricultural University, Tamil Nadu
2. Allwin, L., Ph.D., Assistant Professor (Agricultural Entomology), Agricultural College and Research Institute, Killikulam, Tamil Nadu Agricultural University, Tamil Nadu
3. Arul, S., Agricultural Officer, Department of Agriculture, Tamil Nadu
4. Arunachalam, P., Ph.D., Assistant Professor (PB&G), Agricultural College Research Institute, Madurai, Tamil Nadu Agricultural University, Tamil Nadu
5. Govindaraj, M., Ph.D., Scientist (Small millets), ICRISAT, Hyderabad
6. Kannan, P., Ph.D., Assistant Professor (SS&AC), Dryland Agricultural Research Station, Tamil Nadu Agricultural University, Chettinad, Tamil Nadu
7. Ramesh, T. Ph.D., Assistant Professor (Agronomy), ADAC&RI, Tirchy, Tamil Nadu Agricultural University, Tamil Nadu
8. Menaka, S., Assistant Professor (Agricultural Economics), College of Agricultural Technology, Theni, Tamil Nadu
9. Mooventhan, P., Scientist (Agricultural Extension), National Institute of Biotic Stress Management, Indian Council of Agricultural Research, Raipur, Chhattisgarh
10. Muthukumar, P., (Research Fellow), Division of Vegetable Science, Indian Agricultural Research Institute, New Delhi
11. Nagalingam, M., Ph.D., Scientist (Animal Science), Project Directorate on Animal Disease Monitoring and Surveillance , Hebbal, Bangalore
12. Prabukumar, G., Ph.D., Assistant Professor (Agronomy), National Pulse Research Station, Tamil Nadu Agricultural University, Tamil Nadu
13. Rajesh, M., Ph.D., Assistant Professor (Plant Pathology) Centre of Excellence in Millets, Tamil Nadu Agricultural University, Tamil Nadu
14. Ganeshamoorthy, P., (Agricultural Officer) Department of Agriculture, Tamil Nadu
15. Sridharan, N., Ph.D., Assistant Professor (Crop Physiology), Department of Crop Physiology, Tamil Nadu Agricultural University, Tamil Nadu

16. Thirunavukkarasu, M., M. Ph.D., (Assistant Professor) (SS&AC) Krishi Vigyan Kendra, Gandhigram Rural Institute, Gandhigram, Dindigul, Tamil Nadu

17. Mariappan, G., Ph.D., (Agricultural Officer) Department of Agriculture, Vellore, Tamil Nadu

CROP PRODUCTION

1

T. Ramesh and G. Prabukumar

1.1. WEATHER AND TILLAGE

Weather : A state or condition of the atmosphere at a given place and at a given instant of time'.

Climate : The generalized weather or summation of weather conditions over a given region during comparatively longer period.

Meteorology : Greek word "Meteoro" means 'above the earth's surface' (atmosphere) and "logy" means "indicating science". Branch of science dealing with that of atmosphere is known as meteorology

Agricultural meterology : Branch of applied meteorology or a branch of agriculture as it deals with agriculture

Coordinates of India : 8° N and 37° N latitude; 68° E and 98° E longitude

Coordinates of Tamilnadu: 08° 05' N and 13° 35' N latitude; 76° 15' E and 80° 20' E longitude

Hydrosphere - water portion of the earth

Lithosphere - solid portion of the earth

Atmosphere - gaseous portion of the earth

Atmosphere : It is the colourless, odourless and tasteless physical mixture of gasses, which surrounds earth on all sides.

Vertical Layers of atmosphere

Troposphere:

- The word "Trop" means vertical mixing or turbulence and "sphere" means region.
- This layer is called as "seat of weather phenomena".
- Tropopause : A shallow layer separating troposphere and stratosphere

Stratosphere

- This layer is called as "Seat of photochemical reactions"
- Temperature of this layer increases with altitude
- Ozone layer that it absorbs UV rays present
- The upper boundary of the stratosphere is called the stratopause

Mesosphere

- Temperature decreases with height.
- The upper boundary of this layer is called the "Mesospause".
- Mesosphere is the coldest region in the atmosphere (-95°C at the mesopause (80km))

Thermosphere(Ionosphere)

- Temperature increases in the order of 1000°C.
- Long distance radio communication is possible in this layer.

Exosphere

- ❖ The outer most layer of the earth's atmosphere
- ❖ Lies between 400 and 1000 km.
- ❖ Hydrogen and Helium gases predominate in this outer most region.

Lapse rate: The decrease in air temperature with height is known as the normal / environmental lapse rate and it is 6.5°C/km.

Adiabatic lapse rate: The rate of change of temperature in an ascending or descending air mass through adiabatic process is called as adiabatic lapse rate. (10.0°C/km)

Latitude: The distance from the equator, either south or north. Based on the latitude, the climate has been classified as i) Tropical ii) Sub-tropical iii) Temperate & iv) Polar.

Longitude: The distance from the equator, either east or west

Altitude(elevation): The height from the mean sea level (MSL)

Precipitation: The crop regions are classified on the basis of average rainfall as follow.

Scales of climate

Rainfall(mm)	Name of the climatic region
Less than 500	Arid
500-750	Semi-arid
750-1000	Sub-arid
More than 1000	Humid

Microclimate: It deals with the climatic features peculiar to small areas and with the physical processes that take place in the layer of air very near to the ground.

Meso climate: Study of climate over relatively smaller areas between 10 & 100 km across.

Macro climate: Macro climate deals with the study of atmosphere over large areas of the earth

Type of climate	Horizontal scale (km)	Vertical scale(km)	Time Scale(hrs)
A. Macro climate			
1. Planetary scale	2000-5000 & more	10	200 to 400
2. Synoptic scale	500-2000	10	100
B. Meso climate	1 to 100	1-10	1-10
C. Micro climate	<100m	200 m	6-12 minutes

Climate classification

- ❖ Thornthwaite (1948) classified the climate using precipitation and evaporation /Potential evaporation
- ❖ Mathur (1955) classified using Moisture Index (Im)

  Im = 100 [(P-PE)/PE]

  Where P = Precipitation, PE = Potential evapo-transpiration
- ❖ Troll (1965) classified based on number of humid months. Humid month is one having mean rainfall exceeding the mean Potential evapo transpiration.

Climate	Number of humid months	% geographical area of India
Arid	<2.0	17.00
Semiarid-dry	2.0-4.5	57.17
Semiarid-wet	4.5-7.0	12.31
Humid	>7.0	1.10

❖ **ICAR classification**

The ICAR under All India Coordinated Research Project on Dryland Agriculture adopted classification based Moisture Deficit Index (MDI)

$$MDI = \frac{P - PET}{PET} \times 100$$

Where P is annual precipitation (cm) and PET is Potential Evapotranspiration. Based on MDI the climate is divided into three regions as below.

Type of climate	MDI
Subhumid	0.0 to 33.3
Semiarid	-33.3 to -66.6
Arid	> -66.6

Solar radiation

Sun

- ❖ Sun is the prime source of energy (99% of the energy to biosphere is only from the sun)
- ❖ Diameter of the sun is 1.39 X 10^6 km
- ❖ Sun is on an average 1.5 X 10^8 km away from the earth
- ❖ It rotates on its axis about once every four weeks
- ❖ Surface temperature of the sun is 5462° K
- ❖ Every minute, the sun radiates approximately 56 X 10^{26} calories of energy.
- ❖ **Spring equinox :** March 21st
- ❖ **Autumn equinox :** September 23rd
- ❖ **Summer solstice :** June 21st
- ❖ **Winter solstice :** December 21st
- ❖ **Aphelion day :** July 4th
- ❖ **Perihelion day :** January 3rd

Transfer of heat

Conduction

Heat transfer through matter without the actual movement of the substances or matter. Eg. The energy transmission through an iron rod which is made warmer at one end.

Convection

Processes of transmission of heat through actual movement of molecules of the medium. Eg. Boiling of water in a beaker

Radiation

Transfer of energy from one body to another without the aid of the material medium (solid, liquid or gas). Eg. The energy transmission through space from the sun to the earth.

Radiation flux

The amount of radiant energy emitted, received and transmitted across a particular area is known as radiant flux.

Spectrum of Radiation

Band	Spectrum	Wavelength (μ)	Importance
Ultra violet	Cosmic rays	< 0.005	Shorter wave lengths of spectrum
	Gamma rays and X-rays	0.005 – 0.20	& Chemically active, unless filtered
	Ultraviolet rays	0.20 – 0.39	there is danger of life on earth
Visible	Violet	0.39 – 0.42	Visible spectrum known as Light
	Blue	0.42 – 0.49	essential for all plant processes
	Green	0.49 – 0.54	
	Yellow	0.54 – 0.59	
	Orange	0.59 – 0.65	
	Red	0.65 – 0.76	
Infra red	Infrared rays	> 0.76	Essential for thermal energy of the plant (Source of heat)

Band	Wavelength(nm)	Specific effect on plant
1.	Radiation within 1000 and more	No specific effect on plant activity. Radiation absorbed by plants are transformed into heat. This radiation does not interfere with bio-chemical processes.
2.	1000-720	Radiation in this band helps in plant elongation, can be accepted as an adequate measure of plant elongation activity. The far red region (700-920 nm) has important role on photo-periodism, germination of seeds, flowering and colouration of fruits.
3.	720-510	In this spectral region light is strongly absorbed by chlorophylls. It generates strong photosynthetic and photo-periodic activity.
4.	610-510	This is green-yellow region. Absorption in this spectral region has low photosynthetic effectiveness and weak formative activity.
5.	510-400	It is the strongest chlorophyll and yellow pigment absorption region. In the blue-violet range, photosynthetic activity becomes very strong. This region has very strong effect on formation of tissues.
6.	400-315	Radiation in this band produces formative effects. It has dwarfing effect on plants and thickening effect on plant leaf.
7.	315-280	Radiation in this band has detrimental effect on most plants
8.	Less than 280	Lethal effect most of the plants get killed due to radiation in this band UV ranges have germicidal action.

Solar constant

It is the energy received on a unit area at the outer most boundary of the earth (atmosphere) surface held perpendicular to the sun's direction, at the mean distance between the sun and the earth. Solar constant is 1.94 cal / cm^2 / min

Albedo

It is the percentage of reflected radiation to the incident radiation. **Eg.** Pure water – 5-20%, Vegetation 10-40%, Soils 15-50%, Earth 34-43% and clouds 55%.

Temperature: It is defined as the measure of the average speed of atoms and molecules

Kinetic energy: Energy of motion.

Heat: It is the aggregate internal energy of motion and molecules of a body.

Sensible heat

It is the heat that can be measured by a thermometer and thus sensed by humans.

Latent heat

It is the energy required to change a substance to a higher state of matter. This same energy is released on the reverse process. From water to water vapour takes 600 calories and water to ice takes 80 calories.

Blue colour of the sky is due to Rayleigh scattering

Red Colour of the sky at sunset & sunrise: It is because of increased path length in the atmosphere. Percentage of solar energy in the visible part decreases.

Light

Light is the visible portion of the solar spectrum with wavelength range is from 0.39 to 0.76μ.

The characteristics of light viz. intensity, quality, duration and direction are important for crops.

Light intensity: The intensity of light is measured by comparing with a standard candle.

Sciophytes (shade loving plants): The plants grow better under partially shaded conditions. (eg) Betel vine, buck wheat etc.

Heleophytes (Sun loving): Many species of plants produce maximum dry matter under high light intensities when the moisture is available at the optimum level. (eg) Maize, sorghum, rice etc.

Quality of Light

Visible part of the solar spectrum

Violet	400 – 435 m μ	Blue	435 – 490 m μ
Green	490 – 574 m μ	Yellow	574 – 595 m μ
Orange	595 – 626 m μ	Red	626 – 750 m μ

- Wave length used in photosynthesis: violet – blue and the orange - red regions.
- Infrared: Detrimental to plant growth.
- Red light is the most favourable light for growth followed by violet – blue.
- Ultra – violet and shorter wave lengths kill bacteria and many fungi.

Duration of light:

Photoperiodism : Response of plants to the relative length of the day and night

(i) Long day plants: The plants which develop and produce normally when the photoperiod is greater than the critical minimum (greater than 12 hours). eg. Potato, Sugarbeet, Wheat, Barley etc.

(ii) Short day plants: The plants which develop normally when the photoperiod is less than the critical maximum (less than 12 hours). Rice, Sorghum, cotton, Sunflower

(iii) Day neutral plants / Indeterminate: Those plants which are not affected by photoperiod. (eg) Tomato, Maize

Direction of light

Phototropism: The change of position or orientation of organs of plants caused by light

Eg. The leaves are oriented at right angles to incidence of light to receive maximum light radiation.

Photomorphogenesis: Change in the morphology of plants due to light. This is mainly due to U.V and violet ray of the sun.

Growing Degree Day

A degree day is obtained by subtracting the threshold temperature from daily mean temperature. Summation of the daily values over the growth period gives degree days of the crops.

$$GDD = \Sigma\left\{\frac{T_{max} + T_{min}}{2} - T_b\right.$$

Where

T_{max} – Maximum air temperature of the day

T_{min} – Minimum air temperature of the day

T_b - Base temperature of the crop

The base temperature is the threshold temperature. The minimum threshold temperature is the temperature below which no growth takes place. Usually ranges from 4.5 to 12.5 °C for different crops (Most commonly used value is 6.0°C)

Heat Injuries

'Thermal death point" – the temperature at which the plant cell gets killed when the temperature ranges from 50-60°C.

Sun clad : Injury caused on the barks of stem by high temperature during day time and low temperature during the night time.

Stem griddle : The stem at ground level scorches around due to high soil temperature. It causes death of plant by destroying conductive tissues. Eg. Common in young seedlings of cotton in sandy soil when soil temperature exceeds 60°C.

Cold Injuries

(i) Chilling injury: Plants which are adapted to hot climate, if exposed to low temperature for sometime, are found to be killed or severely injured or development of chloratic condition (yellowing) (eg.) cholratic bands on the leaves of sugarcane, sorghum and maize in winter months when the night temperature is below 20°C.

(ii) Freezing injury: when temperate plants are exposed to very low temperature, water freezes into ice crystals in the intercellular spaces of plants. The protoplasm of cell is dehydrated resulting in the death of cells. (eg.) Frost damage in potato, tea etc.

(iii) Suffocation: In temperature regions, usually during the winter season, the ice or snow forms a thick cover on the soil surface. As a result, the entry of oxygen is prevented and crop suffers for want of oxygen. Ice coming in contact with the root prevents the diffusion of CO_2 outside the root zone. This prevents the respiratory activities of roots leading to accumulation of harmful substances.

(iv) Heaving: This is a kind of injury caused by lifting up of the plants along with soil from its normal position. This type of injury is commonly seen in temperate regions. The presence of ice crystals increases the volume of soil. This causes mechanical lifting of the soil.

Cardinal temperatures: Minimum, optimum and maximum temperatures requirement of a plant

Thermoperiodism

Response of living organism to regular changes in temperature either day or night or seasonal is called thermoperiodism.

Humidity: The amount of water vapour that is present in atmosphere is known as atmospheric moisture or humidity.

Absolute humidity: The actual mass of water vapour present in a given volume of moist air. It is expressed as grams of water vapour per cubic meter or cubic feet.

Specific humidity: Weight of water vapour per unit weight of moist air. It is expressed as grams of water vapour per kilogram of air (g/kg).

Relative Humidity: The ratio between the amount of water vapour present in a given volume of air and the amount of water vapour required for saturation under fixed temperature and pressure. It is expressed as percentage. The relative humidity of saturated air is 100 per cent its and this is expressed as percentage. The relative humidity of saturated air is 100 per cent.

Dew Point temperature: The temperature to which a given parcel of air must be cooled in order to become saturation at constant pressure and water vapour content. In this case, the invisible water vapour begins to condense into visible form like water droplets.

Vapour Pressure deficit: The difference between the saturated vapour pressure (SVP) and actual vapour pressure (AVP) at a given temperature.

Wind: Air in horizontal motion is known as wind.

Mountain winds

a) Blows from mountain up slope to base; Occurs during night time
b) Cooling of air close to slope takes place
c) Adiabatic heating decreases this phenomenon
d) Also known as 'Katabalic winds'

Valley winds

a) Blow from valley base to up slope; Occurs during day time
b) Over heating of air adjacent to slope takes place
c) Adiabatic cooling decreases this phenomenon
d) Also known as 'Anabatic winds'

Atmospheric pressure

The atmospheric pressure is the weight of the air, which lies vertically above a unit area centered at a point. The weight of the air presses down the earth with the pressure of 1.034 gm / cm^2. It is expressed in millibar (mb) equal to 100 N/m^2 or 1000 $dynes/cm^2$.

Low / Depression: When the isobars are circular or elliptical in shape, and the pressure is lowest at the centre, such a pressure system is called 'Low' or 'Depression' or 'Cyclone'. The movement will be anti-clockwise in the Northern hemisphere while it is clockwise in the southern hemisphere. Wind speed hardly exceeds 40 km per hour.

Anticyclone: When isobars are circular, elliptical in shape and the pressure is highest at the centre such a pressure system is called 'High' or 'Anticyclone'. When the isobars are elliptical rather than circular the system is called as

‘Ridge’ or ‘Wedge’. The movement will be clockwise in the Northern hemisphere while it is anti-clockwise in the southern hemisphere.

Storm: Low pressure centre surrounded by winds having their velocities in the range of 40 to 120 km/hour. These storms produce heavy precipitation and bring about a change in the existing weather.

Hurricane: A severe tropical cyclone with wind speed exceeding 120 km per hour.

Thunderstorms: Storms produced by cumulonimbus clouds and always accompanied by lightening and thunder.

Tornadoes: Defined as a violently rotating column of air attended by a funnel-shaped or tubular cloud extending downward from the base of cumulonimbus cloud. Tornadoes are the most violent storms of lower troposphere.

Waterspouts: It is column of violently rotating air over water having a similarity to a dust devil of tornado. In other words, tornadoes are weak visible vortices occurring over water are called waterspouts. They are formed over tropical and subtropical oceans.

Wind systems of the world

The wind belts found on earth’s surface in each hemisphere are:

a. Doldrums

b. Trade wind belt

c. Prevailing westerlies

d. Polar easterlies

Doldrums

Owing to continuous heating of the earth by insolation, pressures are low and winds converge and rise near the equator. This intertropical convergent zone (ITCZ) is known as ‘Doldrums’. The location is 5°S and 5°N latitudes.

Trade winds (Tropical Easterlies)

Reduction in surface pressure on the equator (doldrums) causes an increase in pressure at 35°N and 35°S which are known as horse latitude (sub-tropical high).As a result, the winds flow from the horse latitude to the equatorial region. These winds flow from 35°N to the equator in NE direction in the northern hemisphere and from 35°S to the equator in SE direction in the southern hemisphere. These are known as ‘Trade winds’. These are known as ‘Tropical easterlies’.

Anti-trade winds

The anti-trade winds mostly flow from land to ocean and brings no rain.

Prevailing Westerlies

The winds that flow from sub-tropical high to the low-pressure area about 60-70° latitudes in both the hemispheres are known as ‘Prevailing westerlies’.

Polar Easterlies / Polar winds

A permanent high pressure exists on the poles. From these high pressure polar regions, cold winds flow to areas at about 60-65° latitudes in both the hemispheres.

Wind Direction Winds are always named after the direction they come from. Thus, a wind from the south, blowing towards north is called south wind.

Windward: It refers to the direction wind comes from

Leeward: It refers to the direction it blows to.

Prevailing wind: A wind blows more frequently from one direction than from any other is called prevailing wind

Cloud seeding: It is defined as a process in which the precipitation is encouraged by injecting artificial condensation nuclei through aircrafts or suitable mechanism to induce rain from rain bearing cloud.

Cloud Seeding materials

Cold clouds: Dry ice and Silver Iodide

Warm clouds: Water drops and Common salt

Monsoon : "Monsoon" is derived from Arabic word "MAUSIM" means season

South west monsoon : Contribute more than 70% of the total rainfall of the country SW monsoon is "Grand period of rainfall" in India

North East Monsoon : This is also called as Retreating monsoon (or) Returning monsoon.

The depression in Bay of Bengal is common and it gives rainfall in Andhrapradesh, TN and also Karnataka

Season: Season is a period in a year comprising few month during which the prevailing climate does not vary widely

Seasons of India (IMD classification)

1.South west monsoon (June – September) or Monsoon period or Kharif or Rainy season

2.North east monsoon (October – December) or Post Monsoon or Post Rainy season

3.Winter (Dec/Jan- Feb.) or Cold weather period

4.Summer or Pre monsoon (March / April –May/June) or Hot weather period

Seasons for Agricultural Crops

1.Hot weather (Mid February –Mid June).

2.Kharif or rainy season (Mid June –Mid October).

3.Rabi (Mid October –Mid February).

4.Spring falls between January and March.

Rainfall distribution in Tamilnadu in different seasons		
Rainfall season	**Quantity (mm)**	**Percentage share**
South West monsoon	311.7	32
North East Monsoon	457.8	47
Cold weather period	48.7	5
Hot weather period	155.9	16

DROUGHT

Aridity: It is a permanent climatic feature and is the culmination of a number of long term processes.

Drought: It is a temporary condition that occurs for a short period due to deficient precipitation for vegetation, river flow, water supply and human consumption.

1. **Permanent drought**: This is characteristic of the desert climate where sparse vegetation growing is adapted to drought and agriculture is possible only by irrigation during entire crop season.
2. **Seasonal drought:** This is found in climates with well defined rainy and dry seasons. Most of the arid and semiarid zones fall in this category.
3. **Contingent drought**: This involves an abnormal failure of rainfall.
4. **Invisible drought:** This can occur even when there is frequent rain in an area. When rainfall is inadequate to meet the evapo-transpiration losses, the result is borderline water deficiency in soil resulting in less than optimum yield.

5. **Meteorological drought:** It is defined as a condition, where the annual precipitation is less than the normal over an area for prolonged period (month, season or year).
6. **Atmospheric drought:** It is due to low air humidity, frequently accompanied by hot dry winds. It may occur even under conditions of adequate available soil moisture. It refers to a condition when plants show wilting symptoms during the hot part of the day when transpiration exceeds absorption temporarily for a short period.
7. **Hydrological drought:** Meteorological drought, when prolonged results in hydrological drought with depletion of surface water and consequent drying of reservoirs, tanks etc.
8. **Agricultural drought:** It is the result of soil moisture stress due to imbalance between available soil moisture and evapo-transpiration of a crop.
9. **Early season drought:** It occurs due to delay in onset of monsoon or due to long dry spells after early sowing
10. **Mid season drought:** Occurs due to long gaps between two successive rains and stored moisture becoming insufficient during this long dry spell.
11. **Late season drought:** Occurs due to early cessation of rainfall and crop water stress at maturity stage.
12. **Apparent drought:** What is drought for one crop may not be drought for another crop; what is drought in red soil may not be drought in black soil.
13. **Physiological drought:** Refers to a condition where crops are unable to absorb water from soil even when water is available, due to the high osmotic pressure of soil solution due to increased soil concentration, as in saline and alkaline soils. It is not due to deficit of water supply.

Evaporation: A physical process in which liquid water is converted into its vapour.

Transpiration: The loss of water from living parts of the plant is known as Transpiration. The loss of water through stomatal openings of the leaves is termed as stomatal transpiration. The loss of water through cuticle is known as cuticular transpiration and from lenticels is known as lenticular transpiration.

Evapotranspiration: Evapotranspiration denotes the quantity of water transpired by plants or retained in the plant tissue plus the moisture evaporated from the surface of the soil.

Reference Evapotranspiration (ET_0): This represents the maximum rate of evapotranspiration from an extended surface of 8 to 10 centimeters tall green grass cover, actually growing and completely shading the ground under limited supply of water.

Potential Evapotranspiration (PET): Potential evapotranspiration (PET) for any crop is obtained from reference evapotranspiration and crop factors (Kc) when water supply is unlimited.

$$PET = Kc \times ET_0$$

Type of weather forecast

	Types of forecast	Validity period	Main users	Predictions
1	Now casting	Next Few hours (Max upto 12 hours)	Aviation	Clouds coverage, Rainfall, fog, thunder storms etc.
2	Short range	Up to 48 hours	Farmers, marine agencies and general public	Rainfall distribution, heavy rainfall, heat and cold wave conditions, thunder storms etc.
3	Medium range	Beyond 3 days and upto 10 days.	Farmers and direction	Rainfall, Cloud cover, temperature, RH, wind speed
4	Long range	Beyond 10 days upto a month and a season.	Planners and Farmers	This forecasting is provided for Indian **monsoon rainfall**. The out looks are usually expressed in the form of expected deviation from normal condition.

Methods of Weather forecasting

1. Synoptic or conventional
2. Statistical or empirical
3. Numerical or deterministic

Synoptic chart

Plotting weather data observations on a large outline map using the international code of signals and abbreviations which in popular term is called as "weather map" and technically as, "synoptic chart"

Synoptic climatology

Synoptic climatology is defined as the description and analysis of the totality of weather at a single place or over a small area, in terms of the properties and motion of the atmosphere over and around the place or area.

Synoptic report

Observed weather conditions are marked in brief coded form as a synopsis of the conditions. Such a brief report on weather conditions is known as synoptic report.

Crop Weather Calendar

This is the pictorial form of chart containing the different stages of crops, optimum and unfavourable weather conditions during that stage of crop. This diagram helps weather forecaster to see at a glance what warnings are to be issued to a particular area during given weather situation and during the particular stage of the crop.

Remote sensing: It is defined as the art and science of gathering information about objects or areas from a distance without having physical contact with objects area being investigated.

Remote Sensing platforms: Ground based platforms: Infrared thermometer, Spectral radiometer, Pilot-Balloons and Radars

Air based platforms: Aircrafts

Satellite based platforms:

Polar orbiting satellites: These satellites operate at an altitude between 550 and 1,600 km along an inclined circular plane over the poles. These satellites are used for remote sensing purposes. LANDSAT (USA), SPOT (FRANCE), and IRS (INDIA) are some of the Remote Sensing Satellites.

Geostationary satellites: These have orbits around the equator at an altitude of 36,000 km and move with the same speed as the earth so as to view the same area on the earth continuously. They are used for telecommunication and weather forecasting purposes. INSAT series are launched from India for the above purposes.

Applications in agriculture and forestry:

1. Crop identification.
2. Crop acreage estimation.
3. Crop condition (Vigour and density) assessment and stress detection.
4. Identification of planting and harvesting dates.
5. Crop yield modeling, estimation and early season forecast.
6. *In situ* spectral measurement of crops.
7. Identification of pest and disease infestation.
8. Soil moisture estimation, irrigation monitoring and management
9. Frost damage.
10. Mapping/monitoring of droughts and their prediction.

11. Soil mapping.
12. Land use / land cover and land degreadation mapping.
13. Delineation of forest types and condition of range.
14. Mapping and monitoring of forest resources.
15. Forest fire detection.

Application for water resources:

1. Flood mapping and damage assessment.
2. Snow mapping, extent and depths.
3. Sea surface temperature monitoring.

Applications in meteorology:

1. Tropical storm surveillance.
2. Monsoon studies
3. Water forecasting.

Limitations of Remote Sensing Applications in Agriculture :

1. Small land holdings with smaller cropped fields, Mixed cropping .
2. Different cultural practices, Confusing crops (e.g. wheat and barley)
3. Large areas under rainfed conditions with poor growth, Cloud cover during monsoon season, which is major crop growing season in India.

Crop model: It is a representation of a crop through mathematical equations explaining the crops interaction with both above ground and below ground environment.

Crop weather models may also be defined as a simplified representation of the complex relationships between weather or climate on one hand and crop performance such as growth, yield or yield components, on the other hand by using mathematical and /or statistical techniques.

Types of models

a) Statistical models
b) Mechanistic model
c) Deterministic models
d) Stochastic models
e) Dynamic models
f) Static models
g) Simulation models
h) Descriptive models
i) Explanatory models

Crop Models	**Crops**
Crop Environment REsource Synthesis (CERES)	Rice, wheat, maize and sorghum
CROP GROwth (CROPGRO)	Soybean, groundnut and chickpea
Soil Plant Atmosphere Water (SPAW)	Wheat, maize, pearl millet and chickpea

Climate change

Any permanent change in weather phenomena from the normals of a long period average is referred as climate change. It is global phenomenon.

Climate variability

It refers to variations in the mean state of climate on all temporal and spatial scales beyond that of individual weather events. It is local/regional phenomenon

El-Nino

- El-Nino is a Spanish word meaning **"the boy child"** ('Child Christ') because El-Nino occurs around Christmas time each year when the waters off the Peruvian coast warm slightly. 'El Niño is now used more widely to refer to this **abnormal warming of the ocean** and the resulting effects on weather.

La-Nina

- 'La Niña' is used popularly to signify the opposite of El Niño, occurring when the waters of the **eastern Pacific are abnormally cold**. La Niña episodes are associated with more rainfall over eastern Australia, and continuing drought in Peru.

ENSO

- The El-Nino event is due to decrease in atmospheric pressure over the South East Pacific Ocean. At the same time, the atmospheric pressure over Indonesia and North Australia increases. Once the El-Nino event is over, the atmospheric pressure over the above regions swings back. This sea-saw pattern of atmospheric pressure is called **Southern Oscillation**. Since El-Nino and Southern Oscillation are linked they often termed as ENSO.

Southern Oscillation Index (SOI)

- A measure of the monsoon low pressure is the Southern Oscillation Index (SOI) represented by the difference in sea level pressure over **Tahiti,** an Island in South central pacific and **Darwin** in North Australia, which represents the northern part of the Indian Ocean.

Impact of Climate change

- Sea-Level Rise
- Green house gases emission
- Changes in global temperature
- Erosion and soil fertility
- Changes in pests, diseases and Weeds s
- Chacnges on monsoonal pattern
- Desertss, Land Degradation and Desertification

In the long run, the climatic change could affect agriculture in several ways :

- ***Productivity***, in terms of quantity and quality of crops
- ***agricultural practices***, through changes of water use (irrigation) and agricultural inputs such as herbicides, insecticides and fertilizers
- ***environmental efsfects***, in particular in relation of frequency and intensity of soil drainage (leading to nitrogen leaching), soil erosion, reduction of crop diversity
- ***rural space***, through the loss and gain of cultivated lands, land speculation, land renunciation, and hydraulic amenities.

- ***adaptation***, organisms may become more or less competitive, as well as humans may develop urgency to develop more competitive organisms, such as flood resistant or salt resistant varieties of rice.

Adaptation and mitigation strategies to climate change

1. **Altering agro techniques of crops**
   - Changing date of sowing, spacing and input management
   - Use of alternate crops or cultivars
   - More area under horticulture and agro-forestry
2. **Development of resource conserving technologies**
   - Zero-tillage establishment of upland crops
3. **Increasing income from agricultural enterprises**
   - Reduce the cost of cultivation
   - Increase the input use efficiency
4. **Improved water and nutrient management techniques to reduce the GHGs emission**
   - Rice- SRI and Aerobic rice cultivation
   - Use of nitrification inhibitors & slow releasing fertilizers
   - Split application of N&K
5. **Improved land use and natural resource management policies**
   - Improve soil health using organic farming
   - Incentive for green manuring
6. **Improved risk management through early warning system and crop insurance**
   - Use of ICT tools for early warning
   - Weather based crop insurance

Institutes

Name	**Particulars**	**City**
IMD	India meteorology department	Pune
NCMRWF	National center for medium range weather forecasting	New Delhi
WMO	World Meteorology Organization	Geneva, Switzerland
IITM	Indian Institute of Tropical Meteorology	Pune
UNFCCC	United Nations Framework Convention on Climate Change	Bonn, Germany
IPCC	Intergovernmental panel on climate change	Geneva, Switzerland
FWB	Farmers Weather Bulletin	
AAS	Agro Advisory Services	
AMFU	Agro meteorological field units	
AASU	Agro-met Advisory Service Units	
WWW	World Weather Watch	

TILLAGE

Tillage: Mechanical manipulation of soil with tools and implements to bring favourable soil condition for germination of seeds

Tilth: Physical condition of soil obtained out of tillage

Objectives of tillage

- To prepare a good seed bed, to control the weeds, to absorb more rain water in soil, to mix up the manure and fertilizers uniformly in the soil, to aerate the soil, to provide adequate seed-soil contact, to remove the hard pan and to increase the soil depth.

On-season tillage: Tillage operations made for raising crops or at the onset of season

Preparatory cultivation: Tillage operations to prepare the field for raising crops

Primary tillage:

- Tillage operation is done after the harvest of crop to bring the land under cultivation
- Ploughing is the opening of compact soil with the help of different ploughs.
- Country plough, mould board plough, bose plough, tractor and power tiller drawn implements are used for primary tillage.

Secondary tillage

- Tillage operations performed after primary tillage to bring a good soil tilth
- To clean the soil, break the clods and incorporate the manure and fertilizers. Harrowing and planking is done to serve those purposes.

After cultivation/ Inter cultivation

- Tillage operations carried out in the standing crop
- Includes harrowing, hoeing, weeding, earthing up, drilling or side dressing of fertilizers etc. Spade, hoe, weeders etc. are used for inter cultivation

Off season tillage

- Tillage operations done for conditioning the soil suitably for the forthcoming main season crop
- Post-harvest tillage, summer tillage, winter tillage and fallow tillage

Sub soiling / Chiseling

- To break the hard pan and reduce compaction. Once in four to five years

Clean tillage

- Entire field - no living plant is left undisturbed. To control weeds, soil borne pathogen and pests

Blind tillage

- Tillage done after seeding or planting either before emergence or early stages of crop (sugarcane, potato etc.)
- Extra plants and broad leaved weeds are uprooted.

Dry tillage

- Dry land condition having sufficient moisture (21-23%) for germination of seeds. Suitable for broadcasting crops- rice, jute, wheat, oilseed crops, pulses, potato and vegetable crops. Soil water holding capacity and aeration are increased. More favourable for soil micro-organisms

Wet tillage / pudding

- Tillage operation is done in a land with standing water until the soil becomes soft and muddy and creates an impervious layer below the surface
- To reduce deep percolation losses of water, to provide soft seed bed for planting rice, incorporation of green manures and weeds.
- Cage wheel mounted tractor and power tillers are used

Zero tillage

- Crop is planted in the residues of the previous crop without any prior soil tillage
- Weeds are controlled by the use of herbicides.

Minimum tillage: Reducing tillage operations to the minimum necessity for ensuring a good seed bed.

Conservation of tillage

- To conserve soil and soil moisture; Organic residues remain on surface
- If stubble forms the protective cover - stubble mulch tillage.

Stubble mulch tillage

- Soil is protected by growing a crop or by crop residues on the surface.
- Sweeps or blades are used to cut the soil up to 12 to 15 cm depth in the first operation after harvest

Sowing

Seed rate

- The required number of plants/unit area is decided by seed rate. The seed rate depends on spacing or plant population, test weight, germination percentage. The formula is as follows.

$$\text{Seed rate (kg/ha)} = \frac{\text{Plant population (per ha) x No. of seeds/hill x Test weight (g) x 100}}{\text{1000000 x Germination percentage (\%)}}$$

Methods of sowing

Broadcasting:

- Scattering the seeds, Followed for small sized to medium sized crops, Largest method of sowing followed in India, Easiest, cheapest and requires minimum labours and Skilled person is required

Dibbling

- Inserting a seed through a hole at a desired depth and covering the hole. Line sowing, Practiced on plain surface and ridges and furrows or beds and channels. Seeds are dibbled at $2/3^{rd}$ from top or $1/3^{rd}$ at bottom of the ridge.

- ❖ Sorghum, maize, sunflower, cotton are dibbled on ridges and furrows.
- ❖ Uniform population, better germination and reduced seed rate.

Sowing behind plough

- ❖ Seeds are dropped in the furrows opened by the plough and closed
- ❖ Seeds are sown at uniform distance.
- ❖ Laborious and time consuming process.
- ❖ Redgram, cowpea and groundnut are sown

Seed drilling

- ❖ Dropping seeds in a definite depth covered with soil.
- ❖ Animal drawn Gorus and power operated (seed drills) implements are used
- ❖ Seeds are drilled continuously or at regular intervals in rows.
- ❖ Fertilizer can also be applied simultaneously.

Transplanting

- ❖ Raising of seedlings in nursery and planting in the main filed
- ❖ Rice, finger millet and vegetables crops
- ❖ Age of seedlings is $1/4^{th}$ of the total duration of the crop
- ❖ Area required for nursery normally is $1/10^{th}$ of the total area.

Depth of sowing

- ❖ Optimum depth; Seeds placed at deeper layers, spend more energy for germination
- ❖ Placed on soil surface, taken away by birds/worked away
- ❖ Approximately 3 to 4 times diameter of the seed
- ❖ Optimum depth of sowing - 3 and 5 cm depth

Plant geometry

- ❖ Arrangement of the plants in different rows and columns in an area
- ❖ Area occupied by a single plant
- ❖ Ex. Rice – 20 cm x 15 cm.

1.2. WATER MANAGEMENT

Water covers 75 percent of the earth surface, of which fresh water occupies 2.5 percent but only 30% is available to meet water demands

India facts

- ❖ Total geographical area is 329 Mha
- ❖ Average annual rainfall is low (100 mm) in Western region (Rajasthan)
- ❖ Average annual rainfall is high (11000 mm) in North eastern region (Meghalaya)

- ❖ Average annual potential evapo-transpiration is 1239 mm in North India
- ❖ Average annual potential evapo-transpiration is 2052 mm in south and central India
- ❖ Irrigation potential is 139.5 Mha and it is comprising
    - ☞ Surface water based irrigation potential is 58.5 Mha under major and medium projects and 15 M ha under minor projects
    - ☞ Ground water based irrigation potential is 66 Mha
- ❖ 15 Mha degraded due to development of water logging and salinity
- ❖ Flood prone area is 40 Mha in south west monsoon season
- ❖ Average productivity of irrigated lands is 2.5 t ha^{-1} and in rainfed area is 0.8 t ha^{-1}
- ❖ National water policy (2002) states "water" is prime natural resources, a basic human need and a precious national asset"

  * Central water commission (2000) states estimate of utilizable water as follows :

  Surface : 690 cu.km
  Ground : 432 cu.km
  Total : 1122 cu.km

Water utilization in 2000 is 60.6 Mha of which irrigation purpose (83%), industrial use (3%), domestic use (5%), energy sector use (6%) and other purposes (3%).

Water productivity of major crops

To produce	Water requirement (litre)
1 kg Paddy	2400
1 kg Rice	3400
1 kg Groundnut Pod	3670
1 kg Kernal	5240
1 Litre Groundnut Oil	13000
1 kg Sugarcane	220
1 kg Sugar	1850
One Banana	
✻ Poovan	50
✻ Rasthali	10
✻ Monthan	200
Cotton	
✻ 1 kg Kapas	3000
✻ 1 kg Lint	9000
✻ 1 Saree	19000

Understand about Thousand Million Cubic feet (TMC)

$$TMC = 10^3 \times 10^6 \text{ ft}^3$$
$$= 10^9 \text{ ft}^3$$

$1\ ft^3 = 0.027\ m^3$

$= 10^9\ (0.027\ m^3)$

Water requirement for rice is 1100 mm (or) 1.1m

eg 1. How much rice area can be irrigated with 1TMC of water for entire period?

$$\text{Area to be irrigated by 1 TMC} = \frac{10^6 \times 27}{1.1 \times 10000}\ m^3 = 2450 \text{ hectares}$$

1 TMC of water irrigate 2400 hectares of rice.

eg 2. How much rice area can be irrigated with 1TMC of water to the depth of 5 cm?

$$\text{Area to be irrigated by 1 TMC} = \frac{10^6 \times 27}{0.05 \times 10000}\ m^3 = 54000 \text{ hectares}$$

1 TMC of water irrigate 54000 hectares of rice to the depth 5 cm in single irrigation.

Soil water availability to plants

- ❖ The amount of water held depends upon the porosity and pore size distribution and the capillary pressure of the water in the soil.

  1 bar - 1020 cm of water column / 10^3 pa/100k pa/0.1Mpa

  1 atm = 1030 cm of water column / 10^3 pa/100k pa/0.1Mpa

- ❖ Soil water potential expressed in terms of hydraulic need/ height of water column

  (1 bar = 100 cm, 10 bar = 1000 cm)

- ❖ PF- Logarithm of the (-) ve pressure head expressed in centi moles of water column
  - ✓ A tension head of 10 cm of water would be PF of 1
  - ✓ A tension head of 1000 cm of water would be PF of 3
- ❖ Soil moisture constants are saturation, field capacity, permanent wilting point and hygroscopic coefficient.

Saturation : Soil is a porous medium and when all the pores of the soil are filled with water, it is referred to as saturated soil.

Porosity (P) percentage= [1- BD/PD] x100

BD- Bulk density (Mg m^{-3})

PD - Particle density (Mg m^{-3} =2.65)

Field capacity (FC) : The maximum amount of water that a soil will retain after allowing free drainage. Applied pressure for clays soil is 1/3 bar and sandy & sandy loam - 0.1 to 0.2 bar at field capacity and it is the upper limit of available water to plants.

- ❖ Coarse textured soil- 24 to 36 hrs
- ❖ Medium textured soil- 2-3 days
- ❖ Crop transpire maximum of soil water at field capacity conditions
- ❖ Transpiration rate does not decrease significantly until soil modified falls below 50% of field capacity

Permanent wilting point (PWP) : The moisture content of a soil below which plants wilt to such an extent that they fail to recover even when fully watered and it is the lower limit of available water to plants.

- ❖ FC and PWP will be estimated by using pressure plate or a pressure membrane apparatus

Hygroscopic Coefficient (HC) : It is the water held by soil at 31 bar (3.1pa) which is unavailable to crop plants.

- ❖ The water between saturation and field capacity is called gravitational water, which is unavailable to plants
- ❖ The water between field capacity and permanent wilting point is called available water which is utilizable for crop plants
- ❖ The water between permanent wilting point and hygroscopic coefficient is unavailable water for plants
- ❖ Available water capacity is the difference between soil moisture content of field capacity and wilting point

Water requirement of crops

Consumptive use = Evaporation (E) + Transpiration (T) + plant metabolic needs

Water requirement = Irrigation water (IR) + Effective rainfall (ER) + Soil profile moisture contribution (S)

Irrigation requirement (IR) = Water requirement (WR) – (ER+S)

Measurement of soil water : Soil water is measured by direct and indirect method

1. Direct method : Gravimetric or oven drying method

This method involves of water from soil samples by evaporation, leaching or chemical reaction and samples kept in hot-air oven at 105-110° C

Other direct methods: Alcohol burning method, rapid moisture determination using IR-lamp, calcium carbide and gypsum absorption plugs

2. Indirect method : It is measured by soil water content and soil metric potential

i. *Soil water content* is measured by using electrical resistance blocks, neutron moisture meter, Time Domain reflectometry (TDR) and Depth moisture gauge.

ii. *Soil metric potential* is measured by using Tensiometer (up to a suction of 0.85 bars) and Plaster of paris or gypsum block technique.

Plant and water absorption

- ❑ Plant extracts water from soil with root hairs through osmosis and diffusion process
- ❑ Loss of water from plants through stomata and plant surface known as transpiration
- ❑ Closure of stomata leads to arrest of photosynthesis and reduction in nutrient uptake

Irrigation : Artificial application of water to land or soil. It is used to assist in the growing of agricultural crops. Predominant irrigation systems are Canals, tube well and well irrigation

Methods of irrigation

- *Surface methods*
    - i. Uncontrolled flooding
    - ii. Controlled flooding: check basin, border strip irrigation, furrow and corrugation

Border : Water is applied at the upper end of a strip in which the lateral flow of water is controlled by small earth ridges called border dikes or borders.

Basin : Water is applied rapidly to nearly level plains surrounded by levees or dikes.

Controlled flooding : Water is released at intervals from closely spaced field ditches and distributed uniformly over the field.

Corrugation : Water is applied to small, closely spaced furrows or ditches in fields of close-growing crops or in orchards so that it flows in only one direction.

Furrow : Water is applied in small ditches made by cultivation implements. Furrows are used for tree and row crops.

Wild flooding : Water, released at high points, is allowed to flow onto an area without controlled distribution.

Pressurized irrigation system such as sprinkler and drip irrigation useful in undulating and uneven lands / coarse texture soils.

Sprinkler head irrigation : Water is sprayed over the soil surface through pipes or nozzles from a pressure system.

- ❖ Water use efficiency - 60%
- ❖ Used in flowers, vegetables and horticultural crops
- ❖ Rain guns is the modified form of sprinklers system

Drip Irrigation : Water is applied slowly and under low pressure to the surface of the soil or into the soil through such applicators as emitters, porous tubing, or perforated pipe.

- ❖ Used for wider spaced and high value crops :(Sugarcane,Coconut, banana, grapes and lime, Cotton and Redgram)
- ❖ Water savings - 40 to 70%

Sub-irrigation : Water is applied in open ditches or tile lines until the water table is raised enough to wet the soil.

Scheduling of irrigation : Irrigation scheduling is the process used by irrigation system managers to decide when to irrigate and how much water to apply to a crop in the field

Water budget : It is an estimate of the income and expenditure of soil moisture, which may be consider as a bank account

Irrigation Efficiency : It is defined as the ratio of irrigation water actually utilized by growing crops and water diverted from a source

Water use efficiencies

$$\text{Consumptive use efficiency} = \frac{\text{Consumptive use of water by crop}}{\text{Water depleted from root zone}} \times 100$$

$$\text{Water use efficiency} = \frac{\text{Marketable yield}}{\text{Evapotranspiration}}$$

$$\text{Field Water Use Efficiency} = \frac{\text{Yield}}{\text{Evapotranspiration + Deep percolation}}$$

Evaporation accounts for the movement of water to the air from sources such as the soil, canopy interception, and water bodies.

Transpiration accounts for the movement of water within a plant and the subsequent loss of water as vapor through stomata in its leaves.

Evapotranspiration (ET) is a term used to describe the sum of evaporation and plant transpiration from the earth's land surface to atmosphere.

Reference evapo-transpiration is computed by Penmans's method (or) Pan evaporation method and empirical methods (Blaney-criddle and Hargreaves method).

$$ET_o = K_p \times E_{PAN}$$

Where,

E_{PAN} Measured values of evaporation using US class A open pan evaporation meter

K_P - pan co-efficient

ET_O Reference Evapo-transpiration

- Evapotranspiration (ET in mm) under irrigation: $ET_c = K_c \times ET_o$
- Kc - is the crop co-efficient

Devices used measuring irrigation water : Orifices, weirs and flumes

- Commonest devices is 900 V notch weir and parshall flume
  - **The discharge rate through 90° V notch**

    $Q = 0.0138\ H^{5/2}$

    Where, Q - discharge rate (1/s), H - head (cm)

  - **The discharge rate through rectangular weir**

    $Q = 0.0184\ LH^{3/2}$ (for suppressed rectangular weir)

    $Q = 0.0184\ (L - 0.2H)H^{3/2}$ (for contracted rectangular weir)

    Where, Q - discharge (1/s), L - length of crest (cm), H - head over the weir (cm)

  - **The discharge rate through contracted trapezoidal weir (or) cipoleth weir**

    $Q = 0.0186\ LH^{3/2}$

    Where, Q - discharge (l/s), L - length of crest (cm), H - head over the crest (cm)

  - **The discharge rate of orifice**

    $Q = 0.16 \times 10^{-3}\ a\ \ 2gh$

    Where, Q = discharge (1/s)

    a = area of cross section of orifice (cm^2)

    g = acceleration due to gravity (981 cm/s^2)

    h = depth of water over centre of orifice (cm)

Rain water conservation

In-situ water conservation : It is a process of enhancing infiltration of rain water to fill-up soil pores. eg. ridges and furrow, random tied ridging, broad bed and furrow, raised bed, basin listing, compartmental bunding and mulching

Ex-situ water conservation : It is a method to collect runoff in small storage structures which are called water harvesting ponds, small farm reservoirs, farm ponds and percolation ponds.

Constructing terraces and erecting small bunds along the terrace and breaking the slope length. If the bunds are along a contour, the bund is a contour bund and the terrace is contour terrace. If the bund is constructed between two contours, it is a graded bund and this terrace is graded terrace.

Classification of irrigation water

Salinity hazard

Low salinity water	EC is Less than 0.25 dSm^{-1}
Medium salinity water	EC is 0.25 to 0.75 dSm^{-1}
High salinity water	EC is 0.75 to 2.25 dSm^{-1}
Very high salinity water	EC is More than 2.25 dSm^{-1}

Sodium hazard

Low sodium water	SAR is Less than 10
Medium sodium water	SAR is 10 to 18
High sodium water	SAR is 18 to 26
Very high sodium water	SAR is more than 26

Management of poor quality water

- Gypsum application
- Growing of tolerant crops
- Sowing at the side of the ridges
- Use of drip irrigation
- Provision of adequate drainage
- Mulching with organic residues
- Inclusion of lowland rice in crop rotation

1.3. WEED MANAGEMENT

Weed is the plant out of place or weeds are the plant growing where they are not wanted (Bletchley, 1912). Botanical division of anthophyta (angiosperms) has 250000 species in the world, out of which 250 species (0.1%) are troublesome to be called as weeds.

❖ It compete with crops for water soil, nutrients, light and space and thus reduce crop yields

❖ It acts as alternate hosts for insects, pests, diseases and micro organism

❖ Some weeds release into the soil inhibitors or poisonous substances that may be harmful to crops, human beings and reduce the quality of agricultural produce

❖ Aquatic weeds pollutes water by deoxygenating it and killings the fishes

❖ Average seed production capacity (perennial weeds) 16500 seeds per plant, eg. wild oats (*Avena fatua*) 250 seeds per plants and wild amaranthus (*Amaranthus viridis*) 11 million seeds per plant.

❖ Seed viability of weeds: The seed viability of annual meadow grass (*Poa annua*) and Searlet pimpernel (*Anagallis arvensis*) is 8 years, creeping thistle (*Cirsium arvense*) is 20 years and field bind weed (*Convolvulus arvensis*) is 50 years.

Classification of weeds

1. Based on habitat: Terrestrial and Aquatic
2. Based on duration of life/season
   i. Annual (complete life cycle in a year or a seasons)
      - Summer annuals : *Echinochola crusgalli, E.colona, Panicum repens, Eclipa alba* and *Amaranthus viridis*
      - Winter annuals : *Phalaris minor, Poa annua* and *Chenopodiun album*

   ii. Biennials (live > 1 year but <2 years) - *Cichorium intybus, Raphanus raphanist*run

   iii. Perennials (Live >2 years) - *Cynodon dactylon, Saccharum spontaneum; Cyperus rotundus, Sorghum halepense* and *Convolvulus arvensis*

3. *Based nutritional habits/ nature of competition*

i. Autotrophs: It produces their own food

ii. Heterotrophs: It depends on other plant for food
   - Total root parasite - Orobanche on tomato, tobacco and chickpea
   - Partial root parasite - Striga on sorghum and maize
   - Total shoot parasite - Cuscuta on alfalfa, niger and lintseed
   - Partial shoot parasite - Loranthus on mango and other trees

Aquatic weeds further classified

- Floating weeds are rest upon the water surface - eg. water hyacinth and pistia
- Emergent weeds are found along the shorelines or ponds and canals eg. Typha and scirpus
- Submerge weeds are grow completely under water, few stones and leaves may exist on above water surface.eg. hydrilla, salvinia and vallisneria

Concepts of weed control

Weed management concepts are prevention, eradication, control and management

1. Prevention : It is a measure to inhibit or delay weed establishment. Cultural practices like seed cleaning, sowing, spacing, irrigation, intercropping, crop rotation and cover crops and highly cost effective as establish of new weeds which create problem for many years

2. Eradication : Total elimination of a weed species from a field area or region. Soil sterilants or fumigants (carbon disulphide, methyl bromide) are used particulars in nurseries for eliminating soil pathogens and pests including weeds.

3. Control : It is a measure to reduce or suppress the established weeds.

Physical methods

a. Hand pulling , hoeing, tillage, mowing and slashing, burning, flooding and mulching.

b. Pre-planting or post-planting tillage is an economical methods of weed control.

c. Soil turning plough, disc and spring tooth harrows and the dug foot, blade and bucket- type cultivars are used to keep weed growth under control.

Cultural methods

a. Quick growing and short duration varieties with larger leaf area and good branching / tillering ability are complete with weeds successfully.

b. Proper seed rate, depth, time & method of sowing, method of irrigation and fertilizer application are competing with weeds successfully.

c. Close row crops compete with weeds better than wide row crops.

d. Early and faster growing crops competes better with weed eg: pigeon pea.

e. Crop rotations minimize the dominance of particular weed.

f. Intercropping fast growing crops like cowpea, soybean and special crops such as maize, pigeon pea and soybean to reduce weed competition.

Chemical methods

- ❖ Selective control of weeds by application Bordeaux was first noticed by Millardet, a French scientist in 1896
- ❖ Bordeaux mixture that control fungal disease in grapes and also control weeds
- ❖ 2,4-D, MCPB, MCPA (Selective herbicides) used for wheat, barley, oats, maize, rice and millets
- ❖ Use of herbicides is easier, less time consuming and cost effective

Classification of Herbicides

Based on their chemistry, selectivity mode of action and time of application

1. Based on chemistry - phenoxys, urease and triazines
2. Based on selectivity - selective and non-selective
3. Based on mode of action - contact and translocative

Selective herbicide : It removes particular weeds from certain crops

eg : Atrazine in maize, isoproturon in wheat

Non-selective herbicide : It directly kills all vegetation including crops when it applied

- ❖ Mostly used for vegetation management in non-crop areas and also used for to control weeds in orchards and plantation crops, eg. paraquat and glyphosate.

Contact herbicide : It kills only the plant or portions of the plant that actually comes in contact with herbicide. Both selective and non-selective comes under this category, eg. paraquat

Translocative herbicide: Move within the plant from the place of application to the place where they show their ultimate effect. It is effective in destroying the roots and other underground storage organs of perennial weeds eg. Glyphosate and 2, 4-D.

Herbicides may be applied :

- ❖ Before planting /pre-plant incorporation eg. fluchloralin
- ❖ After planting but before emergence eg. atrazine, pendimethalin (pre-emergence)
- ❖ After emergence of the crop eg. 2,4-D, clodinafop (post-emergence)
- ❖ Pre-planting and pre-emergence herbicides applied on soil (soil-acting) and the post-emergence applied on plant foliage (foliar-acting)

Herbicide formulations

1. ***Emulsion :*** A liquid dispersed in another liquid each maintaining its original identity eg. Ester formulation of 2, 4-D
2. ***Granular herbicides :*** The herbicide is adsorbed, mixed with a generally inert carrier in such a way that the final product consists of granular particles many carriers are used, eg. clay, sand, vermiculite and finely ground plant parts.
3. ***Wettable powders :*** The active ingredient (a.i) of herbicides as wettable powders is nearly insoluble in water and may not be soluble in organic solvents. eg. simazine, atrazine, diuron.
   - The oil is a part of a chemical formulated that is directly responsible for the herbicidal effect. If the chemical is 99% pure considered as 99% a.i
   - The strength of herbicide formulated in expressed in terms of a.i
   - Some herbicides activity is calculated on acid-equivalent basis
   - For manual application, knapsack sprayer is ideal
   - Spray volume of 250-500 l/ha is an optimum for herbicide application
   - Flat -fan nozzle is most preferred nozzle for application of herbicides.
   - Even flat fan nozzle can be used with single nozzle sprayer
   - Tapering fan nozzle is used in boom sprayers with multiple nozzles

Outstanding example of biological control of weed species

a. Cochineal insect *Dactylopius ceylonicus* is used for control of prickly pear (Opuntia spp)
b. Destruction of lantana with the help of *Teleonemia scrupulosa*
c. Water hyacinth is controlled by *Neochetina bruchi, N. eichhorniae*
d. *Parthenium hysterophorus* is controlled by mexican beetle (*Zygogramma bicolorata*)
e. *Salvinia moesta* is controlled by *Cyrtobagous salviniae*
f. Destruction of aquatic vegetation in ditches by using herbivorous fish, eg. *Tilapia mossambica* (sunfish family)

Chemical used for weed control

Crop	Herbicide	Rate kg/ha (a.i.)	Time of application
Direct seeded rice	Butachlor	1.5	Pre-emergence
	Benthiocarb	1.5	Pre-emergence
	Pendimethalin	1.5	Pre-emergence
	Oxadiazon	0.75	Pre-emergence or Early post-emergence
	Pyrazosulfuron-ethyl	0.020	Pre-emergence or Early post-emergence
	Bensulfuron-methyl	0.060	Post-emergence
	2, 4-D	1.0	Post-emergence
	Bentazon	2.0	Post-emergence
	Metsulfuron-methyl	0.004	Post-emergence

Transplanted rice	Pyrazosulfuron-ethyl	0.020	Pre-emergence or Early post-emergence
	Cyhalofop-butyl	0.28	Post-emergence
	Pretilachlor	0.5	Pre-emergence
	Bensulfuron-methyl	0.060	Post-emergence
	Almix (metsulfuron methyl + chlorimuron ethyl	0.004	Pre-emergence
	Alachlor	2.00	Pre-emergence
	Anilofos	0.43	Pre-emergence
	Fluchloralin	1.0	Pre-transplanting
	Butachlor	1.5	Pre-emergence
	Pendimethalin	1.5	Pre-emergence
	2, 4-D	1.0	Post-emergence
	Oxadiazon	0.75	Pre-emergence or Early post-emergence
	Bentazon	2.0	Post-emergence
	Metsulfuron-methyl	0.004	Post-emergence
Wheat	Terbutryne	1.0	Pre-emergence
	Pendimethalin	1.0	Pre-emergence
	Isoproturon	1.0	Post-emergence
	Diclofop-methyl	1.0	Post-emergence
	Difenzoquat	1.2	Post-emergence
	2, 4-D	1.2	Post-emergence
	Metribuzin	0.2	Post-emergence
	Metsulfuron-methyl	0.004	Post-emergence
Maize	Atrazine	2.0	Pre-emergence
	Alachlor	2.5	Pre-emergence
	Metolachlor	2.0	Pre-emergence
	Pendimethalin	1.5	Pre-emergence
	Bentazon	1.5	Post-emergence
	2, 4-D	0.75	Post-emergence
Sorghum	Simazine	1.0	Pre-emergence
	Atrazine	1.5	Pre-emergence
	2, 4-D	0.75	Post-emergence
	Bentazon	1.5	Post-emergence
Bajra	Atrazine	0.5	Pre-emergence
	Pendimethalin	1.0	Pre-emergence
	2, 4-D	0.75	Post-emergence
Greengram, Blackgram and cow pea	Pendimethalin	1.0	Pre-emergence
	Alachlor	2.0	Pre-emergence
	Metolachlor	0.75	Pre-emergence
	Oxyfluorfen	0.2	Pre-emergence

	Fluazifop-P-butyl Bentazon	0.250 1.0	Post-emergence Post-emergence
Redgram	Pendimethalin Alachlor Linuron Fluazifop-P-butyl	1.0 2.0 1.5 0.250	Pre-emergence Pre-emergence Pre-emergence Post-emergence
Bengal Gram	Pendimethalin Fluchloralin Linuron	0.75 1.0 0.5	Pre-emergence Pre-planting Pre-emergence
Groundnut	Oxyflurofen Fluchloralin Pendimethalin Alachlor Metolachlor Oxadiazon Fluazifop-P-butyl Bentazon Diclofop-methyl	0.24 1.0 0.75 1.5 0.75 0.75 0.250 1.0 1.0	Pre-emergence Pre-planting Pre-emergence Pre-emergence Pre-emergence Pre-emergence Post-emergence Post-emergence Post-emergence
Sesame	Fluchloralin Alachlor Pendimethalin Diuron	1.0 1.5 0.75 0.75	Pre-planting Pre-emergence Pre-emergence Pre-emergence
Sunflower	Pendimethalin Alachlor Metolachlor Fluazifop-P-butyl	0.75 1.5 1.5 0.250	Pre-emergence Pre-emergence Pre-emergence Post-emergence
Castor Mustard	Fluchloralin Pendimethalin Alachlor Metolachlor Fluchloralin Pendimethalin Alachlor Oxyflurofen Isoproturon	1.5 0.75 1.5 1.5 1.0 1.0 1.5 0.25 0.5	Pre-planting Pre-emergence Pre-emergence Pre-emergence Pre-planting Pre-emergence Pre-emergence Pre-emergence Post-emergence
Cotton	Fluchloralin Diuron Pendimethalin Metolachlor Alachlor	1.5 1.0 1.0 1.5 2.0	Pre-planting Pre-emergence Pre-emergence Pre-emergence Pre-emergence

	Fluazifop-P-butyl	0.250	Post-emergence
Sugarcane	Metribuzin Simazine Atrazine Alachlor 2,4-D	1.0 2.0 2.0 2.0 2.0	Pre-emergence Pre-emergence Pre-emergence Pre-emergence Post-emergence
Potato	Paraquat Metribuzine	2.5 1.0	Post-emergence Pre-emergence
Tomato	Metribuzine	0.7	Pre-emergence
Onion	Fluchloralin	2.0	Pre-planting
Garlic	Oxyflurofen	0.24	Pre-emergence
Oxadiazon	0.75	Pre-emergence	
Fluchloralin	0.90	Pre-planting	
Cabbage	Fluchloralin	2.0	Pre-planting
Lucerne	Decathal 2,4-D	2.0 1.5	Pre-planting Post-emergence

1.4. FIELD CROPS

CEREALS

RICE : *Oryza sativa* (2n:24) Origin : South East Asia

- ❖ Rice farming implements dates back to 8000 BC
- ❖ Rice is staple diet for 2.7 billion people in the world.
- ❖ It occupies 150 Mha and producing 573 Mt with an average productivity of 3.83 t/ha.
- ❖ India is largest rice growing country with 44.6 Mha production to the tune of 93 Mt.
- ❖ There are 21 species in genus Oryza, of these only 2 cultivated species *O.sativa* in Asia and *O. glaberrima* in Africa.
- ❖ There are 3 varietal group in *O.sativa* viz., *indica*, *japonica* and *javanica*
- ❖ About 4, 20,500 accessions conserved all over the world.
- ❖ India had one of the richest rice germplasm collections 60,000 accessions in the world
- ❖ Rice is an short day C_3 plant
- ❖ The process of tillering start at 10 Days After Planting (DAP)
- ❖ In rice a new leaf produces at 7 days interval.
- ❖ In rice, flowering start from tip of the panicle. The topmost leaf below the panicle is the flag leaf
- ❖ Rice grain is a caryopsis fruit type.
- ✍ The dehulled rice grain is called caryopsis, commonly referred to as brown rice

- ❖ There are 3 important phases are vegetative (Germination to panicle initiation), reproductive (Panicle initiation to flowering) and ripening (50% flowering to maturity)
- ❖ Optimum temperature requirement for rice is 20-35°C
- IR 8 is Miracle rice released in 1966
- Special operations in paddy cultivation : Puddling, Trimming, Plastering and Levelling

Season

- In India
    - *Kharif* – June- Sep
    - *Rabi* – Oct – March
- In Tamil Nadu
    - Kuruvai / kar / Sornavari – dry season – May, June
    - Samba / Pissanum – August
    - Thaladi – after the harvest of Kuruvai – Sep – Feb, March
    - Navarai – December and January
    - Kodai / Summer – February and March
    - Rainfed – September and October
- Seasons in other states
    - Sarwa, Abi, Dalwa, Tabi (AP)
    - Ahu, Sali, Boro (Assam)
    - Haine (Karnataka)
    - Viruuppa, Mundakan, Poonja (Kerala)
    - Beali, Sarad, Dalua (Orissa)
    - Aus, Aman, Boro (WB)
    - Kharif, Rabi – Most Hindi belts

Nursery area: 20 cents (800 m^2) for raising seedlings for one hectare.

Seed rate

- 30 kg for long duration
- 40 kg for medium duration
- 60 kg for short duration varieties and
- 20 kg for hybrids
- 75 kg for rainfed rice

Optimum age of seedlings for transplanting

- 18-22 days for short, 25-30 days for medium and 35-40 days for long duration varieties.

Major system of rice cultivation

1. **Direct seeded/Rainfed :** Dry or pre-germinated seeds are sown under dry- or wetland condition and cultivation mainly depends on rainfall
2. **Semi dry :** Seeds are pre-monsoon sown and put under rainfed condition for a period of 30 - 45 days. Later it was converted into wet situation by irrigating from the tanks.

3. **Wet / transplanted :** A system of planting rice where seeds are raised in a nursery bed for 20 to 30 days. The young seedlings are then directly transplanted into the flooded puddled field
4. **Deep water rice :** Deepwater rice is defined as rice that is flooded deeper than 50 cm for one month or longer during the growing season
5. **Aerobic rice:** Growing of rice under non-flooded and non-puddled situation. It is growing rice just like any irrigated dry (ID) crop.

There are 3 type of nursery

- ✓ Wet nursery - 25-30 days
- ✓ Dry nursery - 20-25 days
- ✓ Dapog nursery - 30 m^2 area for 1ha, 1kg seed/m^2, 14 days seedling. This techniques widely adapted in System of Rice Intensification (SRI) planting.

- Pusa Basmati 1 is a world first high yielding, semi dwarf basmati variety and Neck blast damage is severe in Basmati varieties.
- Father of hybrid rice : Prof. Yuan Longping
- Father of hybrid rice in India: Dr.E.A.Siddiq
- GA3 application enhances panicle and stigma exertion.
- Rice hybrid technology based on three line breeding system (A,B,R Lines)
- Biofertilizers for rice : Azospirillum, Phsphobacteria, Azolla and Blue green Algae
- Blue green algae (Azolla) provide 20-25 kg N/ha
- Chlorophyll meter and IRRI leaf colour chart used for leaf N determination
- 'Zn' deficiency cause '*khaira*' disease in rice
- ❖ False smut and sheath rot are new threat to rice.
- ❖ Paddy harvest at 25% moisture content and reduced 14% for storage

SRI techniques

- ❖ Seed rate of 7.5 kg ha^{-1} and seedlings are rised in the mat nursery
- ❖ Square planting with a spacing of 25 x 25 cm
- ❖ Transplanting of 14 days old young seedling
- ❖ Transplanting with single seedling per hill
- ❖ Weeding by Cono/rotary weeder at 15, 25, 35 and 45 days after planting for incroporate the weeds and aerate the soil
- ❖ Watering by alternate wetting and drying
- ❖ LCC based Nitrogen management (Critical value is 4)

Golden rice: It produces provitamin-A (beta carotene) which is useful to alleviate vitamin A deficiency in the diets of poor people.

WHEAT : (2n:42) Origin: Asia Minor

Wheat is second most important food crop in the country. There are three species are cultivated in India

- *Triticum aestivum* (90% area, bread wheat)
- *T. durum* (10% area, macaroni wheat)
- *T. dicoccum* (emmer wheat)

❖ Wheat revolution started in 1967 with dwarf variety HD-2329, Sonora 64 & Lerma Rojo

✍ Long day plant

❖ Seed rate is 100kg /ha. Optimum sowing time: November 1st fortnight

❖ Seeds sown by hand behind the plough called kera method

❖ Crown root initiation (CRI) stage is critical stage of wheat for irrigation

❖ *Phalaris minor* (foxtail weed, mimic weed, canary weed, satellite weed) and oat are major weeds in wheat

❖ Nematode cause disease

- Ear cockle / tundu (*Anguina tritici*)
- Molya disease (*Heterodera avenae*)

❖ Wheat grains stored well in less than 10% moisture content.

BARLEY : *Hordeum vulgare* (2n:14) Origin: Ethiopia

❖ Lugri is a fermented drink developed from hull less barley grain

❖ Pearl barley is suitable for person suffered kidney disorders

❖ Tolerant to salinity and alkalinity but sensitive to acidity

❖ Seed rate 75-80 kg/ha

❖ Barley has 3 critical stages, such as "active tillering" (30-35 DAS), flag leaf stage (60-65 DAS) and milk stage (80-85 DAS)

❖ Water requirement is 200-300 mm

MAIZE : *Zea mays* (2n:20) Origin: Mexico

❖ Moisture content in maize grain 12-15% is preferable for storage. Abundant starch (65%)

❖ During harvest 25-30% moisture is present in maize.

❖ Quality protein maize variety (QPM) are released by using "Opaque 2 gene" varieties such as Shaktiman 1&2 , HQPM1, Composite Shakti 1 (sakti, rattan, protina are QPM composite)

❖ Tassel : It is the branched inflorescence

❖ Depend on maturity maize can be grouped into four type

- Full season maturity (100-110 days)
- Medium maturity (85-95 days)
- Early maturity (80-85 days)
- Very early maturity (75-80 days)

Maize grain types

❖ Flint corn (*Zea mays indura*)

- Entire outer portion of kernel is hard starch

❖ Dent corn (*Zea mays intendata*)

- 95% of USA is occupied by dont corn. Hard starch is confined to kernel
- The amylose of soft starch in the core contracts when the grain is dried producing characteristic dent in the top of the kernel

❖ Sweet corn (*Zea mays saccharta*)

- Grown for food and harvested at 70% moisture
- Good source of energy and vitamin C & A
- About 20% of dry matter is sugar compared to 3% in dent

❖ Floury corn (*Zea mays amylacea*)

- Kernel is largely composed of soft starch with little or no hard strach
- Kernels are easy to grind

❖ Pop corn (*Zea mays averta*)

- It's kernel is small and extreme form of flint corn
- When heated to 170° C, the grain swells and burst, turning inside out

❖ Waxy corn (*Zea mays ceretina*)

- Due to waxy appearance of the kernel
- The starch is entirely amylopectin whereas dent has 78% and 22% amylose

❖ Recommended seed rate for maize is 20 kg/ha

❖ It is a crop highly sensitive to excess moisture and moisture stress also

OATS : Avena sativa (2n= 6x =42) Origin: Asia minor

❖ Oats rank sixth among cereal production in the world

❖ Optimum temperature for oat cultivation is 21-24°C

❖ Seed rate of oat 70-80 kg/ha for seed crop and 90-100 kg / ha for fodder crop.

MILLETS

❖ Major millets - Sorghum and pearl millet

❖ Minor millets - finger, foxtail, proso, barnyard, kodo and little millets

❖ Millets are belongs to C4 group of plants

❖ Among millets, the productivity higher in fingers millet (1400 kg/ha)

❖ Millets anti-nutritional factor : poly phenol enzyme inhibitors and phytate in seed coat.

Sorghum : *Sorghum bicolor* (2n:20) Origin : Ethiopia

☞ Plant remain practically dormant during periods of drought and resume growth as soon as sufficient moisture - so as called camel crop

☞ 'Adaptation to limited rainfall and high temperature' is due to the leaves and stems are covered with a waxy material that reduces 'transpiration'

- CSV 15 and CSV 13 suitable for forage and grain production
- HCN present in sorghum especially early stage 40-50 days.
- Sorghum poisoning: HCN present in sorghum especially early stage 40-50 days causes poisoning to cattle when they grace young plants.
- Sorghum injury: The after effects of sorghum roots and stubbles on succeeding crop of sorghum is called as 'sorghum injury'. While decaying of sorghum stubbles, lots of microbes are build up (sugar content more) and they temporarily lockup N
- Sorghum injury can be overcome by
  - Additional N fertilizer application
  - Incubating with leguminous green manure
  - Avoiding next crop immediately
- Nursery technique for transplanted crop
  - Seed rate 7.5kg in 7.5cents
  - Age of seedlings 18 days – maximum
- Seed rate (kg /ha)
  - Transplanted – 7.5
  - Direct seeding -10.0
  - Rainfed direct seeding-15.0
- Striga management Witch weed (*Striga asiatica*, S *lutea*, S. *hermonthica*)
  - Post emergence application of 2,4 D Na salt 2.0kg/ha at 25-30 DAS
  - Crop rotation with trap crop like sunflower, groundnut and cowpea – they induce weeds to germinate but not affected
  - Spraying urea 10% solution 25-30 DAS
  - Using germination stimulants like Strigol & Ethylene
- Sweet sorghum yield 30-50 t green cane / ha and 1-2 t grain /ha
- Sugar content in sweet sorghum juice varies from 16 - 23% brix
- Sweet sorghum ideally sown during June

Pearl Millet : *Pennisetum glaucum* (2n:14) Origin: West tropical Africa

- Drought withstanding mechanism
  - Deeper root system
  - Efficient photosynthetic system
  - Rapid transfer of food materials from leaves to grain
- Rainfall of 400-750mm
- Temperature are 28 to 32° C is optimum for vegetative growth
- Rain at grain maturity – ergot disease due to high humidity & low temperature
- X 6, X 7, CO 7, WCC 75(World Cumbu Composite), COH 8, K 3 etc

- CO 9 is a good fodder variety
- Seed rate is 5 kg/ha
- Pearl millet harvest at 20% moisture content & stored with 12% moisture

Finger Millet : *Elusine coracana* (2n:36)

- Can tolerate salinity > pH 11.0
- Spacing & seed rate for rainfed
    - A spacing of 20-25 cm row
    - 22.5cm was seen better than 15 cm
    - Seed rate of 6-8 kg/ha
- For transplanting
    - 5 kg for nursery (12.5 cents, 18-20 days old seedlings)
- Seed rate is 10 kg/ha
- Mottle streak and streak virus is prevalent in Karanataka, Tamil Nadu and Andhra Pradesh
- Mottle disease transmitted by leaf hopper and its generally severe in summer
- Varieties: CO 9, CO 14, CO15 K1 and K2 - Rabi (September -October) TRY 1, Paiyur 1, INDAF 5, GPU 28

Kodo Millet : *Paspalum scorbiculatum* (2n:40)

- Seed rate is 10 kg / ha for line sowing and 15 kg for broad casting
- Varieties: CO 2, CO 3, APK1, Paiyur 1, Vamban 1 - 100-125 days

Foxtail Millet : *Setaria italica* (2n:18)

- Seed rate is 8-10 kg/ha for line sown and 15 kg/ha for broadcasting
- Varieties: CO 3, CO 4, CO 5, CO 6, CO 7 - 90 days

Prosomillet (*Panicum milliaceum*) (2n:36)

- Seed rate is 10 kg / ha for line sowing and 15 kg for broad casting
- Varieties: CO 1, CO 2, CO 3, CO 4, CO 5 - 70 days

Barnyard Millet : *Echinochloa frumentacea* (2n:36&54)

- Important component of tribal agriculture
- Seed rate is 8-10 kg/ ha for line sown 12-15 kg/ha for broadcast
- Varieties : K 1 is highly suitable for Tamil Nadu condition. CO 1 - 110 days; CO 2 - 90 days

Little Millet : *Panicum sumatrence* (2n:36)

- The crop strongly associated with tribal agriculture and grown as catch crop
- Seed rate is 8 kg/ha for line sowing and 12 kg/ha for broadcasting
- Varieties: For Tamil Nadu, CO3, CO4 Paiyur 1 and Paiyur 2 are promising.

PULSES

Pulses are the major sources of dietary protein in the vegetarian diet in our country. Besides being a rich source of protein, they maintain soil fertility through biological nitrogen fixation in soil and thus play a vital role in furthering sustainable agriculture

- Kharif pluses: Redgram, blackgram, greengram, cowpea, horsegram, mothbean etc
- Rabi season pulse or Cool season food legumes: Chickpea, Filedpea, Lentil, Lathyrus, Frenchbean, Fababean

Bengalgram : *Cicer arietinum* (2n:16)

- Varieties
  - Two types *Desi* & Kabuli
  - Desi is small seeded
    - Angular shaped edge
    - Shape like chickens head
    - 90% of the world's cultivated
  - Kabuli, large and round seeded with white pale cream seed coat
  - CO 2, CO 3, CO 4 are 90days
  - All India – many varieties , Vijay, Pusa 391, DCP 91-3 (HYV, High input response, 150days)
- Seed rate is 60-100 kg /ha
- Yield potential is 1500-2000 kg / ha
- Major pest : pod borer (control 1:5 pheromone trap/ha)
- Major disease : *Ascochyta blight*

Pigeonpea (Arhar) : *Cajanas cajan* (2n:22)

- It was consumed extensively as dhal
- India ranks 1st – 90% area, 85% production
- Pigeon pea sensitive to frost and to avoid this go for short durationvariety
- Season: June - July for all the variety and VBN3 and CO(RG)7- throughout the year
- Seed rate 15-20 kg/ha except VBN 2 and CO6 (10 kg/ha) and average yield 1.2-1.5 t/ha
- BSR 1 is a perennial redgram and its seed rate is 5 kg/ha
- Spacing is 60x30 cm except VBN 2, CO6 and BSR1(90x30 cm)
- Varities : VBN3 and APK1(drought resistant) are short duration variety. VBN2 and CO6 are long duration Sterility Mosaic Resistant(SMD) variety and CO5(SMD and pod borer resistant) and CO(RG)7 are medium duration root rot resistant variety.
- Major disease: wilt, sterility mosaic and phytophora stem blight
- Giant African marigold grown as border crop to manage pod borer.
- Foliar application of 40 ppm NAA (4ml /4.5 lit) is spray recommended at the time of flowering (80-90 DAS for short duration and 120-130 long duration) in the morning or evening hours for redgram

Greengram (Mungbean) : *Vigna radiata* (2n: 22)

- ✓ Highly nutritious (23-24% protein)
- ✓ It is very sensitive to water togging
- ✓ Seed rate is 20-25 kg/ha
- ✓ Spacing is 30x10 cm
- ✓ Average yield is 6-8 q/ha
- ✓ Major disease:Yellow mosaic virus, Cercospora leaf spot, powdery mildew and Macrophomina blight
- ✓ Varieties: VBN 3 suited for all the season with duration of 70-80 days. Multibloom nature and tolerant to YMV. Other YMV tolerant varieties are CO (Gg) 6, KM (Gg) 2. ADT 3 preferred for rice fallow situation.

Blackgram (Urd bean) : *Vigna mungo* (2n: 22)

- ❖ Black gram rich in phosphoric acid
- ❖ Seed rate is 20-25 kg/ha and average yield is 10-12 q/ha
- ❖ Spacing is 30x10 cm
- ❖ Varieties : T9, CO 4, CO 5, KM 2, VBN 1, VBN 2, VBN 3, VBN 4, ADT 3, ADT 5
- ❖ VBN 5, VBN (Bg) 6 and VBN (Bg)8 suited for through the year with the duration of 70 days. Moderately resistant to yellow mosaic disease (YMV). ADT 3 suited for rice fallow situation.

Rice fallow pulses

- ✓ It is also called relay crop /zero tilled crop /residual crop
- ✓ Crops suitable blackgram & greengram
- ✓ Deltaic rice belts of AP & TN. In TN – Cauvery delta and Tamirabarani
- ✓ Season: Dec-Jan or end of Samba /Early Thaladi or end of Rainy season
- ✓ Seed rate 25 kg for greengram & 30 kg for blackgram
- ✓ Broadcast the seeds 7-10days before harvest rice at waxy condition of soil
- ✓ No irrigation is needed if sown in optimum season; No inter cultural operation is needed
- ✓ Foliar nutrition by DAP 2% solution just at the time of flower initiation and 15 days later is recommended

Improved technology for pulses

- ❖ To overcome the flower shedding, increase the drought tolerance and increase the yield TNAU pulse wonder 6.0 kg ha^{-1} is mixed with 500 lit of water and spray at the time of flowering (30-35 DAS) in the morning or evening hours.
- ❖ Foliar application of 2 % DAP (4 kg of DAP dissolved in 10 lit of water and keep for overnight and decanted solution mixed with remaining 190 lit of water) spray is recommended at the time of flowering (30-35 DAS) in the morning or evening hours for blackgram and greengram.
- ❖ Basal application gypsum 250 kg ha^{-1} is recommended for red soil to overcome the surface hardness and crusting problems.

Field pea : *Pisum sativum*

- ❖ Optimum temperature requirment is 15-25°C
- ❖ Seed rate is 70-80 kg/ha
- ❖ Average yield is 15-20q/ha

French bean : *Phaselolus vulgaris* (2n:22)

- ❖ Seed rate is 100-120kg/ha and average yield is 15-20q/ha
- ❖ Bean mosaic is a major disease of french bean

Cowpea / labia : *Vigna ungiculata* (2n:22,24)

- ❖ More tolerant to heavy rainfall than the other pulses
- ❖ Seedrate is 20-25 kg/ha and grain yield 10q/ha

Horsegram/Kulthi : *Macrotyloma uniform* (2n:24)

- ❖ Seed rate is 22-30 kg/ha and average yield is 7.8q/ha
- ❖ Variety AK42 having 30% protein and CO1, Paiyur 1 and Paiyur 2 are promising variety for Tamil Nadu.

COMMERCIAL CROPS

Sugarcane : *Saccharum officinarm* (2n= 80)

Origin

- ✓ *Saccharum officinarum* (2n=80); South Pacific islands-New Guinea.
- ✓ *Saccharum sinense* (2n=110-120) - China
- ✓ *Saccharum barberi* (2n=82-124) – India
- ✓ Sugar cane contribute nearly 75% to the total sugar pool at the global level
- ✓ It occupies 3% of the total cropped area in the country
- ✓ UP ranks first in area and production, Maharashtra ranks first in sugar recovery and Tamil Nadu ranks first in productivity
- ✓ India Productivity is 70t/ha and Average sugar recovery of 11%
- ✓ Tamil Nadu Productivity is 100t/ha
- ✓ Inflorescence is a silky panicle called 'arrow'. Spikelets open from 3 a.m. to 8 a.m. with peak between 6 a.m. to 8 a.m. Fruit is a caryopsis and used as seed.

Sugar cane contribute nearly 75% to the total sugar pool at the global level. It occupies 3% of the total cropped area in the country and Tamil Nadu recorded high productivity in India.

Climate

- ❖ Warm humid condition continue its growth and flowering
- ❖ Temperature above 50°C arrests its growth
- ❖ Sugarcane best in Tropical region
- ❖ For ripening it needs cool and dry season
- ❖ Maximum & Minimum temperature range between 10-15°C
- ❖ Essential climatic factors for sugar accumulation and cane ripening are bright sunny days and cool nights with a mean temperature of 15-18°C, lower RH of 50-60% and frost free days.
- ❖ Ideal climate for sugarcane cultivation in India prevailing in the states of Maharashtra, parts of Karnataka, Telengana region of Andhra Pradesh

Planting season

1. Eksali (1 Year) - Dec - Feb (Tamil Nadu and Andra Pradesh)
2. Pre seasonal - Oct - Nov (15-16 Months)
3. Adsali - Jul- Aug (18 Months)

❖ Maharastra and Parts of Karanataka main cropping season is December to April

❖ Planting period is 8 months from July to February

There are four main seasons in TN

- Early – Dec-Jan
- Mid-season – Feb-Mar
- Late – April – may
- Special season – June-July

Seed Cane

❖ 35,000 - 40,000 Setts (3 budded) / ha

❖ In TN

- 50,000 three budded setts /ha
- 75,000 two budded setts /ha
- 187,500 single budded setts /ha

❖ Length of 3 budded setts is 30-40 cm long

SETT TREATMENT

❖ The setts should be soaked in 100 litres of water dissolved with 50g Carbendazim, 200ml

❖ malathion and 1 kg urea for 15 minutes.

❖ Treat setts with Aerated steam at 50°C for one hour to control primary infection of grassy

❖ shoot disease.

❖ Dipping with carbendazim 0.2% improves the germination, prevent rotting & infection

❖ Chlorpyriphos - keeps off termites & checks shoot borer

Methods of planting

✍ Flat planting : Most Parts of northern India , Bihar and Karanataka

✍ Furrow planting : Peninsular India and Eastern U.P

✍ Trench planting : Coastal regions

✍ Deep Trench system

✍ Paired - row system

✍ Ring or pit system etc.

✍ **Transplanting technique (STP technique)**

Water requirement : 2000 - 3000 mm

- Important cultural operations in sugarcane are
  - Earthing up
  - Detrashing
  - Propping and
  - Flowering control

Earthing-Up

- The first earthing-up is known as "partial earthing-up (45 DAP)
  - To cover the fertilizers
  - To provide anchorage to root system
- The second operation is "full earthing-up".
  - It is done after final manuring (90-120 days coinciding with peak tillering)

Detrashing

- Removal of dry and lower leaves on 150th and 210th day. It helps in clean cultivation, easy movement of air within the crop canopy, reduce certain pests like scales, mealy bugs, white fly etc.,

Propping

- Tying the canes by using the lower bottom leaves to check lodging of cane at the age of 210 days of the crop

Control of flowering

Spraying of ethrel @ 500 ppm in twice or 1000ppm once at floral initiation

Assessing maturity

- Use of hand refractometer - Brix reading
- Assess the maturity by HR meter survey – 18-25% indicates optimum maturity
- When the reading between top and bottom is 1:1 – is right time to harvest

Fehlings test is used to estimate the glucose content (< 5% in juice)

Sugarcane Ripeners

- Spray Sodium metasilicate 4kg in 750 litres /ha 6 months after planting
- Repeat at 8th & 10th months and not at declining phase
- Polaris and Ethrel are most extensively used sugarcane ripener in Hawaii
- Polaris @ 5 kg in 600 l /ha

Harvesting : Sugarcane will be harvested with the sugar content of 12 - 15% and sucrose content of 10-18% and this proper maturity will give sugar recovery from 9.5 to 11.5%. Stalks are cut in to 3-5 cm above the ground level.

Crushing and gur making

- ❖ 30-35% sugarcane used for manufacturing gur and khandsari
- ❖ In North India juice is purer and neutral pH
- ❖ In South India juice is acetic (Need lime)

Ratooning

- ✍ The crop raised from planting cane sett is called plant crop
- ✍ After the harvest of plant crop stubble sprouts and gives rise to succeeding crop called as ratoon crop. The practice of taking up ratoon crop is called 'ratooning'
- ✍ Advantages:
  - Reduction in the cost of field preparation, planting material and operation cost
  - Saving in field duration – ratoon matures earlier than plant crop
  - Ratoon may give equal yield that of plant crop
- ✍ Ratoon occupies 50-55% total cane area
  - 30-35% of the total cane production
  - Optimum temperature for quick sprouting of stubble is 27 ^{0}C

Ratoon management

- ✍ Plant crop should be harvested at right maturity
- ✍ Harvesting close to the ground with sharp cutting
- ✍ Stubble shaving to 4-6cm is recommended if no uniform cut at harvest
- ✍ Remove the trash but do not burn it
- ✍ Shoulder breaking or off-barring to remove decayed stubbles
- ✍ Gap filling with sprouted setts or seedlings
- ✍ Ratoon is less efficient in N utilization hence 25% additional N from 5-7 days after ratooning is desirable
- ✍ Spraying of FeSO4 @ 2.5kg/ha in 150 litres on 15th day if chlorotic symptom is noticed
  - If persists repeat twice at 15 days interval
  - Ratoon requires more plant protection for grassy shoot disease and ratoon stunting
- ✍ Application of growth regulators such as ethrel 500 ppm or GA 100 ppm increase sprouting
- ✍ Insect pests: White wolly aphids - *Certavacuna lanigera* was major outbreak in 2003

Sustainable Sugarcane Initiative (SSI)

- ❖ Raising nursery using single budded chips in pro-tray
- ❖ Transplanting young seedlings (25-35 days old)
- ❖ Maintaining wide spacing (5x2 feet) in the main field
- ❖ Providing sufficient moisture through water saving efficient irrigation technologies viz., skip furrow, alternate furrow and subsurface drip irrigation and fertigation with water soluble fertilizers

Sugarbeet *(Beta vulgaris)*

- Sugar beet is the day neutral plant of the temperate regions
- Optimum temperature for seed germination is 15°C
- Optimum temperature for sugar accumulation is 21°C
- Sugar beet is high tolerance to soil salinity & alkalinity
- Ramonskaya 06 - diploid multigerm variety and Maribo - polyploid variety.
- Seed rate is 10 kg / ha for unprocessed and multigerm (1 gromervie contains 3-4 seeds)
- Fruit type is gromervie
- Roots sugar content is 15-16%
- Major pest is cut worm (affect young stage)
- Major disease is root rot (*Sclerotium rolfsii*) and it is controlled by soil drenching with chlorphyriphos

Cotton (commonly called white gold)

- It is an allopolyploidy crop
- Asiatic cotton - *Gossypium arboreum* (2n - 26 (desi cotton)
- Desi cotton *G. herbaceum* (2n = 26)
- Upland cotton - *G. hirsutum* (2n =52)
- Egyptian cotton - *G. barbadense* (2n =52)
- India ranks first in cotton area and third in the production in the world
- India is the only country in the world where an the 4 cultivated species of cotton are cultivated on commercial scale
- Maximum area covered by hybrids (45%) followed by *Gossipium hirsutm* (35%)
- Cotton is semi xerophyte crop
- Cotton is moderately tolerant to salinity
- Seed rate hybrids is 2 kg/ha and water requirement is 700-1200 mm
- 1 bale of cotton is 170 kg
- **Herbicide injury in cotton:** Cotton is sensitive to 2, 4-D herbicide

Varietal improvement

- MCU 5 first extra long stable variety released in Tamil Nadu
- First cotton hybrid -H4 (or) Sankar (1970) at Surat station in Gujarat
- First world interspecific hybrid was developed between *G.hirsutum* and *G. barbadense* and released 1972 in Dharwad
- World first diploid hybrid DH 7 was released in Surat (1985) Gujarat

Major Cotton growing states :

- *G. hirsutum* and *G. arboreum* - All 9 cotton growing states
- *G. herbaceum* - Gujarat and Karanataka
- *G. barbadense* - Tamil Nadu, Karnataka, Andhra Pradesh

- *G. barbadense* - only one variety is Suvin
- Most of the varieties of *G. arboreum* belong to short a medium stable.

Donors for resistance :

- *Helicoverpa* - *G. somalense*
- White fly - *G. armourianum*
- Frost - *G. thurberi*
- Ginning is a process of separating fibre from seed cotton

Type	Staple length (mm)
Superior long stable	27
Long stable	24.5 -26.5
Superior medium stable	22-24.5
Medium stable	20-21.5
Shorts table	< 19.5

Acid delinting of cotton seeds: Put the required quantity of seeds in the container and add commercial conc. H_2SO_4 @ 100 ml per kg of fuzzy seed. Stir vigorously and continuously with a wooden stick for 2 to 3 minutes till the fuzz sticking to the seeds is completely digested and the seed coat attains a dark brown colour of coffee powder.

Earthing up: Digging and earthing up is essential for soil aeration

Nipping Topping: Topping in cotton is done to intermediate types/varieties. If monopodia continues more than 15 nodes, topping is done to breaks apical dominance. Topping is done to prevent lodging, reduce boll rot, increase flowering and fruiting, branches and thereby increases boll production. Topping is done between 75 and 90 days.

Rice fallow cotton

- It is cotton cultivated in rice fields immediately after the harvest of rice when the field is in waxy condition when the season is optimum.
- No field preparation is needed to sow the crop.
- Season: Jan 15th –Feb 15th
    - Field : Immediately after the harvest of rice; Sowing at waxy soil condition
      Varieties: MCU 7, SVPR 3, ADT 1 and LRA 5166
    - Spacing: 60 x 30 and (75 X 30 for LRA)
    - Fertilizers: 60:30:30 kg NPK /ha
    - Digging the rows and earthing up: From 21st day:
    - Earthing up: Around 40th day after the moisture dried and soil loosened for second time

Insect pests

- Boll worms: Spotted bollworm (*Earias vittella*), American boll worm (*Helicoverpaarmigera*) and pink boll worm (*Pectinophora gossypiella*).
- Red cotton bug : *Dysdercus singulatus*
- Dusky cotton bug : *Oxycarenus hyalinipennis*

Quality parameters: Ginning percentage, Color, Trash, Fibre quality, Fibre length, Fibre fineness, Fibre strength, Spinning performance and Oil content (14-26%) of the seed etc.,

Boll shedding in cotton: Premature shedding of bud , flower and boll is a serious problem in cotton. Two foliar sprays of planofix (NAA) at 40 ppm (40 mg NAA in one litres of water) at one-month interval at square and peak flowering stage prevent the shedding of squares and flowers. Spray of gibberllins @ 100 ppm at the opening of flowers and young bolls helps in retention of bolls with plants (GA 61 g per ha). Spray of CCC @ 40 ppm twice in cotton is recommended.

JUTE AND ALLIED FIBRES

Jute (White Jute)**:** *Corchorus capsularis* (2n: 14), Origin: China

- Seed rate : Tossa jute 6 - 7 kg/ha
  : White jute 8- 10 kg/ha
- Jute requires a warm and humid climate
- Harvest at 50% flowering that is early pod stage of the crop
- *Retting* : It is the process through which the fibres in the bark are being separated from the plants with removal of pectins, gums and other mucilaginous substances. Jute bundles are kept in standing water of 30 cm deep and 32 °C temperature
- *Mesta* : (Madras hemp): *Hisbiscus sabdariffa* var. *altissima* 2n =36 (sp)
- 1 bale of Mesta = 181 kg

Sisal : *Agave sisalana*

- Introduced from Mexico, agave semi perennial plant and it is also known as century plant. it propagated through suckers bulbils

Sun hemp : *Crotalaria juncea* (2n =16)

- Bombay hemp (or) banaras hemp; Seed rate for broadcasting 25kg/ha and for direct sowing 15 kg/ha

Green manure Sun hemp

- Seed rate is 60kg/ha, it produces total green matter of 20 t/ha. It contains 50-60 kg of N /ha and fixes : 50-60 kg of N/ha

Flax or linseed : *Linum usitatissimum* (2n =30, 32)

- Flax is a dual purpose temperate zone crop, stem used as fibre and seeds used for oil purpose. Flax fibre contains 80 - 90% of cellulose. Seed rate requirement is 40 kg/ha

OILSEED CROPS

Status in oilseed production in the world

✓ First : Castor, Safflower, Sesame and Niger
✓ Second : Groundnut, Rapeseed and Mustard
✓ Third : Linseed

Status in oilseed production in the India : Groundnut, Rapseed- Mustard, Soybean occupies 77% oilseed area, 88% oilseeds production, 13% of gross cropped area, 3% of gross national product and 10% of value of all the agriculture commodities

Groundnut : (*Arachis hypogaea*) 2n : 40, Origin : Brazil

- ❖ It is a self pollinated crop and kernel contains 44-50% oil with the Calorific value - 349/100g.
- ❖ Shelling Percentage : 70-75%
- ❖ The optimum temperature for vegetative growth is between 20° C to 30 ° C and for reproductive growth is 24°C to 34°C.
- ❖ Bunch type : TMV 7, CO 2, CO3, CO4, BSR 1, JL 24, VRI 2, TMVGn 13
- ❖ Semi Spreading : CO Gn 5, TNAU CO 6, TMV 10, VRI Gn 7
- ❖ Seed rate for bunch type is 120 kg /ha and for semi spreading type is 100-110 kg/ha
- ❖ Seed Hardening: Selected seeds are soaked for 6 hrs with 0.5% CaCl2 solution in 50% by volume and shade dried.
- ❖ **Earthing up:** Earthing up can be done during second hand weeding to provide medium for the peg.
- ❖ **Gypsum application:** Application of gypsum at 200 kg/ha as basal and 200 kg/ha at 45 DAS reduces pop pods and improve the oil content.
- ❖ Foliar Spray : 2 sprays of groundnut rich @ 5.0 kg/ha (for each spray) at 35 DAS (50 % flowering) and 45 DAS (Pod developing stage) in 500 litres of water is recommended to increase flower retention, improve pod filling and to induce drought tolerance apart from yield improvement.

Botanical varieties :

Self pollinated : *Hypogaea* : Virginia group: var. hirsuta

- ❖ Spreading to upright and seeds are dormant

Self pollinated : *fastigata* : Valencia group var: Vulgaris (Spanish group)

- ❖ Upright growth habit, little fresh seed dormancy, shorter duration than sub sp. hypogaea, include Valencia & Spanish group, Peg - Positively geotropic and best suited soil is sandy loam and loamy soils.

Dormancy : It is a condition of plant seeds that prevents germination when the seeds are under optimal environmental conditions for germination

- ❖ Virginia types have seed dormancy up to 60-75 days and dormancy broken by using ethrel (500 ppm)
- ❖ Spanish bunchy type has lack of seed dormancy and it can be used immediately after harvesting for sowing.

Rapeseed- Mustard (Brassicaceae or Cruciferae)

Brown mustard (or) India Mustard

- ❖ *Brassica juncia* - Leaf mustard, mustard greens
- ❖ *Brassica nigra* - Garden land crop and used as a spice
- ❖ Rapeseed and mustard- second most important oil seed crop after groundnut in India
- ❖ 25-30% of the total oil seed production in the country
- ❖ Indian mustard introduced from china to North Eastern India
- ❖ Largest producer of rapeseed mustard is Rajasthan
- ❖ Oil content is 30-48%
- ❖ White mustard oil content is 25-33%
- ❖ Inflorescence type is corymb or flowers are gathered in raceme
- ❖ Fruit type is siliqua
- ❖ Seed rate recommended is 4-5 kg/ha

- Harvesting at 75% of pods turn yellow and moisture content at 30%
- Threshing of seed done at moisture content of 12-20%
- Safe moisture content for storage is 8%

Soyabean : *Glycine max* (2n : 40) origin : China

- Seed rate for kharif season @ 75-80 kg/ha and spring @100 kg/ha
- Oil content is 20%
- Protein content is 40-42% quality protein
- Protein - Mostly globulin (partly albumins)
- Oil - poly unsaturated fatty acids (oleic & lionoleic)
- Soya bean - Major oil seed crop in the world.
- 50% of the total acreage as well as production
- Provide 60% of vegetable protein 23% of oil in the world
- Soyabean ranks 3rd in vegetable oil economy in India after groundnut and rapeseed - Mustard
- Madhya Pradesh produces 70% of total soybean production
- Harvesting at seed moisture content of 15-17%
- Ideal threshing moisture content is 13 -14%
- Safe Storage of seeds at 10% moisture content

Sunflower (*Helianthus annus*), Asteraceae or Compositae (2n: 34) Origin: USA

- Cross Pollination due to Protoandry
- Sunflower was introduced in to India in 1969
- Next to soybean and groundnut, sunflower is the 3rd important oil seed crop of the world
- Seed rate is 5 kg/ ha
- Seed oil content is 48-53%
- Oil contains unsaturated fatty acids (oleic and linoleic)
- Flower type is capitulum
- Flower movement is heliotropic
- Movement related to auxin level not due to the sun light
- Productivity highest in Punjab and Haryana
- Storage moisture content is 9-10%
- **Growth regulator:** Spray of NAA (Napthelene Acetic Acid)) / Planofix @ 20 ppm (280 gm in 625 lit of water) on 30 and 60 DAS to increase seed setting.
    - Artificial pollination:
    - Keeping beehives in the field @ 5 per ha improves seed setting and yield.
    - Hand pollination may also be resorted to induce cross pollination.
    - Mild rubbing of the capitulam with the hand covered with soft cloth or rubbing two flowers, face to face gently on alternate days for two weeks.
    - Foliar spray of borax at 0.2%

Sesame : *Sesmum indicum*, Pedaliaceae (2n -26)

- Sesame is the oldest oilseed crop in the world.
- India is the largest producer of sesame in the world.

- ❖ India holding first places in the world with respect to sesame growing area
- ❖ In India, Gujarat has major producer of sesame (20%)
- ❖ Sesame seed is albuminous in nature
- ❖ Sesame Seed oil content is 46-52%
- ❖ Sesame seed protein is 20-26%
- ❖ Sesame is a self pollinated crop
- ❖ Fruit type is capsule
- ✍ **Varieties:** CO 1, TMV 3, TMV 4, TMV 5, TMV 6, Paiyur 1, SVPR 1 (white), VRI 1
- ✍ Seed rate is 5 kg /ha; Spacing : 30cm x 30cm (11 plants m^{-2})
- ✍ Critical stage for moisture requirement :Flowering phase (35th to 45th DAS).
- ✍ Phyllody disease : Phytoplasma ; Vector : *Orosius albicinctus*

Safflower : *Carthamus tinctorius*, Asteraceae or Compositae (2n : 24)

- ❖ India is the largest producer of safflower in the world.
- ❖ Seed oil contains (PUFA) polyunsaturated fatty acids (linoleic acid content 78%)
- ❖ Safflower oil Reduce the blood cholesterol level
- ❖ Oil content of seed is 28-32%
- ❖ It is often cross pollinated crop
- ❖ Safflower is day neutral, thermo sensitive and salt tolerant crop
- ❖ Maharashtra is the highest producer of safflower (72%) in India

Niger : *Guizotia abyssinica*, Asteraceae or Compositae (2n : 30)

- ❖ Niger is a temperate region cross pollination crop
- ❖ It is resistant to drought and resistant to poor oxygen supply in soils because it has ability to develop aerenchymas
- ❖ Fruit type is achene
- ❖ Seed rate is 5-8 kg/ha
- ❖ Seed oil content 35-40%

Castor : *Ricinus communis*, Euphorbiaceae (2n : 20) Origin: Eastern Africa (Ethiopia)

- ❖ India is the largest producer of castor and also a largest castor growing acreage in the world
- ❖ It is a cross pollinated crop especially through wind
- ❖ Flower type is panicle racemes
- ❖ Seed rate is 10-15 kg/ha for rainfed and 5 kg/ha for irrigated condition
- ❖ Seed Oil content is 36-58%
- ❖ Fruit type is capsule
- ❖ Highest area and productivity in Gujarat
- ✍ Varieties : SA 2, TMV 5, TMV 6, CO1
- ✍ Hybrids: GCH-4, TMVCH -1, YRCH-1

1.5. FORAGE CROPS AND GRASSES

India supports nearly 20 percent of the world's livestock and 16.8 percent human population with the 2.3 percent world's geographical area.

Animal population contribution to world by India

- India leader in cattle and buffalo population
- Cattle - 16%
- Buffalo - 55%
- Goat - 20% (world second largest)
- Sheep - 5% (world 4th largest)
- India stands 1st in milk production

WHO Recommendations:

- Availability of milk is 281g/head/day but recommended is 250g/head/day
- Availability of egg is 35 eggs/head/year but recommended is 180 eggs/head/year
- Availability of meat is 800g/head/year but recommended is 11kg/head/year

Forage : It is a plant material (mainly plant leaves and stems) eaten by grazing animals.

Fodder : Food given to animals (Including plants cut and carried to them) rather than that which they forage for themselves. It includes hay, straw, silage and legumes.

Grazing Resources

- In India grazing intensity is 12.6 Adult cattle unit (Acu) /ha
- In developed countries intensity is 0.8 Acu /ha
- Total potential area of grazing land in India 85.9 Mha
- Main grazing resources in India pasture lands (12 Mha)
- Pasture lands high in Himachal Pradesh (36.44%)

Fodder Cholam

- **Varieties :** CO 27, COFS 29 & COFS 31 (Multicut Cholam)
- For COFS 29 & COFS 31 (multicut variety) : Apply 45 : 40 : 40 kg N,P, K/ha as basal and 45 kg N as top
- Dressing on 30 DAS followed by the application of 45 kg N/ha after every cut. After 4th cut, apply 40 kg P and 40 kg K along with 45 kg N to sustain the fodder yield and quality.
- Seed rate : CO-27 -50 kg/ha and COFS 29 & COFS 31– 5 kg/ha
- Spacing : 30 x 10 -15 cm ; Harvest at 50% flowering for fodder

Fodder cumbu

- **Seed rate:** 10 kg/ha. **Spacing:** 25 x 10 cm.
- **Harvest:** Harvest at boot leaf stage, 40 - 45 DAS. Green Fodder Yield: 25-30 t/ha
- Note: CO 8 fodder cumbu can be intercropped with CO 5 or COFC 8 cowpea that can be harvested together to provide a nutritious fodder to cattle.

Fodder Maize

- Composite/hybrids: African Tall, Ganga 5
- 30 : 40 : 20 kg N, P_2O5 and K_2O / ha. Apply 30 kg N/ha at 30 DAS as top dressing.
- Spacing : 30 x 15 cm, Seed rate : 40 kg/ha; Harvest the crop when the cob is in the milky stage.

Neelakolukattai (Blue Buffel Grass) - (*Cenchrus glaucus*)

- **Variety:** CO-1, Seed rate: 6 - 8 kg/ha or 40,000 rooted slips/ha. Spacing: 50 x 30 cm.
- Fresh seeds have dormancy for 6 - 8 months.
- To break dormancy, soak the seeds in 1 % potassium nitrate solution for 48 hours prior to sowing.
- A pure crop yields 40 t/ha/year in 4 - 6 cuts.

Guinea grass (*Panicum maximum*)

- Variety: CO 2, seed rate: 2.5 kg/ha seeds or 40,000 rooted slips. Spacing : 50x 50 cm
- First cut 75-80 days after germination or 45 days after planting of slips. Subsequent cuts at
- interval of 45 days. Green Fodder Yield: 260 t/ha per year in 8 cuts.
- Intercropped with Desmanthus (Velimasal) at 3:1 ratio

Deenanath Grass (*Pennisetum pedicellatum*)

- Variety **:** CO 1, Seed Rate: 2.5 kg/ha, Spacing, 35 x 10 cm or solid sowing in lines 30 cm

Cumbu - Napier Hybrids

- Hybrids : CO-3 and KKM-1, Spacing 50 x 50 cm and 40,000 planting material /ha.
- Fertilizer : 150:50:40 of NPK in kg/ha.
- Apply full dose of P,K and 50% N basally. Top dressing of 50% N on 30 DAS.
- Repeat the basal application of 75 kg N/cut for sustaining higher yield
- As a mixed crop, 3 rows of Cumbu Napier Hybrid and one row of Desmanthus
- First harvest is to be done on 75 to 80 days after planting and subsequent harvests at intervals of 45 days and yield is around 400 t/ha.
- Note :To alleviate the ill effects of oxalates in this grass,
- Feeding 5 kg of leguminous fodder per day per animal along with these grasses or
- Providing calcium, bone meal or mineral mixture to the animal or
- Giving daily half litre of supernatant clear lime water along with the drinking water or sprinkling this water on the fodder.

Lucerne – Kudirai Masal (*Medicago sativa*)

- **Variety** : CO 1, Seed Rate: 20 kg/ha Spacing: 25 cm x solid sowing
- Good quality seeds without the seeds of Cuscuta should be used.
- Apply the full dose NPK of 25:120:40 basally before sowing, Yield : 70-80t/ha/yr

Hedge Lucerne - Velimasal (*Desmanthus virgatus*)

- Pure crop: Sow the seeds at 20 kg/ha in solid stand on the side of the ridges
- Mixed Crop: Velimasal + Cumbu- Napier hybrid grass at 1:3 ratio
- First cut on 90th day after sowing at 50 cm height and subsequent cuts at intervals of 40 days at the same height. Green fodder yield is 125t/ha/yr

Fodder Cowpea

- Varieties : CO 5, CO FC 8 & COFC 9, Seed rate - 25 kg/ha. Spacing : 30 x 15 cm
- **Harvesting:** Harvest 50 - 55 days after sowing (50% flowering stage).
- **Green Fodder Yield:** Irrigated : 20-30 t/ha, Rainfed : 10-15 t/ha

Muyal Masal (STYLO) - (*Stylosanthes scabra*)

a. *S.hamata* (Annual) avd b. *S.scabra* (Perennial)

- For line sowing (30 x 15 cm) the seed rate is 6 kg/ha and for broadcasting 10 kg/ha.
- First harvest can be taken 75 DAS at flowering stage and subsequent harvests depending upon the growth.
- Low yield during the first year, Later on due to self seeding, it yields about 30 to 35 t/ha/year from the third year onwards.

Leucaena – Soundal (*Leucaena leucocephala*)

- June - July :Hawaiian giant (Ivory coast), CO 1
- Rainfed (Sep - Oct): K 8, Giant Ipil - Ipil, CO 1
- Scarification of seeds with sand in mortar. Acid scarification: concentrated sulphuric acid for three minutes or hot water treatment by soaking the seeds in hot water (80° C) for 4 minutes

Fodder preservation

Silage : It is a preserved under anaerobic condition used as a lean season feed material

Hay : Dehydrated green fodder used as a lean season feed material

Silage : It is a controlled fermentation of crops under anaerobic condition

- The process of making silage is known as ensilage
- Nutrient losses are less in silage than hay making
- Lactic acid fermentation is desirable for silage making
- Avoid butyric acid fermentation by *clostridium spps* under cold condition by adding molasses, cereal grains and citrus pulp

Techniques in silage making

- A dry matter should be added >30% in silage pit
- Green fodder should be used at specific stage of different crops
  - Sorghum at flowering to dough stage
  - Maize at milk to dough stage
  - Oat at 50% flowering to dough stage
  - Grasses at early flowering stage

Silage quality

- pH ranges from 4.5 - 5.0
- Ammonical N: <10% of total N
- Butyric acids: <0.2%
- Lactic acid: 3-12%

Hay making : It is conservation of high quality forages by drying process is called hay making and a process of reducing the moisture content of fresh fodder crops to 10 to 20 percent.

Cultivated grasses and legumes fodder

Description	Subabul (*Leucaena* Sp.)	Cowpea (*Vigna unguiculata*)	Cumbu Napier Hybrid (*Pennisetum purpureum*)	Guinea Grass (*Panicum maximum*)	Rainfed Grass Kolukattai Grass (*Cenchrus glaucus*)	Irrigated Legume Leucerne (Medicago sativa) (*Kudirai masal*)	Rainfed legumes Stylosanthes Stylosanthus scabra (*Muyal masal*)	Cereal Maize (*Zea mays*)	Fodder Sorghum (*Sorghum bicolor*)	Desmodium (*Aattu masal*)
Sowing time	Feb -Mar Nursery Jun -July Transplanting June - Aug Direct sowing	Jun -Jul Sep-Oct Feb-Mar	Feb-Aug	Feb-Aug	Feb-Mar	Oct-Nov	Jun -Jul Sep-Oct	Jun- Aug Jan-Mar	Jun- Aug Jan-Mar	Jun- Aug Through out the year under irrigation
Variety	Hawayan, Giant, Co 1, K 8	CO 5	NB21, CO1, CO2, BN2	CO 1	CO 1	CO 1	Stylo-Santhes Namaka	African Tall	CO11, CO27, COFS 29 K7,K10	-
Seed rate (kg/ha)										
i) Slips(Nos.) ii) Seed (kg.)	- 20	- 35	40000 -	66000 2.5	- 4	- 20	- 10	- 40	- 40	- 20
Spacing (cm)	100x20	30x10	50x50	50x30	50x30	25x20	30x15	30x10	25X10	30x30
Nutrient kgha^{-1}										
Basal N P_2O_5 K_2O Top dressing N	25 100 30 -	25 40 20 -	50 50 40 100 after Each cutting	50 50 40 25 after Each cutting	25 40 20 25 after Each cutting	25 120 40 -	20 60 15 -	30 40 20 30 kg on 30th day	30 40 20 30 kg on 30th day	25 12.5 62.5 30 kg on 30th day
Irrigation interval	10 days	10 days	10 days	10 days	1 week	20 to 30 days	7 to 10 days	7 to 10 days	7 to 10 days	20 days In summer
No. of cuts First cutting	6 3 to 4 months	1 50% flowering	5 to 6 50 DAP	7 to10 60 days	4 60 to 65 days	10 60 days	5 75 days	1 70-75 days	1 50% flowering	6 to7 4 months onwards
Fodder Yield (T/ha) Green Dry	75 to 100 22 to 33	20 to30 7.5 to 10	250 to 300 60 to 75	150 to200 25 to 35	40 12	70 to 80 10 to 12	35 to 45 8 to 10	35 to 40 12 to 15	40 to50 10to12.5	18 to20 10 to 12

1.6. CROPPING SYSTEMS

1. **Cropping system :** It is an important component of farming system. It represents cropping pattern used in a farm and their interaction with farm resources, other farm enterprises and available technology which determine their makeup.
2. **Farming system :** It represent an appropriate combination of farm enterprises, cropping systems, livestock, fisheries, forestry, poultry and the resources available to the farmers to raise them for food and profitability.
3. **Cropping pattern :** The proportion of area under various crops at a point of time in a unit area. It indicates the yearly sequence and spatial arrangement of crops and fallow in an area.
4. **Mono cropping :** The land may be occupied by one crop during one season year after year is called mono cropping.
5. **Double cropping :** The land may be occupied by two crops in a year is called double-cropping.
6. **Triple cropping** : Growing of three crops in a year sequence
7. **Quadruple cropping**: Growing of four crops in a year sequence
8. **Multiple cropping :** The land may be occupied by more than two crops in a year is called multiple cropping. It includes inter cropping, mixed cropping and sequence cropping. Intensification of cropping in space and time dimensions.
9. **Intercropping :** Growing two or more crops simultaneously on the same piece of land with a definite row pattern. Intensification of cropping in space and time dimensions
10. **Row intercropping:** Growing two or more crops simultaneously where one or more crops panted in rows
11. **Strip intercropping :** Growing two or more crops simultaneously in different strips wide enough to permit independent cultivation but narrow for the crops to interact agronomically.
12. **Mixed cropping :** Growing of two or more crops simultaneously intermingled without any row pattern.
13. **Sequential cropping :** Growing of two or more crops in sequence on the same piece of land in a farming year.
14. **Relay cropping :** Planting of the succeeding crop before harvesting the preceding crop.
15. **Ratoon cropping :** Raising a crop with regrowth coming out of roots or stalks after harvest of the crop.
16. **Parallel cropping :** Cultivation crops which have different natural habit and zero competition.
17. **Companion cropping :** It is an intercropping system where the production of both inter crops is equal to that its solid planting.
18. **Multitier cropping :** Cultivation of two or more than two crops of different heights simultaneously on a certain piece of land in any certain period. (eg. Coconut + pepper + pine apple + grass)
19. **Synergistic cropping :** The yields of both crops are higher than their pure crops on unit area basis
20. **Additive series intercropping :** One crop is main crop or base crop and another crop is intercrop. The geometry of base crop is altered to introduce intercrop, in this type the main crop population is same that of recommended population.
21. **Replacement series :** Plant population of both component crops is less than recommended population in pure stand.
22. **Land equivalent ratio**: Ratio of area needed under sole cropping to one of intercropping at the same management level to give an equal amount to yield.

    Length of growing period (LGP): It is the period (in days) during a year when precipitation exceeds half of the potential evapotranspiration plus period required to evapotranspire 100 mm of water from precipitation stored in the soil profile.

LGP	Cropping system
<75 days	Perennial vegetation and mono cropping of short duration varieties
75-140 days	Mono cropping
140-180 days	Intercropping
>180 days	Double cropping

Cropping system evaluation indices

- Relative yield total
- Land equivalent ratio
- Relative crowding coefficient
- Aggressivity
- Area time equivalent ratio
- Income equivalent ratio

Cropping season

In India three seasons are followed

Rainy season	-	Kharif (July to October)
Winter season	-	Rabi (October to March)
Zaid season	-	Summer (March to June)

1.7. DRY FARMING AND AGROFORESTRY

Dry Farming

Dryland agriculture: Growing of crops entirely under rainfed conditions

Dry farming:

- It is cultivation of crops in regions with annual rainfall less than 750 mm.
- Crop failure is most common due to prolonged dry spells during the crop period.
- These are arid regions with a growing season less than 75 days.
- Moisture conservation practices are necessary for crop production.

Dryland farming:

- It is cultivation of crops in regions with annual rainfall more than 750 mm.
- Crop failure is relatively less frequent.
- These are semiarid tracts with a growing period between 75 and 120 days.
- Moisture conservation practices are necessary for crop production.
- Adequate drainage is required especially for Vertisols.

Rainfed farming

- It is crop production in regions with annual rainfall more than 1150 mm.
- Crops are not subjected to soil moisture stress during the crop period.
- Emphasis is often on disposal of excess water.
- These are humid regions with growing period more than 120 days.

Significance of Dry farming in Indian Agriculture:

- About 70% of rural population lives in dry farming areas
- 55 per cent of the net sown area will be still in rainfed.
- About 45 per cent of the total food grain, 75 per cent of oilseeds, 90 per cent of pulses and about 70 per cent of cotton is obtained from rainfed agriculture in India.
- More than 90 per cent of the area under sorghum, groundnut and pulses is rainfed.
- Dryland crops oscillates with the quantity, onset, and progress, spatial and temporal distribution of monsoon rains.
- Offers good scope for development of Agroforestry, Social forestry, Horti-Sylvi-pasture and such other similar systems

Development of Dry farming in India:

1970-23 Research Centres established under AICRPDA

1972-Establishment of ICRISAT

1985-Central Research Institute for Dryland Agriculture at Hyderabad

1986-Launching of NWDPRA programmes by Government of India in 15 states.

Aridity

Aridity refers to a condition of deficiency of water due to either insufficient precipitation or excess water loss over supply. The term "arid" is derived from a Latin word, "arere" which means 'dry'.

Semi arid tropics: This region lies between 10° and 30° north and south latitudes.

Constraints for crop production in dry farming regions:

- Climatic constraints
- Soil related constraints
- Cultivation practices and
- Socio economic & political constraints.

a. Climatic constraints

1. Vagaries of monsoon
    - Variable Rainfall: Annual rainfall varies greatly from year to year
    - Intensity and Distribution
    - Late Onset of Monsoon
    - Early Withdrawal of Monsoon
    - Prolonged Dry Spells

2. High atmospheric temperature
3. Low relative humidity
4. Hot dry winds
5. High atmospheric water demand (potential evapo-transpiration) exceeding precipitation during most part of the year.

b. Soil Constraints

1. Inadequate soil moisture availability
2. Poor organic matter content
3. Poor soil fertility
4. Soil deterioration due to erosion (wind and water)

c. Cultivation practices

- Ploughing along the slope
- Broadcasting seeds/gorru sowing / sowing behind the country plough leading to poor as well as uneven plants stand
- Selection of traditional varieties
- Monsoon sowing
- Choice of crop based on rain fall
- Application FYM in limited quantity
- Hand weeding
- Mixed cropping
- Use of conventional system of harvesting
- Traditional storage system

d. Socio-economic constraints

Soil erosion

Soil erosion is the process of detachment of soil particles from the topsoil and transportation of the detached soil particles by wind and water.

- Detaching agents are falling rain drop, channel flow and wind.
- Transporting agents are flowing water, rain splash and wind.
- Out of 328 mha of India's geographical area 175 m ha (53.3%) subject to soil erosion and all kind of land degradation.
- Recent estimates indicate that about 5,333mt (16.35 t/ha) of soil is detached annually.

Types of erosion

Geological erosion: This long time slow process has developed the present topographic features like stream channels, valleys, etc. through weather abnormalities such as intense rainfall and biotic interference.

Accelerated erosion: It is due to disturbance in natural equilibrium by the activities of man and animals through land mismanagement, destruction of forests, overgrazing, etc.

Water erosion: Loss of soil from land surface by water, including runoff from melted snow and ice is usually referred to as water erosion.

Forms of water erosion

Sheet, rill, gully, ravine, landslide and stream bank erosion.

Wind erosion: Erosion of soil by the action of wind is known as wind erosion.

Saltation: Movement of soil particles by a short series of bounces along the ground surface

Suspension: Movement of fine dust particles, smaller than 0.1mm dia floating in the air

Surface creep: Rolling and sliding of soil particles along the ground surface due to impact of particles descending and hitting during saltation is called surface creep

Erosion control

Control measures are grouped in to three agronomic, mechanical and forestry measures

Agronomic – Choice of crops, land preparation, contour cultivation, strip cropping, mulching, application of manures and fertilizers and appropriate cropping systems

Mechanical – Contour bunding, Graded bunding, Bench terracing, Contour trenching, Gully control, Vegetative barriers

Forestry – Perennial trees and grasses

***In-situ* Moisture Conservation Techniques**

Storage of rainfall in soil at the place where it falls is termed as *"in-situ"* soil moisture conservation. It aims at increasing infiltration of rainfall into the soil and reducing runoff loss of rainwater.

In-situ soil moisture conservation can be accomplished through.

- Cultural /agronomic methods
- Mechanical methods
- Agrostological / biological methods

Cultural/ agronomical	Mechanical	Agrostological
Addition of organic matter Summer ploughing mulching cultivation strip cropping	Basin listing, Bunding, Ridges and furrows, Tie ridging, Random tie ridging, Broad bed furrow, Dead furrow and Furrows after crop establishment	Pasture, Strip cropping with grasses, Ley farming and Vegetative barriers

Contour farming: Ploughing along the contour and sowing reduce soil erosion and reduce runoff.

Cover crops: Erosion reduced by fully covering the land surface with foliage

(Eg.) Black gram, green gram, groundnut, fodder grasses like *Cenchrus ciliaris*, *Cenchrus glaucus*, Dinanath grass, marvel grass. Both contour cropping and cover cropping can be practiced when the slope is less than 2 per cent.

Strip cropping: Strip intercropping involves erosion resistant crops and erosion permitting crops in alternate strips of 2-3 m width across slope and along the contour. Erosion resistant crops include grasses and legumes with rapid canopy development. (e.g) *Cenchrus glaucus* + *Stylosanthes hamata*

2. Mechanical methods of soil moisture conservation

i) Basin listing: Formation of small depressions (basins) of 10-15 cm depth and 10-15 cm width at regular intervals using an implement called basinlister.

ii) Bunding: Formation of narrow based or broad based bunds across slope at suitable intervals depending on slope of field. The bunds check the free flow of runoff water, impound the rainwater in the inter-bund space, increase its infiltration and improve soil moisture storage.

Contour bunding

- Bunds of 1 m basal width, 0.5 m top width and 0.5 m height are formed along the contour. The distance between two contour bunds depends on slope.
- The inter-bund surface is leveled and used for cropping.
- It is suitable for deep red soils with slope less than 1 %.
- It is not suitable for heavy black soils with low infiltration where bunds tend to develop cracks on drying.

Graded/ field bunding

- Bunds of 30-45 cm basal width, and 15-20 cm height are formed across slope at suitable intervals of 20-30 m depending on slope.
- The inter-bund area is leveled and cropped.
- It is suitable for medium deep to deep red soils with slopes upto 1%.
- It is not suitable for black soils due to susceptibility to cracking and breaching

Compartmental bunding

Small bunds of 15 cm width and 15 cm height are formed in both directions (along and across slope) to divide the field into small basins or compartments of 40 sq m size (8 x 5 m). It is suitable for red soils and black soils with a slope of 0.5 to 1%.

Ridges and furrows

Furrows of 30-45 cm width and 15-20 cm height are formed across slope. The furrows guide runoff water safely when rainfall intensity is high and avoid water stagnation. They collect and store water when rainfall intensity is less. It is suitable for medium deep to deep black soils and deep red soils.

Tie ridging: It is a modification of the above system of ridges and furrows where in the ridges are connected or tied by a small bund at 2-3 m interval along the furrows.

Broad bed furrow (BBF)

Here beds of 1.5 m width, 15 cm height and convenient length are formed, separated by furrows of 30 cm width and 15 cm depth. Crops are sown on the beds at required intervals. It is suitable for heavy black soils and deep red soils. The furrows have a gradient of 0.6%.

Dead furrow

At the time of sowing or immediately after sowing, deep furrows of 20 cm depth are formed at intervals of 6-8 rows of crops. No crop is raised in the furrow. Sowing and furrowing are done across slope. It can be done with wooden plough in both black and red soils.

Agrostological methods

The use of grasses to control soil erosion, reduce run off and improve soil moisture storage constitutes the agrostological method.

Ley farming: It is the practice of growing fodder grasses and legumes and annual crops in rotation.

Vegetative barriers

Vegetative barrier consists of one or two rows of perennial grasses established at suitable interval across the slope and along the contour. It serves as a block to free runoff and soil transport. Vettiver, Cenchrus etc are suitable grasses for this purpose.

Farm ponds

Farm ponds are small storage structures constructed at the lowest point of a farm to collect and store runoff water.

Mulching: Mulching means covering the soil surface with any material such as organic wastes, plastic, polythene, etc. The organic wastes used for mulching include crop stubbles, straw, coir pith, groundnut shell, husk etc. These wastes at 5-10 t ha^{-1} are spread on the soil surface to a thickness of 5-10 cm.

Dust mulch: When surface soil is stirred by tillage, the continuity of capillary pores is broken and the rise of water through capillary movement is obstructed. This process is called dust mulching.

Vertical mulching: It is a technique where in trenches of 40 cm wide, 15 cm deep are dug at 2-4 m interval across slope and filled with stubbles or organic wastes to a height of 10 cm above soil surface. Runoff is checked, collected in the shallow trenches and redistributed to adjoining soil layers.

Live mulching: It is the term used to describe the covering of soil surface through the plant canopy in intercropping system. (Eg) sorghum + forage cowpea, sorghum + sword bean

Stover mulch or straw mulch: It refers to covering the soil surface with cumbu / sorghum straw, sugarcane trash reduces the evaporation and increases soil moisture efficiency.

Stubble mulch: It is referred to the stirring of the soil with implements that leave considerable part of the vegetative material or crop residues or vegetative litter on the surface as a protection against erosion and for conserving moisture by favouring infiltration and reducing evaporation.

Pebble mulch: Small pebbles like stone are placed on the soil surface. This mulching will be successful in dry land fruit tree culture.

Shelter belt: It is a practice of growing one of multi rows of trees / shrubs or crop plants across the wind direction either in the field or field boundaries to reduce the wind effect and to reduce the wind velocity.

Wind break: It is also a form of shelter belt, but only one row of tall trees having good leaf canopy are grown in North - South direction in order to reduce wind velocity and there by reduce soil erosion. Tall trees like eucalyptus, casuarinas and wood apple are grown as wind breaks.

Antitranspirants: These are substances or chemicals applied on plant-foliage to control rate of transpiration.

a) Stomatal closing type: Eg. Phenyl mercuric acetate (PMA), alkanyl succinic acid (ASA).

b) Film forming type: Eg. Paraffin and wax emulsions, folic 2% and power oil 1 %.

c) Reflectant type: Eg. Kaolin (3-5%) and lime solution

d) Growth retardant type: Chemicals like cycocel (ccc-chloro choline chloride, chlor mequat)

Alley cropping: This practice refers to raising perennial shrubs or tall crops as hedge rows up to 1-2 m height at 48 m intervals and raising short statured annual crops in the alleys (inter space between hedge rows). Eg. Leucaena / Agathi + Blackgram / cowpea / sunflower / groundnut

Establishment of optimum population

It depends on

- Seed treatment
- Sowing at optimum soil moisture

- Time of sowing
- Depth of sowing
- Method of sowing
- Crop geometry

Seed treatment

Seed treatment is done for many purposes such as protection against pests and diseases, inoculation of bio-fertilizers and inducing drought tolerance.

Seed hardening

It is the process of soaking seeds in chemical solution and drying to induce tolerance to drought. Seeds are subjected to partial hydration followed by dehydration before sowing.It is done to induce drought tolerance in emerging seedlings.

Seed hardening in various crops

Crop	Chemical	Concentration	Soaking time	Volume of solution per kg seed
Rice	Potassium chloride	1 %	Water-10hrs chemical-10hrs	1 litre
Sorghum	Potassium di-hydrogen phosphate	2 %	6 hrs	350ml
	Potassium chloride	1 %	5 hrs	1 litre
Pearl millet	Potassium chloride	2 %	16 hrs	1 litre
	Sodium chloride	3%		
Ragi	Calcium chloride	0.5%	Until visibility of embryo growth	1 litre
Sunflower	Zinc sulphate	2 %	12 hrs	1 litre
Cotton	CCC	1000 ppm	6 hrs	1 litre
	KCl	2%	5 hrs	1.6 litres
	DAP	2%	5 hrs	1.6 litres

Crop geometry

Crop geometry refers to the shape of land occupied by individual plants as decided by spacing between rows and between plants.

Contingency crop planning and midterm corrections for different aberrant weather situations

Rainfall abnormality	Contingency plan and midterm correction
1. Delayed onset of rainfall	
a) Delay exceeding3-4 weeks	Alternate crops of short duration to be sown
Delay in South west monsoon Normal - June Delay - July Delay - August Delay in North East monsoon	 Groundnut Ragi / pearl millet Samai / Cowpea
Normal - October Delay - Early November Delay - Late November	Cotton / Sorghum Sunflower / Pearl millet / Ragi Coriander / Senna
b) Delay of 1-2 weeks	Alternate varieties of short duration of same crop Eg. Sorghum Co 19 (150 days) Co 25 (110 days) Red gram VBN3 (130 days) Co 5 (130 days)
2. Early withdrawal of rainfall	Antitranspirant spray, harvesting for fodder (millets) and harvesting at physiological maturity
3. Intermediary dry spell a. Immediately after sowing	Gap filling with subsequent rains if stand reduction is less than 20%. Re-sowing if stand reduction is more than 20%, mulching between crop rows. Stirring soil surface to create dust mulch to reduce evaporation
b. At vegetative phase	Mulching, antitranspirant spray, spraying potassium chloride and thinning of 33-50% population
c. At flowering	Antitranspirant spray, harvesting for fodder and ratooning with subsequent rains in millets (e.g) sorghum
d. At ripening	Antitranspirant spray, harvesting for fodder and harvesting at physiological maturity

Dry land horticulture

Fruit trees with drought tolerance potential can substitute annual crops in many dry land tracts.

Rainfall (mm)	Fruit trees suitable
560 – 700	Ber, pomegranate, cashew sapota, pomegranate, jamun, amla
700-900	Mango, cashew, custard apple, guava, fig,

Soil type	Fruit trees suitable
Black soils	Ber, sapota, pomegranate, jamun, amla, wood apple
Red soils	Mango, cashew, custard apple (Annona), pomegranate, sapota, amla.

Pure horticulture: Plantations of mango, cashew, guava etc.

Agri horticulture: Annual crops intercropped in between fruit trees. Eg: Mango + Groundnut / samai / horsegram, ber + cowpea / green gram

Hortipasture: growing pasture grasses and legumes between fruit trees. Eg: Ber / guava + *Cenchrus cilliaris* + *Stylosanthes*

Integrated Farming Systems in Drylands

Integrated farming system (IFS) refers to the adoption of allied agricultural enterprises along with crop production in a mutually beneficial manner in the same farm holding. Eg. Crop + sheep / goat, crop / sericulture, Crop + poultry, crop / tree + forage + livestock.

Integrated dryland technologies (IDLT)

- *In-situ* soil moisture conservation
- Choice of suitable crops and crop substitution
- Selection of high yielding drought tolerant varieties
- Cropping system to suit rainfall quantity, duration of rainy season and soil moisture storage
- Tillage to conserve moisture
- Establishment of optimum population
- Soil fertility management
- Crop protection against weeds, pests and diseases.

Alternate Land Use System in Drylands

The choice of an alternate land use system depends on the land capability. Most of the lands under dry farming tracts come under the land, capability classes of III and above.

Land capability class	Alternate land use recommended
Class II	Dry land horticulture
Class III & IV	Agro-forestry / leyfarming
Class V	Pastures / silvipasture / tree farming
Class VI	Range lands / wood lots

Watershed: It is defined as a natural hydrological unit that covers a specific land surface area from which runoff passes through a common outlet.

Watershed management: It is the integration of technology within the natural boundaries of a drainage area for optimum development of land, water and plant resources to meet the basic minimum needs of the people in a sustained manner.

Objectives of watershed management

P - Production of food -fodder - fuel - fruit - fibres - fish and milk combine on a sustained basis

Pollution control

Prevention of floods

O - Over-exploitation of resources to be minimized by controlling excessive biotic interference like overgrazing

Operational practicability of all on-farm operations and follow- up programmes including easy approachability to different locations in watershed

W - Water storage at convenient locations for different purposes

Wild animal and indigenous plant life conservation at selected places

E - Erosion control

Eco - system safety, Economic stability and Employment generation

R - Recharge of ground water

Reduction of drought hazards, Reduction of siltation in multi - purpose reservoirs and Recreation

Components of watershed approach

1. Water -resource improvement
2. Soil and moisture conservation in cultivated lands

Hardware	Software
Permanent / Semi-permanent	Temporary/Recurring
Contour builds	Compartmental bunding
Bench terracing	Ridging after crop establishment
Conservation ditches	Contour cultivation
Land leveling	Mulching
Runoff collection structures	Vegetative barriers

3. Land treatment in non-arable lands
4. Improved cropping
5. Alternate land use system and integration of livestock in farming system.

Aims of watershed development

1. Increased land productivity through improved technology
2. Sustaining the resource base thorough improved conservation measures
3. Augmentation of resource base *viz* soil productivity and water availability

AGRO-FORESTRY

Agro-forestry : It is a production technique that combines agriculture and forestry, horticulture or animal husbandry on the same piece of land in order to fully utilize the natural resources.

Functions of Agro-forestry

i. *Productive functions :* The five 'Fs' such as fuel wood, food, fodder, feed, fibre and fertilizer are the most important besides timber and Non Timber Forest Produce.

ii. *Protective functions :* It includes shade, reduction in wind speed, control of erosion and maintenance and improvement of soil fertility, recreation, carbon sequestration and climate change.

Processes of soil fertility improvement

Addition Processes

- Maintenance of soil organic matter, through carbon fixation in photosynthesis and its transfer via decay of litter and roots

- ❖ Nitrogen fixation by many leguminous and a few non-leguminous trees like casuarina
- ❖ Nutrient pumping and recycling by trees
- ❖ Increased water infiltration through better soil physical properties and effects of roots

Reduction Processes

- ❖ Protection from erosion and thereby from losses of organic matter and nutrients
- ❖ Reduction in the rate of organic matter decomposition by shading and mulching
- ❖ Reduction of water loss from evapotranspiration by shade and litter

Alternate Processes

- ❖ Maintenance of soil properties through organic matter and effects of roots
- ❖ Penetration of compact soil layers by roots
- ❖ Modification of extremes of soil temperature, by shade and litter

Chemical Processes

- ❖ Reduction of acidity through bases in litter
- ❖ Reduction of salinity and sodicity, by trees in association with other management measures
- ❖ Reduction of soil toxicities caused by pollution

Biological processes and effects

- ❖ Production of leaf litter of high quality, containing a balanced nutrient supply and its transfer to the soil by litter decomposition
- ❖ Improved activity of soil fauna
- ❖ Mycorrhizal associations for increased phosphorus and zinc availability
- ❖ Root nodulation and rhizosphere effects
- ❖ Release of plant growth-promoting substances by the rhizosphere

Types of agro forestry systems

1. ***Agrisilviculture system :*** In this system agricultural crops are intercropped with tree crops in the interspaces between the trees. Under this system agricultural crops can be grown up to 2-3 years.
2. ***Silvipasture system :*** Trees are planted in the poor, degraded soils and fodder grass and legume mixtures are being raised in the interspaces of trees.
3. ***Agrisilvipastoral system :*** In this system forest trees are grown along with fruit trees and in the interspaces grass and legume mixtures are raised.

Agro-forestry systems

Agrisilviculture	:	Agricultural + forest trees
Silvipastoral	:	Forest trees + Grasses
Agrisilvipastoral	:	Agricultural crops + Forest trees + Grasses
Agrihorti system	:	Agricultural crops + Fruit cops
Hortisilvicultural	:	Fruit crops + Forest trees
Agrihortisilviculture	:	Agricultural crops + Horticultural crops + Forest crops
Aquasilvicultural	:	Fish + Forest trees
Agrisilviaquacultural	:	Agricultural Crops + Forest trees + Fish

Characteristics of trees suited for agro forestry system

The properties, which are likely to make a woody perennial suitable for agro forestry system are

- Should be a multipurpose tree with five functions
- Preferably a nitrogen fixing tree
- A thick and dense crown in conical shape
- A well tailored root system with deep roots, dense fine roots and abundant mycorrhizal association.
- Resistance to severe pruning and high leaf biomass production.
- A reasonable balanced nutrient content in the foliage.
- Rapid and faster litter decomposition to enhance nutrient release.
- Absence of toxic substances in the litter or root residues.
- Resistance to pests and diseases.
- Tolerance to protracted moisture stress.
- Low invasiveness.
- Satisfy Productive functions or service functions other than soil improvement.

Aqua forestry : Cultivating fish and prawn in saline water and growing coconut and other trees on bunds of ponds

- Leguminous trees
- *Glyricidia sepium*
- *Leucaena leucocephala*
- *Moringa oleifera*

Suitable tree species for agro forestry

I. Fodder - Cum Fuel wood species : *Albizzia spp, Ailanthus spp, Leucaena leucocephala* and *Sesbania spp*

II. Softwood and pulpwood species : *Ailanthus exelsa, Bombux ceiba, Anthocephalus cadamba, Bamboo spp, Populus deltoides*

Fuel wood and Timber species

Acacia nilotica,
Albizzia lebbeck,
Albizzia. Procera,
Pongamia pinnata,
Melia azadirachta,
Thespesia purpurea,
Casuarina equisetifolia,
Dalbergia sissoo,
Tectona grandis

Multipurpose trees and shrubs:

Bahunia variegate,
Erythrina variegata,
Leucaena leucocephala,
Sesbania grandiflora,
Pongamia Pinnata,
Samanea saman,
Moringa oleifera,
Pithecellobium dulce

Traditional Agro forestry systems in India

- Shifting cultivation/ slash and burn system
  - Traditional follow period : 10-20 Years
  - Recent times follow period : 2-5 Years
- Taungya cultivation (or) Hill cultivation
  - Taungya system practised in Kerala , West Bengal and Uttar Pradesh

Bio diversity conservation : According to the government of India report the biological diversity estimate as follows

Plant species - 45,000

Animal species - 8,10,000

It representing 7% world fauna

It representing 6.5% world flora

Nature of Interactions

- ❖ **Commensalistic :** Positive effect on species one and no effect on species two.
- ❖ **Amensalistic :** Negative effect on species one species and no effect on species two.
- ❖ **Monoplastic :** (Predatory (or) parasitic): Positive effect on species one and negative effect on species two.
- ❖ **Inhibitory :** Negative effect on both species.
- ❖ **Synergisitic :** Positive effect on both species.

Negative interactions

- ❖ Allelopathy : Refers to inhibition of growth of one plant by chemical compounds that are released into the soil from the neighbouring plants.
- ❖ Allelochemicals present in all the plant parts.

Example

Tree species	Effect on
Casuarina	Cowpea, sorghum, sunflower
Eucalyptus	Cowpea, sorghum, potato, sunflower
Glyricidia	Maize, Rice, Tropical grasses.

Naturally occurring toxicants in feed stuffs :

Feed stuff	Toxicants
Unripe Mango	Amylase
Unripe banana	Amylase
Senna leaves	Anthraquinone
Subabul	Mimosine (or) leucenine
Prosopis, Glyricidia, Luecana	Tannin

Soil conservation (or) amelioration:

Suitable grass species for stabilizing ravines and gullies

- *Dicanthium annulatum*
- *Cynodon dactylon*
- *Senima narvosum*
- *Bothriachroa perpusa*

Shelter belt :

- *Crotalaria burhia*
- *Leptadenia pyrotechnica*
- *Aerva pesudomentosa*

- ❖ National forest policy was started in1988.
- ❖ National forest policy has set a goal that 33% of the country's geographical area should be under forest and tree cover.
- ❖ Forest survey of India (2001), Forest cover in the country - 675,538 km^2 and constituting total geographical area is 20.55%
- ❖ Dense forest occupies an area of 12.68%
- ❖ Open forest occupies an area of 7.87%
- ❖ ICRAF - International centre for Research in Agro forestry (or) World Agro Forestry Centre Nairobi, (Kenya)
- ❖ National Research Centre for Agro forestry was established at Jhansi in 1988

Demographic pressure

- Population density is 32.4/Sq.km
- Maximum of 904/Sq.km in West Bengal
- Livestock density is 109 Sq/km

Terminologies

- ❖ Cultivation of crops in areas receiving rainfall above 750 mm is known as dry land farming.
- ❖ Cultivation of crops in areas where rainfall is less than 750 mm per annum is called dry farming.
- ❖ Duty is the volume of irrigation water required to mature a crop. It is the ratio of irrigated area to the quantity of water used (ha/l).
- ❖ In low rainfall regions (<750 mm / annum) mono-cropping is practiced, where as in high rainfall regions (>750 mm / annum) inter cropping is practiced.
- ❖ Mixed farming is a system of farming on a particular farm which includes crop production, raising livestock, poultry, fisheries, bee keeping etc., to sustain and satisfy as many as needs of the farmer as possible.
- ❖ Nipping: Act of removal of ear heads from plant stalk.
- ❖ Ratio of total cropped area in different seasons to the total land area is known as cropping intensity.
- ❖ Rationing is a practice of taking the second crop from previous one is known as rationing.
- ❖ The term "Green revolution" was coined by William Gaude.

1.8. CONSERVATION AGRICULTURE

Conservation Agriculture (CA) is a concept for resource-saving agricultural crop production that strives to achieve acceptable profits together with high and sustained production levels while concurrently conserving the environment.

Principles of Conservation Agriculture

1. **Minimum tillage / no tillage :** CA is practicing minimum mechanical soil disturbance which is essential to maintaining minerals within the soil, stopping erosion, and preventing water loss from occurring within the soil.
2. **Residue retention :** The principle of managing the top soil to create a permanent organic soil cover can allow for growth of organisms within the soil structure. This growth will break down the mulch that is left on the soil surface.

3. **Crop rotation :** Rotational crops will act as a natural insecticide and herbicide against specific crops. Not allowing insects or weeds to establish a pattern within fields will help to eliminate problems with yield reduction and infestations within fields. Crop rotation can also help build up a soils infrastructure.

Components of Conservation Agriculture

1. **Direct Seeding or Planting :** Direct seeding involves growing crops without mechanical seedbed preparation and with minimal soil disturbance since the harvest of the previous crop.
2. **Residue management :** A permanent soil cover is important to protect the soil against the deleterious effects of exposure to rain and sun; to provide the micro and macro organisms in the soil with a constant supply of food and alter the microclimate in the soil for optimal growth and development of soil organisms, including plant roots.
3. **Crop Rotations :** The rotation of crops is not only necessary to offer a diverse diet to the soil micro organisms, but as they root at different soil depths, they are capable of exploring different soil layers for nutrients. Nutrients that have been leached to deeper layers and that are no longer available for the commercial crop can be recycled by the crops in rotation.
4. **Weed Management :** The general practice in reduced tillage systems is to substitute herbicides, mowing, or flaming for cultivation to kill the existing vegetation before the crop is planted, but few options exist for weeds that germinate after the crop is planted. Effective materials for pre-emergence weed control are currently available for all the commercial crops.

MECHANISATION IN AGRICULTURE

Farm Mechanisation: It is the process of using tools, implements and machines for agricultural land development, crop production, harvesting, preparation for storage, storage and on-farm processing. (or)

Hand tool technology is the simplest and most basic level of agricultural mechanization: the use of tools and simple implements using human muscle as the main power source.

Draught animal technology refers to implements and machines utilizing animal muscle as the main power source.

Mechanical power technology is the highest technology level in agricultural mechanization. It embraces all agricultural machinery which obtains its main power from other sources other than muscular power

Need of mechanization

Agricultural labour paucity that cripples the agriculture operations, increase the cost of cultivation and hiders the timely operations that is most essential for agriculture. Current farm power availability hovers around 1.7 kw/ha which is much lower than that of Korea (7+ kw/ha), Japan (14+kw/ha) and the USA (6+kw/ha).

The basic requirements for mechanization in irrigated and rainfed situations are

- Leveled field to an area of minimum of half acre to ensure uniform depth of sowing followed by crop establishment.
- Tropiculator / Seed drill cum fertilizer drill is multipurpose machinery used for simultaneous operations such as broad bed formation, sowing on bed and fertilizer applications
- Power weeder ensures quick weeding using available moisture

1. Tillage and Land shaping implements

i. Improved iron plough

ii. Chisel plough

iii. Sub soil coirpith applicator
iv. Terracer cum leveler
v. Multipurpose tool carrier
vi. Power tiller drawn basin lister

2. Sowing implements and Transplantor

i. Seed / Fertilizer broadcasting unit
ii. Improved gorru
iii. Bullock drawn seed planter
iv. Power tiller operated cup feed seed planter
v. Tractor drawn seed planter
vi. broad bed former cum seeder
vii. cultivator cum seeder
viii. Ridger seeder
ix. Rice transplanter

3. Weeding equipments

i. Cono weeder in rice
ii. Power weeders for different spacing

4. Harvesting Equipments

i. Groundnut harvester, Groundnut Sheller and decorticor
ii. Fodder sorghum harvester
iii. Hand cum pedal operated chaff cutter
iv. Power tiller operated auger digger
v. Combined harvesters

Tropicultor: Tropicultor is developed by ICRISAT and used for the agricultural operations like primary and off season tillage, cultivator operation, blad harrowing, sowing, fertilizer and seed covering operation, inter-row cultivation and harvesting.

Advantages of Tropicultor are different seed plate's available for sowing of different crops, easy fertilizer application by changing arm lever, seed and fertilizer placement in moisture zone, sowing of pure crop and intercrop and intercultural operations cost is low, no compaction in the cropped area, formation of ridge and furrow, Broad Bed and Furrow and can be used for harvesting of turmeric, Potato, carrot and groundnut.

1.9. SALIENT INFORMATIONS

Special purpose crops

Names	Crops
Truck crops	Potato, Onion
Cover crops	Groundnut, Methi
Nurse crops	Sunhemp in Sugarcane
Silage crops	Maize, Sorghum
Catch crops	Sunflower, Radish, Semi-Rabi Sesame
Soiling crop	Berseem
Pasture crops	Zinzvo
Row crop	Cotton
Support crops	Castor, Shevri
Mixed crops	Cowpea, Tur
Companion crop	Maize in Cowpea
Intercrop	Jowar in fruit crop
Green manure crops	Sunnhemp, Cluster bean
Trap crops	French marigold around Tobacco, Castor around Sugarcane
Cash crops	Cotton, Sugarcane, Castor

Agro ecological regions		Agro climatic zones	
ICAR	8	Planning commission	15
NBSS & LUP	20	NARP of ICAR	131
		Classification of Soils by ICAR	8

Terms and associated crops

Term	Crops
Curing	Tobacco, Tea
Stripping	Jute
Nipping	Cotton
Wrapping	Sugarcane
Propping	Banana
Trashing	Sugarcane

Oil percentages in selected crops

Sesamum	46-52%
Groundnut	44-50%
Castor	35-58%
Safflower	24-36%
Niger	37-43%
Rapeseed and Mustard	33%
Soybean	20%
Coconut	60%

Important varieties of selected crops

Variety	Remarks
Rice	
GEB 24, BCP 1, NP 130	Superior quality rice variety
Basmati 370, Type 3	Aromatic
Pusa RH-10	First Super fine aromatic basmati hybrid
TKM, Pattambi	Disease pest resistance
SR 26 B	Saline tolerant
FR 13 A/43 B, Madhukar	Flood
Jalmagna, Jaisuria	Deep water ecology
Jaya	High yield
Vikramarya	Tungro-virus resistant
Ajaya	Bacterial blight resistant
Phalguna	Gall midge resistant
Pusa Basmati 1	Worlds first high yield dwarf export quality aromatic rice
IR-8	First dwarf rice variety introduced in India
Lunishree, Yamini, Krishna, Hansa, Vasumati, Pusa Sugandh-2, Pusa Sugandh-3	Worlds first super rice variety for saline/alkaline conditions
Wheat	
Lermoroja, Sonora-64	Direct introductions (imported)
Kalian sone, Sonalika	Selections
HD 2285, WH 542, UP 2338	North west plains
K 8840, K 9006, HP 1633	N-E Plains
HI 1077, GW 190	Central Zone
Sujata, C 306, K 8027, Malviya, Wheat-510, HD 2733, HW 2045	Rainfed region
Pigeon pea (Red Gram)	
ICPH-8	First hybrid in the world
Mustard	
Pusa Jai Kisan, Kalyani, Pusa Bold, Varuna	First somaclonal selection
Sugarcane	
CoC 671, Gandak, Rajbhog	Wonder Cane: Recovery=10%
Cotton	
Sujata	First Egyptian variety

Optimum temperature and water requirement of crops

Crop	Optimum Temperature (°C)	Water Requirement (mm)
Rice	30-32	900-2500
Wheat	25	450-650
Tobacco	28	500
Sugarcane	32-37	1500-2500
Sorghum	32-35	450-650
Maize	32-37	500-800
Cotton	20-30	500-700
groundnut	20-25	750-850

Critical growth stages of cereals

Rice	1	Seedling stage	
	2	Maximum tillering stage	
	3	Panicle initiation stage	
	4	Flowering stage	
	5	Soft dough or milky stage	
Wheat	1	Crown root initiation stage	
	2	Tillering stage	
	3	Joining stage	
	4	Boot leaf stage	
	5	Flowering stage	
	6	Milky and dough stage	
Maize	1	Seedling stage	2 weeks after sowing
	2	Tasselling stage	8 weeks after sowing
	3	Silking stage	10 weeks after sowing
	4	Dough stage	12 weeks after sowing
Sorghum	1	Seedling stage	2-4 weeks after sowing
	2	Pre-flowering stage	12-14 weeks after sowing
	3	Flowering stage	12-16 weeks after sowing
	4	Grain formation stage	17 weeks after sowing
Barley	1	Tillering stage	
	2	Flowering stage	
Pulses	1	Early growth stage	
	2	Post flowering stage or pod filling stage	
Mustard & Rapeseed	1	Pre-flowering stage	
	2	Pod filling stage	
Sunflower	1	Buttoning stage	
	2	Grain filling stage	
Safflower	1	Branching stage	65 days after sowing
	2	Flowering stage	105 days after sowing
Cotton	1	Pre-flowering stage	
	2	Flowering stage	
	3	Fruit setting stage	

Criteria for harvesting and storage of important crops

Crop	Criteria for harvesting	Moisture content for safe storage of grains
Rice	Moisture content of grains < 20%	14
Wheat	Moisture content of grains 15%	12
Sorghum	Moisture content of grains < 28%	12
Maize	Moisture content of grains < 22-25%	12
Groundnut	-	6
Sugarcane	Brix 18-20%	-
Cotton	Bolls fully opened	-

Agro climatic zones

Zone	India	Zone	Tamil Nadu
1	Western Himalayan Zone	1	North eastern
2	Eastern Himalayan Zone	2	North western
3	Lower Gangetic Plains	3	Western
4	Middle Gangetic Plains	4	Cauvery Delta
5	Upper Gangetic Plains	5	Southern zone
6	Trans-Gangetic Plains	6	High rainfall zone
7	Eastern Plateau and Hills	7	Hilly zone
8	Central Plateau and Hills		
9	Western Plateau and Hills		
10	Southern Plateau and Hills		
11	Eastern Coast Plains and Hills		
12	West Coast Plains and Hills		
13	Gujarat Plains and Hills		
14	Western Dry Regions		
15	Islands Zone		

Agro-climatic zones Vs Agro-ecological regions

Agro-climatic zones	Agro-ecological regions
It is a land unit in terms of major climates suitable for a certain crops and cultivators	It is the land unit cut out of agro climatic region when superimposed on land form and the kind of soils and soil condition that act as modifiers to climate and length of growing period

Crop rotation Vs Intensive Cropping

Sl. No	Crop Rotation	Intensive Cropping
1	Raising different crops one after another i.e. succeeding crop is a different species from the previous crop	Same or different crops are raised sequence or simultaneously
2	Intensification over time only	Intensification over time or space
3	A cycle may last for more than one year	A cycle covers only one year
4	Fallowing may be part of the rotation	Fallowing is avoided or kept to minimum duration
5	Soil fertility maintenance and soil health care are the primary objectives	Maximization of yield per unit area per unit time is the primary aim
6	Resource requirement is normal	Resource requirement increases

Dry farming Vs Rain-fed farming

Dry Farming	Rain-fed Farming
Rainfall < 750 mm	Rainfall > 750 mm
Rainfall distribution is erratic	Rainfall is well distributed
PET exceeds rainfall for most part of the year	Rainfall is equal to (or) more than PET for long period
Arid and Semiarid Climate	Humid / Sub humid and temperate climate
Crops are subjected to severe moisture stress	Cops face little (or) no moisture stress
Soil moisture conservation and safe disposal of excess water are important	Run off and erosion control is of primary importance

Protein Content

Crops	Percentage
Cereals	
Rice	6-7
Wheat	11-12
Maize	10
Sorghum	10-12
Bajra	11-12
Barley	11.5
Pulses	
Gram	21.1
Lentil	25
Pea	22.5
Arhar	21-25
Mung	25

Urad	24
Cowpea	23.4
Soybean	42
Oilseeds	
Groundnut	26
Sesame	120-36
Linseed	36
Safflower	40-45% (Oil cakes)

$$\text{Agronomic efficiency (AE)} = \frac{\text{(Yield from fertilized plot in kg)-(Yield without fertilizer in kg)}}{\text{Amount of fertilizer nutrient applied in kg}}$$

Calibration of spray fluid

$$Q = \frac{V1 - V2}{a} \times A$$

Where, Q - Quantity of water required for given area to be sprayed (litres)

V1 - Quantity of water taken initially

V2 - Quantity of water remained after spraying

A - Area to be sprayed (m^2)

a- measured area sprayed (m^2)

Calculation of herbicide product

$$\text{Herbicide product (kg)} = \frac{100 \times R \times A}{P}$$

R- Rate of application in kg active ingredient (ai) per ha

A- Area to be applied in hectares

P - Percent active ingredient in the commercial product

$$\textbf{Crop Water Use Efficiency } (E_{cu}) = \frac{Y}{ET}$$

$$\textbf{Field Water Use Efficiency } (F_u) = \frac{Y}{WR}$$

Where, Y- Yield, ET - Evapotranspiration and WR - Total amount of water used in the field

Commercial Cane Sugar Percent (CCS) = CCS (%) = (10.5 x S) - (0.3 x B)

Where, S - Sucrose percent, B - Brix percent

$$\text{Cropping Intensity} = \frac{\text{Total Cropped Area}}{\text{Net Cultivated Area}} \times 100$$

$$\text{Ginning Percentage} = \frac{\text{Weight of lint (Cotton fibre)}}{\text{Weight of seed cotton taken for ginning}} \times 100$$

$$\text{Harvest index} = \frac{\text{Economic Yield}}{\text{Biological Yield}}$$

Irrigation requirement (IR) = Water requirement - (Effective Rainfall + Soil Moisture Contribution)

$$\text{Irrigation Efficiency} = \frac{\text{Water Consumed by crops}}{\text{Total water applied to the field}} \times 100$$

$$\text{Seed Rate (kg/ha)} = \frac{\text{Area to be sown x Test weight of seed x 100 x 100}}{\text{1000 x 1000 x Germination \% x Purity \% x Spacing}} \times 100$$

$$\text{Shelling Percentage} = \frac{\text{Weight of grains}}{\text{Weight of pods or maize cobs}} \times 100$$

Points to remember

- Total geographical area of India is 328 Mha
- Net sown area India is 141 Mha
- Area under forest in India is 69.02 Mha (22.6%)
- Average annual rainfall of our country is 1194 mm.
- Percentage forest area to total geographical area more in Sikkim (81%) and Manipur (78%)
- National forest policy (1952) proposed that area under forests to be raised steadily to 33% of total area
- Barren and uncultivable lands to the reporting areas are higher in Manipur (64%) and Sikkim (24%)
- National land use policy -1988
- Wind erosion was predominant in Rajasthan and Jammu Kashmir
- Areas under water logging more in Madhya Pradesh, Kerala, Bihar and Andra Pradesh
- Irrigation schemes are classified as
  - Major - more than 10,000 ha
  - Medium - 2000-10,000 ha
  - Minor - 2000 ha
- Indian national water policy - 2002
- Water technology centers located in Bhuvaneshwar, New Delhi and Coimbatore
- Indian Institute of Legal Methodology- Kanke, Ranchi of Utharkand.
- Dr. Varghese Kurien is the pioneer of operation flood (Milk revolution) in India.

SOIL SCIENCE

2

P. KANNAN and M. THIRUNAVUKKARASU

Soil

Soil is a natural body of mineral and organic constituents differentiated into horizons usually unconsolidated, of variable depth which differs among themselves as well as from the underlying parent material in morphology, physical makeup, chemical properties and composition and biological characteristics (Whitney, 1982; Hilgard, 1892; Dokuchaiev, 1900 and Joffe, 1936).

According to Soil Science Society of America (SSSA, 1970) soil is the unconsolidated mineral matter on the surface of the earth that has been subjected to and influenced by genetic and environmental factors of parent material, climate (including moisture and temperature effects), macro and microorganisms and topography, all affecting over a period of time and producing a product, that is "SOIL" that differs from the material from which it is derived in physical, chemical, biological and morphological properties and characteristics. The unconsolidated mineral material on the immediate surface of the earth, that serves as a natural medium for the growth of plants.

Approaches of Soil Study

Pedological Approach : The origin of the soil, its classification and description are examined in Pedology. (From Greek word *pedon* means soil or earth). It is the study of soil as a natural body and does not focus on the soil's immediate practical use.

Edaphological Approach: Edaphology (from Greek word *edaphos* means soil or ground) is the study of soil from the stand point of higher plants. Edaphologists consider the various properties of soil in relation to plant production.

Soil Science : The science dealing with soil as a natural resource on the surface of the earth, including pedology (soil genesis, classification and mapping) and the physical, chemical and biological and fertility properties of soil and these properties in relation to their management for crop production.

1. ***Soil fertility :*** Inherent capacity of soil to supply the plant nutrients in adequate amount without any toxicity to the crop plants
2. ***Soil chemistry :*** Study of chemical constituents, chemical properties and the chemical reactions in the soil
3. ***Soil physics :*** Study of physical properties in relation to plant growth
4. ***Soil microbiology :*** It deals with micro organisms, its population, classification and its role in transformations of nutrients and plant growth
5. ***Soil conservation*** **:** Dealing with protection of soil against physical loss by erosion or against chemical deterioration i.e. excessive loss of nutrients either natural or artificial means.
6. ***Pedology*** **:** Study of soil genesis, survey and classification

Soil can be compared to various systems of human body

Digestive	-	matters decomposition
Respiratory	-	air circulation & exchange of gases
Circulatory	-	water movement within the soil system
Excretory	-	leaching out of excess salts
Brain	-	soil clay
Colour	-	soil colour
Height	-	soil depth

Components of soil (volume basis)

Pore space (50%)

Solid space (50%)

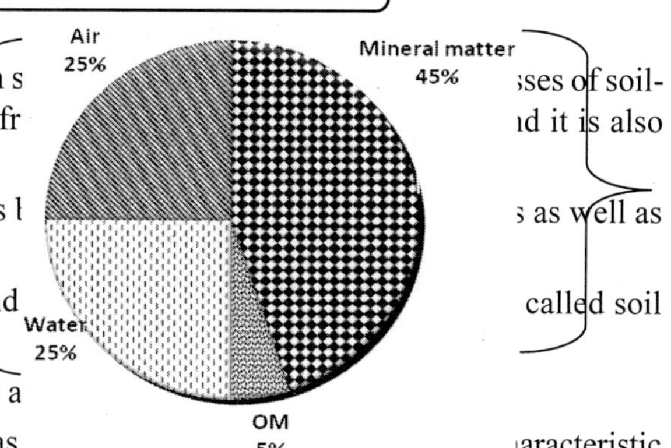

2.1. SOIL GENESIS AND CLASSIFICATION

Soil genesis : The sub-discipline of soil science that deals with s … sses of soil-forming factors responsible for the development of true soil fr … d it is also known as pedogenesis.

Soil classification : It is the systematic categorization of soils … s as well as choices of use.

Soil survey : It is the systematic examination, description and … called soil survey

Rocks : It is a naturally occurring solid aggregate of minerals a …

Minerals: It is a naturally occurring inorganic compound that has … aracteristic geomatric form

Rocks and Minerals

Sedimentary rocks

Shale	-52%	74 %
Sand stone	-15%	
Limestone& Dolomite	-7 %	

Igneous rocks

Granite -15 %
Basalt -3 %
Other rocks -8 %

18 %

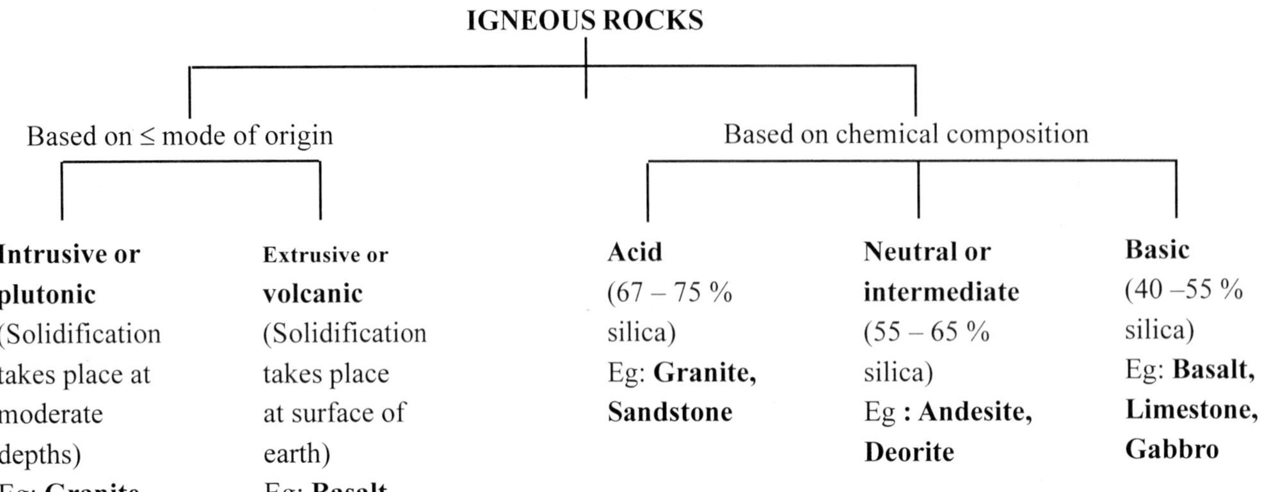

Sedimentary rocks

Mechanically formed rocks : Sand stone and Shale

Inorganically formed rocks : Halite and Gypsum are formed by evaporation

Precipitation and flocculation : Limestone and Dolomite

Organically formed rocks : Lignite and Anorthite

Igneous rocks contain

- Feldspar 60%
- Pyroxene and amphiboles 18%
- Quartz 12%
- Mica 4%
- Others 6%

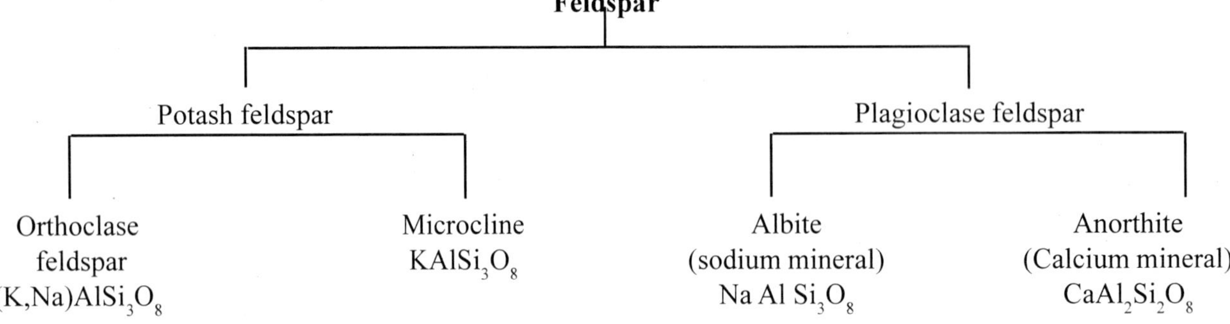

- Pyroxenes, olivine, biotite and amphiboles are called as *ferromagnesian minerals*. These groups are easily weatherable and main source of ***micro nutrients*** in soil.
- Non-ferromagnesian mineral groups are plagioclase feldspar, orthoclase feldspar and quartz

Pyroxene : ($R_2(Si_2O_6)$

- Augite - Ca $Mg(SiO_3)_2$
- Diopside - $Ca(Mg,Fe)(SiO_3)_2$
- Hypersthene - $(Fe,Mg)SiO_3$
- Enstatite - $MgSiO_3$

Amphiboles : ($RsiO_3$) - Hornblende -(K_2,Na,Mg,Ca,Mn)SiO_3

Olivines : Forsterite -(Mg_2SiO_4) and Fayalite -(Fe_2-SiO_4)

Micas :

- Biotite (black mica) - $H_2K(MgFe^{2+})_3(AlFe^{3+})(SiO_4)_3$
- Muscovite (white mica) - H-$_2$ K $Al_3(SiO_4)_3$

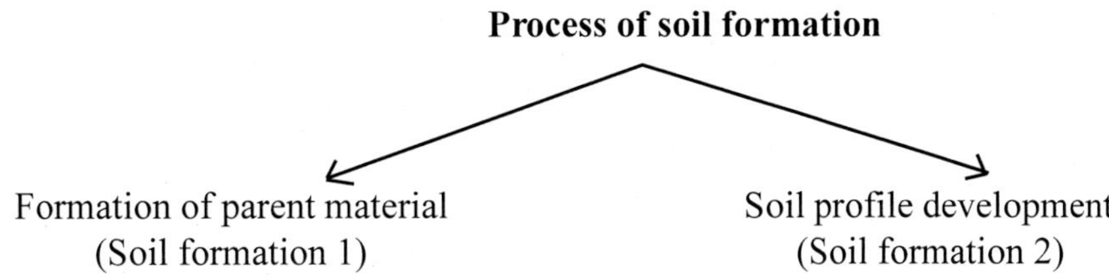

- Weathering is a geological process.
- Weathering is a destructive process.
- Soil formation is a constructive process.
- Rocks are mixture of minerals.
- Dark colored minerals (gabbros and basalt) are easily weatherable than light colored minerals (granite).
- Granite is an acidic rock.
- Basalt is a basic rock.
- Sedimentary rocks are formed from igneous rocks by deposition and recementation
- Primary minerals (Micas, quartz, feldspar, pyroxene, amphiboles) are present in coarse fraction of soil
- Secondary minerals (clay minerals serpentine) are present in fine particles of soil
- Clay minerals are synthesized from weathering of primary minerals
- Accessory minerals are apatite, pyrite and magnetite
- Chemical weathering is important one among the weathering type
- Allophane is the volcanic clay mineral structure less mineral
- Calcium carbonate is absent in acid soil
- Calcareous soils are deficient in Fe, Zn and B
- Calcium carbonate is abundant in calcareous soil.
- Live weight of microorganism is 4000 kg/ha and constitute about 0.01 –0.4% of total soil mass.
- The finer fraction of soil is separated by sedimentation process and coarser fraction is separated by sieving
- The original mineral component of rock is called as primary mineral
- Those primary mineral which form the chief constituents are called as essential mineral
- Those primary mineral occur in small quantities are called as accessory mineral.
- Feldspar comprises 60% by weight of igneous rock.

Deposition (sand stone)

Weathering

Granite Quartz sand

Recementation (shale)

Weathering

Weathering : It is the process of disintegration and decomposition of rocks and minerals.

Parent material : Unconsolidated and more or less chemically weathered mineral material from which soils are developed.

Different type of weathering and weathering agent

Physical	Chemical	Biological
Physical condition of rock	Hydration	Man and animals
Change in temperature	Hydrolysis	Highest plant & their roots.
Action of water, action of freezing and action of glaciers	Solution	Microbes
Action of wind & sand	Carbonation	
Atmospheric electric phenomenon	Oxidation & Reduction	

Example for hydration

- Limonite from Hematite
- Serpentine from Olivine
- Gypsum from calcium sulphate

Example for hydrolysis

- Kaolinite from Potash feldspar
- Kaolinite from Biotite

Example for biological weathering

The growth of plant roots and their exudates broken the rocks or the burrowing of animals make holes and gallaries and speed up the decomposition of rocks.

- ☞ Alluvial, glacial, lake deposits and Aeolian or loess deposits are called as unconsolidated sedimentary rocks.
- ☞ Sand stone contains 78.64% SiO_2.
- ☞ Limestone contains 42.74% CaO.
- ☞ Basic rocks weather more than acidic rocks.
- ☞ Basalt (easily weatherable) >Gneiss >Granite>Hornblende>Andesite
- ☞ Lime stone weather easily than sand stone.
- ☞ Plagioclase feldspar weather easily than orthoclase feldspar.
- ☞ Pyroxene is found in basic rocks.
- ☞ Pyroxene weather easily than amphiboles.
- ☞ The more easily weatherable minerals are mobile.

Weathering ability of minerals

- Very low – Quartz, zircon, rutile, Magnetite and Illmenite
- Low –Orthoclase, Albite, Muscovite and Tourmaline (Boron bearing mineral)
- Soluble – Calcite
- Very high – Anorthite, Olivine and Biotite

- When water freezes its volume increases by 9% and force exerted is150 ton/sq.foot
- Transporting power of moving water varies as the sixth power of its velocity.
- Soil evolved mainly as a result of physical weathering is skeletal soil.
- For chemical weathering water is essential.
- In arctic and desert region physical weathering dominates than others.
- Intensity of weathering increase from humid region to tropic region.
- Physical weathering is dominant in low rainfall area
- Chemical weathering is very fast in humid tropical area
- Hydration occurs in humid region.
- Formation of limonite (yellow colour) from hamatite (red colour) is example of hydration
- Hydration is a non-pedogenic process.
- In humid tropical climate the removal of silicic acid is more than in temperate region.
- Truncated profile is formed when erosion is active.
- Disturbed profile is formed in soil, which is under cultivation.
- Development of mineral profile is mainly due to the movement of water in soil.
- In mechanical eluviations finer particles are washed down results in light texture.
- Chemical eluviations lead to development of podsols.
- A_1 horizon is darker and more porous than A_2 horizon.
- B horizon is store house of calcium, magnesium, potassium, NH_3, manganese
- In truncated soil C-horizon is nearer to the surface.
- Topography influences the water relation in soil.
- Joffe identifies 2 kind of soil formation (i.e) Active and passive formation.
- Hilly and mountainous topography is called as macro topography. It is conducive for endodynamorphism.
- Surface accumulation of organic matter is called as duff or mor.
- Removal of silica and accumulation of sesquioxide is called as Laterisation.
- Removal of sesquioxide and accumulation of silica is called as Podzolisation.
- Sand and silt are called as skeleton of soil and clay is called active portion of soil.
- Light minerals having specific gravity <2.85 where as heavy minerals having >2.85 specific gravity
- Sulphur bearing minerals are Pyrites ($Fe S_2$ & $Cu S_2$)
- Carbonate bearing mineral is calcite
- Weathering sequence are given by Goldich (1938)

All the essential minerals are classified based on oxygen atoms shared between the silica tetrahedrons

- **Neso silicates** – None of the oxygen is shared-eg: Olivine, Garnet and Zircon
- **Sorosilicates** – One oxygen is shared -eg: Epidote, Hemimorphite
- **Inosilicates** – Single chain-Two oxygen is shared - eg:Pyroxene

  Double chain–2.5 oxygen is shared - eg: Amphiboles
- **Phyllosilicates** – Three oxygen is shared- eg: Biotite, Serpentine, clay minerals, Talc
- **Tecto silicates** – All oxygen are shared-eg: Feldspar &Quartz
- **Cyclo silicates** :e.g. Tourmaline and Benitotite

Types of soil formed in different climatic zones

- Podzol –Temperate humid
- Chernozem- Semi arid & Sub humid
- Laterite- Subtropics & Tropics
- Desert –Arid & Semi arid

- Surface soil is called as furrow slice layer (0 –15 cm depth)
- A horizon is called as **elluvial** horizon.
- B horizon is called as **illuvial** horizon.
- **Solum** is A+B horizon.
- **Regolith** is A+B+C horizon.
- Earth crust is upto 16 Km.
- **Lithophile** elements- which ionize or form stable oxyanions.
- **Chalcophile** elements –which tend to form covalent bond with sulphide.
- **Siderophile** elements –which readily form metallic bonds.
- **Atmosphile** elements –which tend to remain in atmospheric gases.
- **Biophile** elements –which tend to associated with living organisms.
- Eart crust contains mainly lithophile elements.
- Rocks are mixture of two or more minerals.
- Sedimentary rocks are known as stratified rocks

Soil survey methods

Type	Detailed soil survey	Reconnaissance soil survey
Base map	Cadastral map (1:4000) & (1: 8000)	Toposheet (1:15000)
Scale	8" =1 mile, 16"=1 mile	1" =1 mile
Mapping unit	Soil type & soil phase	Soil series & soil association
Intensity of observation	640 Acre (Profile sampling) 0.25 –0.5 km (auger sampling)	4 –6 km (profile sampling) 1 –2 km (Auger sampling)

Classification of soil

Order	Concept
Gellisols	Permafrost with in 100cm depth, gelic material with permanent frost with in 2m of the surface, show cryoturbation
Histosols	> 30% organic matter at 40cm depth
Spodosols	Spodic horizon within 2m but have no plaggen epipedon or an argillic or a kandic horizon above the spodic horizon
Andisols	Andic properties in 60% or more thickness
Oxisols	Highly weathered soil with an oxic horizon within 1.5m and 40% clay within 18cm depth
Vertisols	Soil with swell shrink type clays having 30% or more clay upto 50cm depth, deep cracks when dry and have slickensides
Aridisols	Soils have ochric or anthropic epipedons and either have a salic or calcic, gypsic, cambric and duripan
Ultisols	Low base status soil that have an argillic or kandic horizon but with base saturation of <35%
Mollisols	Dark coloured base rich >50% soils of grassland vegetation with a mollic epipedon
Alfisols	High base status >35% soils of the humid & sub humid region with ochric epipedon and an argillic/natric or a kandic horizon
Inceptisols	Soils that have no spodic, argillic, natric, oxic, petrocalcic, plinthite, but have an altered or cambric-B or umbric mollic or plaggen epipedon
Entisols	Recent soil with no diagnostics horizon other than ochric or anthropic epipedon

- Gelisols is the recently formed soil order
- India has 10 soil order out of 12 order
- CEC of Black soil ranging from 30-60 cmol(p+) kg^{-1} and red soil has 25-40 cmol(p+) kg^{-1}
- Laterite soil has CEC of <16 cmol(p+) kg^{-1}
- The term Laterite was first used by Buchanan in 1807 for the highlu ferruginous unstratified materials obsered in malabal hills of south India
- Laterite soil are formed in tropical climate experiencing alternate wet and dry seasons
- Laterite soils are rich in sesquioxide but devoid of bases & silicate minerals
- Black soils are swell-shrink soil, developed from basaltic material (alluvium) under semiarid- sub humid climate conditions.
- Black soil contain smectite type expanding clay.
- Black soil are forming cracks of >50cm depth and width of >1 cm
- Red soil are formed from igneous (Acidic) rocks and contain non-expanding clay of kaolinite
- Red soil are generally poor in N, P, K and OM and also lime content

Soil moisture regime

- ☞ Aquic regime
- ☞ Aridic & torric regime
- ☞ Udic regime (Humid)
- ☞ Ustic regime
- ☞ Xeric regime

Soil temperature regime

- Pergilic regime –Mean Annual Temperature (MAT) is < 0°C
- Cryic regime – MAT is 0 –8°C
- Frigid regime – MAT is 8°C
- Mesic regime –MAT is 8 –15°C
- Thermic regime –MAT is15 – 22°C
- Hyperthermic regime –MAT is > 22°C

Soil forming process

Fundamental

- Addition of water, organic matter and mineral matter to soil
- Losses of these materials from the soil
- Translocation of material from one and place to another place
- Transformation of organic matter and mineral in the soil

Specific pedogenic process

Humification : It is the process of decomposition of organic matter and synthesis of new organic substances.

Eluviation : Movement of soil constituents such as clay, humus, sesquioxides, $CaCO_3$ and soluble salt from upper horizon to lower horizon

Illuviation : Accumulation of elluviated compounds.

Calcification and Gypsification : Formation and accumulation of $CaCO_3$ and $CaSO_4$, respectively

Argillation : It is the process by which the dispersed clay particles are migrated from upper to the lower horizons result a textural horizon.

Leaching : It is the process of removal in solution of a constituent from soil.

Lessivage : Removal of clay particularly the fine clay in suspension from the upper part of soil profile and accumulate in the lower part.

Braunification or Ferruginisation : It is the process of release of iron from primary minerals and their dispersal as coatings on soil particles or as complexes with organic matter or as discrete aggregates to impart a brown to red colour to the soil.

Laterization : It is the process of removal of silica from soil and accumulation of sesquioxides (Fe & AlO_3) with or without the formation of ironstones and concretion.

Podzolization : It is the process of removal of iron and aluminum from the upper part and accumulation of silica.

Gleization : The reduction of iron under anaerobic condition with the production of bluish to greenish grey colour with or without mottles

Salinization : It is the process of accumulation of soluble salts in soil.

Alkalisation : It is the process of accumulation of sodium in exchange complex.

Transported Parent material

Agent	Deposited by (or) in	Name of the deposit
Water	River / Lake / Ocean	Alluvium / Lacustrine / Marine
Wind	Wind	Loess –silt ,Aeolian – sand
Gravity	Gravity	Colluvium
Ice	Ice	Till , Moraine

Soils of India

- Alluvial soil - 78 M ha
- Black soil - 51.8 M ha
- Red soil - 51 M ha
- Laterite - 12.6 M ha
- Desert soil - 37 M ha
  - Acid soil - 49 Mha (In which 26 Mha <5.6 pH and 23 Mha (5.6 –6.5 pH))
  - Saline soil - 7 Mha
  - Alkali soil - 2.8 Mha
  - Acid sulphate soil - 50000 ha
  - Tarai soil - 56000 m^2
  - Coastal alluvium - 85000 m^2
  - Deltaic alluvium - 18000 m^2

Soils of Tamil Nadu

Soil type	Area (Lakh ha)	% area
Alluvial soil	30.8	23.8
Red soil	79.8	61.7
Laterite soil	3.7	2.9
Black soil	15.1	11.6

2.2. SOIL PHYSICS

Soil physics is the study of soil physical properties and processes. It is applied to management and prediction under natural and managed ecosystems. Soil physics deals with the dynamics of physical soil components and their phases as solids, liquids and gases.

Soil texture : Relative proportion of the different particle size fraction specifically referred to as sand, silt and clay.

Textural Classes

Sandy soil (>70% of sand of by weight) (<15% clay)

- Sandy and Loamy Sand

Loamy soil group (40 – 45% sand)

- ❖ Sandy loam, Loam, Silt loam, Silt, Sandy clay loam, Silty clay loam and Clay loam

Clay soil (> 35% clay)

- ❖ Sandy clay (35% clay)
- ❖ Silty clay (40 clay)
- ❖ Clay (>40% clay)

Size of soil particles

Soil particles	USDA	ISSS
Clay	<0.002 mm	0.002 mm
Silt	0.002 –0.05 mm	0.02 mm
Very fine sand	0.05 –0.1 mm	-
Fine sand	0.1 –0.25 mm	0.2 mm
Medium sand	0.25 –0.5 mm	-
Coarse sand	0.5 –1.0 mm	2 mm
Very coarse	1.0 –2.0 mm	-

Size of soil separates >2 mm indicate as gravel in both USDA and ISSS category

Soil texture determination by feel method

- ✍ Smooth, talc like and no grittiness - silt
- ✍ Smooth and talc like and gritty - loam or silt loam
- ✍ No smoothness and gritty - sand

Soil texture determination by mechanical analysis (Robinson International Pipette method)

Stocks Law (1851): Terminal Velocity of a spherical particles settling under the influence of gravity in a fluid of a given density and viscosity is proportional to the square of its radius.

$$V = \frac{2}{9 ç} gr^2 (ps\text{-}pf)$$

Where

V- Settling velocity r – Radius η – Fluid viscosity g – Acceleration due to gravity, ps – Density of solid and pf – Density of liquid

Sedimentation Velocity	
Clay — 347×10^{-6}	Fine sand — 347×10^{-2}
Silt — 347×10^{-4}	Coarse sand — 347

- ❖ Rate of sedimentation is directly proportional to the square of the diameter of the particle.
- ❖ Soil texture determination by hydrometer method - rapid but less accurate and also give erroneous results under conditions of high $CaCO_3$, OM and high salinity.

- ❖ Hydro meter works based on principle of continues decrease in the density of soil suspension over time at the rate the particles fall below the level of the hydrometer

Specific surface of soil particles & clay minerals

Soil separates	Specific surface area (cm^2/g)	Clay minerals	Specific surface area (cm^2/g)
Coarse sand	45	Kaolinite	37-45
Fine sand	446	Illite	120-170
Silt	4458	Montmorillonite	580 – 750
Ultra clay	9.8 millions	Chlorite	130 – 180
Loam soil	50 – 100m^2/g	Vermiculite	780-900
Clay	150 - 200 m^2/g		
Sandy loam	10-40 m^2/g		

Cation exchange capacity (CEC) of soils

Soil texture	CEC (meq/100gm)	Soil texture	CEC (meq/100gm)
Sand (light colour)	3-5	Silt loam	15-25
Sand (dark colour)	10-20	Clay and clay loam	20-50
Loam	10-15	Organic soil	50-100

Properties of clay minerals

Clay minerals	d space (A^0)	Shape	Sio_2 / R_2O_3 Ratio	Y Index
Kaolinite	7.2	Plate like & Hexagonal	<2	0 –5
Halloysite	10.25	Elongated tubules	-	5 –25
Illite	10	Irregular flake	6-10 % K_2O	-
Montmorillonite,	17-18	Irregular flakes	>2	>25
Chlorite	14	Irregular flakes	>2	-
Vermiculite	10-15	Irregular flakes	>2	

Cation exchange capacity (CEC) of different clay minerals

Clay minerals	CEC $cmol(p^+)kg^{-1}$	Surface Area m^2 /kg	Clay minerals	CEC $cmol(p^+)kg^{-1}$	Surface Area m^2 /kg
Kaolinite	3-15	37-45	Saponite	69-81	
Muscovite	11	-	Monmorillonite	80-120	580-900
Illite	30 –40	120-170	Vermiculite	120-150	780-900
Attapulgite	18-22	-	Chlorite	-	130-180
Nontronoite	57-64	-	Humus	200-300	-
Chlorite	20-40				

- ✍ Optimum bulk density requir for plant growth is 1.33 Mg/m^3
- ✍ Optimum particle density requir for plant growth is 2.65 Mg/m^3

- Clay soil has low bulk density than sandy soil.
- Compact soil has high bulk density.
- Organic matter application lowers the bulk density.
- Intensive cultivation increases the bulk density.
- Cultivation of crops decreases the bulk density.
- Muck soil contain OM content 20-50%, where as Peat soil contain OC content >50%
- Most mineral soil show particle density between 2.6 and 2.7 Mg/m^3
- Bulk density of coarse textured soil, ranging from 1.4 – 1.75 Mg/m^3
- Bulk density of fine textured soil, ranging from 1.1 -1.4 Mg/m^3
- Specific volume of agricultural soils may vary from 0.55-0.70 m^3/Mg for coarse textured soil and from 0.70-0.90 m^3/Mg for fine textured soil
- Accumulation of clay in B-horizon under sub humid region show the **kankar layer**
- **Procter moisture content:** during compaction, the density of soil under a load increases with the increase in soil water content up to a certain limit, beyond which further addition of water does not increase compaction because of incompressible nature of soil solids and water.

Porosity

- Porosity is the total volume of pore spaces to the total volume of soil
- Porosity values usually varies from 30 – 60%
- Two types of pore spaces:

  Macro pores larger than about 0.06 mm in diameter

  Micro pores smaller than 0.06 mm in diameter
- Capillary pore spaces are important since water retained in theses pore spaces are used by the plant for their growth
- As the soil dries, non-capillary pores are emptied first

Void ratio:

- It is the total volume of pores to the total volume of soil solids
- Values of void ratio varied between 0.3 and 2.0

Inorganic colloids

- Thomas Graham coined the term colloids and crystalloids.
- Gelatin, gum and organic substances are called as hydrophilic colloid or lyophilic colloid or reversible colloid.
- Metal, metal sulphide and inorganic substances are called as hydrophobic colloid or lyophobic colloid or irreversible colloid.
- Movement of colloidal particles are called as cataphoresis or electrophoresis
- Movement of dispersion medium is called as electro osmosis.
- Precipitation of colloid is prevented by the addition of hydrophobic colloid.
- Clay, organic colloid and silica possess negative charge.
- Sesquioxide possess positive charge.
- Clays are crystalline in nature.

- Kaolinite is called as china clay.
- In kaolinite oxygen-hydroxyl linkages hold the structural units.
- In montmorillonite oxygen-oxygen linkages hold the structural units.
- Kaolinite has better binding power than montmorillonite
- Kaolinite particle is larger than montmorillonite.
- In illite 15% of the Si is replaced by aluminium.
- Soils containing more than 15 % clay exhibit plasticity
- Soil with expanding lattice clay minerals such as smectites have high liquid limits and plasticity indices and on other hand clay minerals such as non-expanding type clay minerals have low liquid limit values.
- 1:1 type clay minerals eg. kaolinite, halloysite, anauxite, dickite -non expanding type,
- 2:1 type clay minerals eg. montmorillonite -expanding type,
- 2:1 type clay minerals eg. Illite- non expanding type,
- 2:2 type clay minerals eg. chlorite - non expanding type
- 2:1 type clay minerals eg. Vermiculite – partial expanding type
- Soil particles less than 0.001 mm size called as soil colloids and its has Brownian movement
- Ca-flocculation and Na-Deflocculation
- Cohesion – attraction of similar particles (soil particle & soil particle) whereas adhesion- attraction of dissimilar particles (Eg. soil &water)
- Negative charges of clay are aroused from isomorphous substitution
- Effective CEC: Sum of milliequivalent of Ca, Mg, K and Na, plus Al and H.
- Percent Base Saturation: (Milliequivalent of Basic cations per 100g of soil/ total CEC Meq/100g of soil)*100

Soil structure : The arrangement of primary and secondary particles into a certain structural pattern is called as structure.

Classification

- Type (Shape and arrangement of peds)
- Class (Size of peds)
- Grade (Distinctiveness and durability)

Type

Prism like - Vertical axis is more developed than horizontal axis

- Structure having flat tops are called as prismatic
- Structure having rounded top are called as columnar.
- Found in subsoil horizons of arid and semiarid region.
- Prism like structure are associated with Swelling type clay and also occur in poor drained soils of humid region
- Mostly occur in sub soils of salt affected/sodic soils

Plate like - Horizontal axis is well developed than vertical axis

- If it is thick it is called as platy.

- If it is thin it is called as laminar.
- This structure is inherited from parent material.
- This soil structure formed due to compaction by heavy tractor on clayey soils

Block like- Structure having angular corners are called as angular blocky.

- Structure having rounded corners are called as sub angular blocky
- Blocky structures are called as nutti or nuci form.
- It is found in soil which is poorly drained. Ped size ranging from 1-10 cm
- Mostly found in sub soil

Spheroidal

- Crumb: It is found in pasture land and more porous in nature
- Granular: It is found in cultivated soil and less porous in nature
- Crumb & Granular structure found in surface soils high in OM, especially in grassland

Classes of structure

- Very fine, Fine, Medium, Coarse and Very Coarse

Grade : Weak, moderate, strong and structure less (Single grain, massive)

Structure less : Massive eg: puddled soil and single grained eg: sand

- ✍ Natural aggregates are called as ped.
- ✍ Artificial aggregates are called as clods.
- ✍ Broken peds are called as fragment.
- ✍ Salt dissolved in water precipitate is called as concretion.
- ✍ Type of the structure is characterized by shape and arrangement of peds.
- ✍ Class of the structure is characterized by size of peds.
- ✍ Grade of the structure is characterized by degree of distinctness and durability of the visible aggregates or peds.
- ✍ Outside of an aggregate is dark and called as clay skin.
- ✍ Soil developed from shales may inherit platy structures.
- ✍ Alkali soil has columnar or prismatic structure.
- ✍ The larger the aggregate present in the soil which show low stability
- ✍ Krillium is a synthetic soil conditioner. Pusa Hydrogel @5kg/ha recommended for soil water conservation

Component of soil colour

Hue	-	Denotes the dominant spectral colour (yellow, red, blue, brown)
Value	-	Denotes the lightness and darkness of colour
Chroma	-	Denotes purity of colour

Factor influence soil colour

- ❖ Mineralogy and chemical constituent
- ❖ Organic matter, moisture content, soil structure

- ✍ Red colour of soil is due to unhydrated ferric oxide and it indicates good drainage.
- ✍ Yellow-goethite & limonite, Red – hematite, Brown maghemite
- ✍ Green-glauconite, black- manganese oxide , white – calcite ($CaCO_3$)
- ✍ Under prolonged anaerobic conditions, soil colour changed into gray/ bluish due to coating of reduced iron oxide
- ✍ Munsell colour chart contains 175 different coloured papers or chips.
- ✍ Value of 0 indicates black colour.
- ✍ Value of 10 indicates white colour.
- ✍ Chroma increases with decreasing greyness.
- ✍ Chroma 0 for neutral grey and 20 for high purity.
- ✍ Normally chroma above 8 is rarely found in soil.
- ✍ Achromatic colours have neutral hue and zero chroma and differ only in value.
- ✍ Munsell colour chart consist of with 7 cards, each has constant hue.
- ✍ In each card value is constant with in horizontal row and increases towards top
- ✍ In each card chroma is constant with in vertical column and increases to the right

SOIL WATER

Functions of water

- ❖ It is the major constituent of plant protoplasm (85 – 95%)
- ❖ It is essential for photosynthesis and conversion of starches to sugars
- ❖ It is the solvent to make available all essential nutrients to crop plants.
- ❖ It regulates physical, chemical and biological property of soil.

Heat of solution: when ions are hydrated the large amount of energy is released, that is known as heat of solution.

Heat of wetting: when clay particles are hydrated a certain amount of energy must be released at this phenomenon is known as heat of wetting.

Suction: The force with which water is held is also termed as suction.

Surface tension: It may be defined as the forces in dynes acting of right angle to any line of 1 cm length in the surface.

Soil water energy concept: It defined as the work the water can do when it moves from its present state to a pool of water in the reference state.

Solute or osmatic potential (y_s) : The effect of attraction of ions and other solutes for water.

Matric Potential (y_m) : The effect of the surface area of soil particles and small soil pores that adsorb water.

Pressure Potential (y_p) : Atmospheric or gas pressure effect

Soil water potential (y_w) = Matric (y_m) + Solute potential (y_s) + Pressure potential (y_p)

Total water potential = Soil water potential (y_w) + Soil gravitational potential (y_g)

$$Yt = yw + yg$$

- Matric potential is most important and dominant for most of the soil.

Classification of soil water

- Physical classification
- Biological classification

Physical	Pressure	Biological	Pressure
Gravitational water	>1 / 3 atm	Superfluous water	1/3 atmosphere or less
Capillary water	1 /3 - 31 atm	Available water	1/3 to 15 atmosphere
❖ *Field capacity*	1 /3 atm		
❖ *Wilting point*	15 atm		
Hygroscopic co-efficient	>31atm	Unavailable water	15 and above atmosphere

Height of water column (cm)	P_F value	Atmospheric pressure
1	0 (Saturated soil)	1/1000
10	1.0	1/100
100	2.0	1/10
346	2.53 (field capacity)	1/3
1000	3.0	1
10000	4.0	10
15849	4.2 (Wilting point)	15
31623	4.5 (Hygroscopic point)	31
100000	5.0	100
10000000	7 (Oven dry soil)	10000

Measurement of soil moisture

- ✓ Gravitational method (Laborious)
- ✓ Gypsum block method
- ✓ Tensiometer measures from 0 to 0.85 atmosphere used for scheduling of irrigation
- ✓ Neutron scattering (Radium and Beryllium, Amerecium and Beryllium as neutron source)
- ✓ Isotope tritium (^3H) is largely used for water movement and ground water recharge & pollution studies
- ✓ Water has dielectric constant of 80 where as dry soil has dielectric constant of 3-6 and from 20-30 for wet soil
- ✓ Time domain reflectometry (TDR) useful techniques is measuring soil salinity in terms of EC.
- ✓ Micro wave remote sensing techniques gives moisture content of only surface soil up to 5-10 cm over larage area.

- ✓ Total soil water potential is determined by soil psychrometer
- ✓ The vapour potential of of soil is estimated by thermocouple psychrometer
- ✓ Piezometer used to measure water level in saturated or flooded soil

Soil Air : It is defined as the gases occupying the free pore space in soil

Total pore space of soil

Soil	Pore space (%)
Sandy	20-30
Loamy	30-50
Clay	50-60

- ❖ Manganese is susceptible to changes due to change in soil aeration condition.
- ❖ Due to the usage of heavy machines bulk density increases and macrospore decreases.

Soil Temperature : The temperature measured near the surface of the soil.

Specific heat (Cal. /gm) : It is defined as the amount of heat in calories required to rise the temperature of 1 g of soil solid by 1^0C

Water	1
O.M	0.462
Clay	0.22
Air	24
Humus	0.4
Ice	0.5
Ice	0.5
dry soil	0.2
mineral (Fe)	0.11

Heat Capacity (Cal.cm^{-3}) : The ratio of of heat supplied to its corresponding rise in temperature is the heat capacity of the soil

Heat Capacity	Specific heat X Specific gravity
O.M	0.462*1.3 = 0.60
Water	1*1 = 1
Air	24*0.000125 = 0.003
Mineral Matter	0.45

- ❖ Favourable temperature for nutrient absorption is 27 –32° C
- ❖ Speed of chemical reaction is doubled by every 10° C rise in temperature.
- ❖ The soil temperature on south slope is higher than that on North Slope.
- ❖ For every 900 feet of height atmospheric temperature falls by 1° C
- ❖ Thermal conductivity: Sand > Loam> Clay >Peat

2.3. SOIL FERTILITY

Soil fertility : It is the inherent capacity of soil to supply the all essential plant nutrients in adequate amount without any toxicity

Soil Productivity : It is defined as the capability of the soil for producing a specified quantity of plant produce per unit area and the ability to produce sequence of crops under a specified system of management.

Plant Nutrition : It is the process of absorption and utilization of essential element for plant growth and reproduction.

Essentiality concept - Arnon and Stout (1939)

- A deficiency of element makes it impossible for the plant to complete its life cycle
- The deficiency is specific for the element is question
- The element directly involves the plant metabolic reaction.

Lie begs law of minimum - Justus van liebig (1812)

Even if all elements present but one of the essential elements be absent the absence of that one constituent will render the crop barren.

Mitscherlich's law -E.A. Mitscherlich (1909)

Growth law : Increase in yield of a crop as a result of increasing a single growth factor is proportional to the decrement from the maximum yield obtainable by increasing particular growth

$$dy/dx = (A-y)\,C$$

Where, dy-yield resulting from an increments of growth factor dx, A-Maximum possible yield, y- yield obtained after any given quantity of the factor x has been applied and C is a proportionality constant.

Spillmans equation : $Y = m\,(1-R^x)$

Mitscherlich's - Spill man equation : Log (A-Y) = log A – Cx

Where, *Y = Yield produced by a factor x, A – Maximum possible yield and C- Constant*

Bray nutrient mobility concept : As the mobility of a nutrient in the soil decreased the amount of nutrient needed in the soil to produce maximum yield increased from a variable net value

$$\text{Log (A-Y)} = \text{Log A} - C_1 b - Cx$$

A, y, x, - maximum yield, optimum yield & fertilizer added

B - amount of immobile, C_1 - Efficiency factor of b and C- Efficiency factor of X

Baule unit : The unit of fertilizer or growth factor to be taken as that amount necessary to produce a yield that is 50 percent of the difference between maximum possible yield and the yield before that unit was applied

1 baule N = 1 baule P_2O_5 = 1 baule K_2O

223 lbs of N + 45 lbs P and 76 lbs of K_2O

Log A / A-Y = cx

A = maximum yield, Y = 50% maximum yield and X = baule unit

Available forms of nutrients

Nitrogen	NO_3^-, NH_4^+
Phosphorus	$H_2PO_4^-$, HPO_4^{2-}
Sulphur	SO_4^{2-}, SO_3^{2-}
Boron	H_3BO_3, $H_2BO_3^-$, HBO_3^{2-}, BO_3^{3-}
Molybdenum	MOO_4^{2-}

Copper : Cu^{2+}, Iron : Fe^{2+}, Zinc:Zn^{2+}, Manganese:Mn^{2+}, Calcium:Ca^{2+}, Magnesium:Mg^{2+}, Potassium:K^+

Average nutrient concentration in plant tissue

Nutrient	Average conc. in plant tissue (%)	Nutrient	Average conc. in plant tissue (mg kg^{-1})
N	1.5	Zn	20
P	0.2	Cu	6
K	1.0	Mn	20
Ca	0.5	Fe	100
Mg	0.2	B	20
S	0.1	Mo	0.1
Cl	100 mg kg^{-1}	Ni	0.1

- Essential element for plant growth and development: 17 elements
- In India, Sulphur has been described as fourth major nutrient after NPK

Mobility of Nutrient in soils	Mobility Nutrient in plants
Mobile - NO_2/ NO_3, SO_4, BO_3, Mn^{2+}, Cl Less mobile - NH_4, K^+ Ca, Mg, Cu Immobile - $H_2PO_4^-$, HPO_4^{2-}, Zn^{2+}	High mobile - N, P, K Moderate - Zn Less mobile - S, Fe, Cu, Mn, Mo & Cl Immobile - Ca & B

Metal elements: K, Ca, Mg, Cu, Zn, Mn & Fe

Non metal elements: N, P, S, B, Mo & Cl

C:N ratio	C:P Ratio	Process
<20: 1	< 200:1 or 200:1	Mineralization
20 – 30:1	>200:1& <300: 1	No mineralization & Immobilization
>30: 1	300:1& above	Immobilization

Ability of cation to flocculate the soil: Al > Ca & H >Mg >K >Na

Power of replacement of cation: H & Al > Ca > Mg >K & NH_4 >Na

Power of replacement of anion: OH > H_2PO_4 >SO_4> NO_3

↑ Mg	–	↓ Fe, ↓ Ca, ↓ Cu
↑ Ca	–	↓ B ,↓ Fe
↑ K	–	↓ Mg ,↓ Mn
↑ P	–	↓Zn , ↓ Fe

Phosphate fixing capacity: Smectite >Vermiculite >Kaolinite> Muscovite

Pesticide absorbing capacity: Montmorillonite > Illite > Kaolinite

Antagonistic effects in nutrients

Excess of nutrient	Causes deficiency	Excess of nutrient	Causes deficiency
Ca	P	N,P,K	Cu
Ca, Mg	K	NO_3	Fe
K, NH_4	Mg	Zn, Al	Cu
Fe, SO_4	MO	N, K, Ca	B
P	Zn		

Elements	Function	Deficiency symptoms
Nitrogen	- Essential component of amino acids, protein, nucleic acid & purines - Integral part of Chlorophyll - Improves the quality of leafy vegetables & Fodder - Improves protein quality of food grains - N imparts vigorous vegetative growth	- Stunted growth - Symptoms first appear on older leaves as light to pale yellow - Reduction in flowering, crop yield and lower protein content
Phosphorus	- P is involved in energy transfer, photosynthesis, transformation of sugar and starch - Constituent of nucleic acid, phytin & phospho-lipids - P act as energy currency (ATP & ADP) within the plants - Stimulate early root growth and development - role in Cell division & development & decrease lodging	- first appear on older leaves become bronzed or develop reddish-purple tips and leaf margins - Severe restriction occurs in growth of plant tops & roots. - P deficient plants are thin, erect and spindly with restricted foliage. - Suppressed development of lateral bud
Potassium	- Regulates permeability of cell membrane (stomata opening & closing) - Resistance to disease, drought tolerance, winter hardiness & reduce lodging - Necessary for sugar crops - Activates enzymes, enhance crop quality and shelf life of fruits& vegetables - Essential for photosynthesis, water & nutrient transport	- first symptom appear on older leaves - Dieback - Leaf scorch - Marginal chlorosis - Rosette in sugar beet, celery carrot - Seeds& fruits are shrivelled - Plant lodge easily and slow & stunted growth.
Calcium	- Stability of cell membrane - Present as Calcium pectate in cell wall - Neutralizes organic acids - Prevent lodging of plants	- Die back - Tip hooking in cauliflower, sugar beet - Blossom end rot in tomato - Black heart of cabbage - Bitter pit of apple

Elements	Function	Deficiency symptoms
Magnesium	- Constituent of chlorophyll & chromosome - Essential for Carbohydrate metabolism	- Cotton reddening - Bronzing of citrus - Interveinal chlorosis - Sand drown disease in tobacco - Woolly flesh of tomato
Sulphur	- Increase oil content in oil seed crops - Responsible for Pungent odour of onion - Promotes nodules formation - Promotes root growth - Essential for amino acids synthesis viz. methionine, cysteine & cystine - Synthesis of glucosides in mustard oil	- first symptom appear on young leaves - chlorosis is common symptom - Tea yellow disease - Upward curling of leaves - Premature shed of older leaves
Boron *(Tourmaline is a source of boron)*	- Cell development - Translocation of sugars - Polysaccharide synthesis	- Cracking is common symptom of B - Sterility & malformaion - Decrease the rate of water absorbtion - Rosetting in Lucerne - Snakehead in walnut - Dieback& corking in apple - Corking & pitting of tomato - Hallow stem & bronzing of cauliflower - Brown heart disease in table beets, turnip - Heart rot of sugar beet - Canker & internal black spot of garden pea - Top sickness in tobacco - Hard fruit of citrus - Hen & chicken disease of grapes
Iron	- Constituent of cytochrome - Catalytic role in enzymatic activity - Essential role in nucleic acid metabolism - Necessary for synthesis & maintenance of chlorophyll in plants	- Interveinal chlorosis - Mottle leaf in sugarcane - Lime induced chlorosis - Growth cessation occurs with whole plant turning necrotic
Manganese	- Constituent of chlorophyll - Oxidation reduction function - Helps in iron movement	- Gum pockets in citrus - Interveinal chlorosis - Grey spec in oats, barley , wheat, rye, maize - Pahala blight of sugarcane - Marsh spot of peas - Speckled yellow of sugarbeet
Zinc	- Essential for chlorophyll synthesis - Necessary for production of IAA(auxin) - Necessary for production of cysteine - Oxidation reduction function - Synthesis of growth hormones	- Kaira disease of paddy - White bud of maize - Rosette of fruit trees - Mottle/little leaf or Frenching of citrus - Sickle leaf of cocoa - Little leaf of apple & cotton

Elements	Function	Deficiency symptoms
Copper	- Necessary for synthesis of Vit A - Catalyst in respiration - Enhances the fertlity to male flowers - Impart disease resistance to plants	- Multiple bud formation - Gum pockets & dieback in citrus - Exanthema of fruit trees - White tip, necrosis, chlorosis of younger shoot tissue, Reclamation disease - Male flower sterility, delayed flowering and senescence
Molybdenum	- Constituent of nitrate reductase, nitrogenase, dehydrogenase & sulphite oxidase - Enhances Biological Nitrogen fixation - Mo- requirement high for nodulated legume - Involved in protein biosynthesis - Affects formation & viability of pollen and development of anthers	- Whip tail in cauliflower - Scald of legumes - Yellow spot of citrus - Down ward cupping in raddish - Affects nodulation

Indicator plants for different nutrients

Elements	Indicator plants
N & Ca	Cauliflower & cabbage
P	Rapseed
K & Mg	Potato
Fe	Cauliflower
Na	Sugarbeet
Mn	Sugarbeet & Oats
Boron	Sunflower

Fixation of nutrients:

Nutrients	Clay minerals
Ammonium fixation	Vermiculite>fine graded micas>smectites
Phosphorus fixation	**In acid soil:** Iron & aluminum react with P to form insoluble aluminum hydroxyl phosphate **In neutral to alkaline soil:** P react with CaCO3 to form insoluble Tri calcium phosphate
Potassium fixation	Largely in vermiculite dominant soilConsiderable in Illite dominant soilSlight in Montmorillonite dominant soilabsent in kaolinite & chlorite dominant soil

FERTILIZERS

Fertilizers are materials containing one or more chemical compound of the organic or inorganic nature, which supply the required plant nutrients to the crop plants.

Classification of fertilizer

Straight: It is a single compound containing one primary nutrient only

Compound and complex: It is a single compound but contains two or more nutrients of which at least two primary nutrients are in chemically combined form

Mixed Fertilizer: It is a physical mixture of fertilizer material containing two or more major plant nutrients.

Fertilizer grade: It refers to the guaranteed minimum percentage of nitrogen, phosphate and potash contained in the mixed fertilizer. eg. 12 –6 – 6 grade for 100kg mixture

Fertilizer ratio: It refers to the ratio of the percentage of nitrogen, phosphate and potash in the fertilizer mixture. eg: 2: 1:1 Ratio

Conditioners: These are materials which are added to the fertilizer mixtures changing their preparation for reducing hygroscopity and improve their physical condition.

Filler: It is a weight makeup material added to fertilizer ingredients to produce a mixture of the desired grade. eg. clay, sand, coal powder, & ground plant parts etc.

- In world, First commercially successful venture of fertilizer production is super phosphate in 1843 by John B. Lawes (treating coprolites with sulphuric acid to produce super phosphate).
- In India, first fertilizer was produced in 1906 that is single super phosphate by EID Pary (India) ltd., Ranipet
- In 1933,ammonium sulphate was produced as by a product of steel industry
- In 1959, Urea and ammonium sulphate nitrate - manufactured by FCI, Sindri and ammonium chloride – manufactured by New Central Jute Mill ltd., Varanasi
- India has no potassium fertilizer is produced due to lack of potassium mineral deposits
- FAI (Fertilizer Association of India) was set up in 1955 – registered as company – is the national representative body of all fertilizer manufacturers in India, comprising public, private, joint and co-operative sectors.
- Ammonium is taken by only certain plants such as rice, tomato. etc.
- *Nitrosomonas* sp. convert ammonium into nitrite and these nitrite is converted into nitrate by *Nitrobacter* sp. (Nitrification-Nitrifying bacteria)
- Denitrifying bacteria are Pseudomonas, Paracoccus, Bacillus and *Thiobacillus* sp.
- In 1905-1908, Haber - Bosch process is the technology for synthetic ammonia synthesis
- Producer gas is N_2 and H_2 and natural gas is a mixtures of alkanes with chain length 1 (Mehtane) to 5 (Pentane)
- Naphtha: Distillate obtained during fractional distillation of petroleum in a temperature range of 200-250 ^{0}C (Benzene, Toluene & Xylene)

Fertilizers Conversions

P to P_2O_5	: 2.29		$ZnSo_4$ 7 H_2O to Zn	:	0.227
P_2O_5 to P	: 0.436		MgO to Mg	:	0.603
K to K_2O	: 1.204		$MgSo_4$ $2H_2O$ to Mg	:	0.098
K_2O to K	: 0.830		N to NH_4	:	1.215
KCL to K	: 0.524		NO_3 to N	:	0.225
SO_2 to S	: 0.321		N to $(NH_4)_2SO_4$	:	4.716
Zn to $ZnSo_4$ 7 H_2O	: 4.39		Ca to CaO	:	1.40

Fertilizer production techniques

Ammonia+ CO_2 ———— NH_2 CO_2 NH_4 ———— CO (NH_2) (Urea - Amide group)

Ammonia + H_2SO_4 ———— $(NH_4)_2$ SO_4 (Ammonium sulphate)

Ammonia + HCl ———— NH_4Cl (Ammonium chloride)

RockPhosphate+HNO_3 ____ Nitro phosphate

Rock Phosphate +H_2SO_4 ____ Single super Phosphate

Rock Phosphate +H_3PO_4 ____ Triple super Phosphate

Rock Phosphate +HCL ———— Dicalcium Phosphate

Rock Phosphate +Elemental Sulphur + *Thiobacillus thiooxidans* ———— Biosuper

Rock Phosphate +Basic slag+H_2SO_4 ____ Pelophos

Nitrogen fertilizers and their content (%)

Fertilizers	N content (%)	Fertilizers	N content (%)
Sodium nitrate	15-16	Calcium nitrate	15
Ammonium nitrate	33-35	Potassium nitrate	13
Ammonium sulphate nitrate	26	Urea formaldehyde	38
Ammonium sulphate	20.5	Crotonylidene diurea	30
Ammonium chloride	25	Thio urea	36.8
Urea (biurate content <2.0 %)	46	Urea Z(urea+ Acetaldehyde)	33-38
Calcium ammonium nitrate	26	Oxamide	31.8
Anhydrous ammonia	80	Sodium nitrate	15

- Citrus crops are more sensitive to biuret , more than 2% is damage to plant

Slow releasing N fertilizers

- Urea Formaldehyde (UF) - 38% N
- Urea Z - 33-38% N
- Oxamide - 32% N

- Crodolidene Di Urea (CDU) - 30% N
- Isobutylene Diurea (IBDN) - 31-32% (2:1 Urea and Isobutyl aldehyde)
- Magnesium Ammonium Phosphate
- GUS – Gunayl urea sulphate
- GUP – Guanyl urea phosphate
- Sulphur Coated Urea (SCU) - 36-38% N

Urea inhibitors	Nitrification Inhibitors
Phenyl Phosphoro Diamidate (PPDA) N – Thio phosphopheric triamine Hydroquinone Nitrapyrin (N-Serve) - Half life 28day	Potassium azide (KN_3), Dicholorophenyl succinic amide Terazole and Dicyondiamide AM (2 amino – 4 choloro 6 trimethyl pyrimidine) ST (2 – sulphanilamdie thiazole)

Phosphatic fertilizers- P_2O_5 Content (%)

✍ SSP	16 and 12 % S; 24 % Ca
✍ Double super Phosphate	32
✍ Triple super phosphate	48
✍ Mono ammonium phosphate	48
✍ Di ammonium phosphate	46-54 and 16-18 % N
✍ Di calcium Phosphate	34-39
✍ Bone meal	25-28
✍ Basic slag/Thomas slag	10 –20
✍ Rock Phosphate	20 –40

Potassium fertilizers (K_2O content in %)

- ✍ Potassium chloride - 60
- ✍ Potassium sulphate - 48
- ✍ Potassium bearing minerals are sylvite, langebeinite,kainite and carnallite

Equivalent acidity and basicity for some fertilizers

Fertilizer	Equivalent acidity	Equivalent basicity
Ca CN	-	62
$NaNO_3$	-	29
KNO_3	-	29
$Ca(NO_3)_2$	21	
Anhydrous ammonia	148	
Nitrate of soda Potash	-	26
NH_4Cl	128	-
$(NH_4)_2SO_4$	110	-
$(NH_4)_2SO_4$ NH_4NO_3	93	-
Urea	80-84	-
DAP	70	
Ammonium Nitrate	60	
Mono Ammonium Phosphate	55	

Fertilizers sources	Formula	Content %
Sulphur sources		
Single super PO_4	$(Ca\ H_2\ PO_4) + CaSO_4\ 2H_2O$	12 – 14
Ammonium Sulphate	$(NH_4)_2\ SO_4$	24
Potassium Sulphate	$K_2\ SO_4$	18
Gypsum	$CaSO_4\ 2H_2O$	16
Pyrite	FeS_2	20
Micronutrient sources		
Zinc sulphate heptahydrate	$ZnSO_4 + 7H_2O$	21
Zinc sulphate monohydrate	$ZnSO_4 + H_2O$	33
Manganese sulphate	$MnSO_4 + H_2O$	30.5
Ferrous sulphate	$FeSO_4 + 7H_2O$	19
Copper sulphate	$CuSO_4 + 5\ H_2O$	24
Borax	$Na_2\ B_4 + O_7 10\ H_2O$	10.5
Ammonium molybdate	$(NH_6)MO_7\ O_{24} + 2H_2O$	52

Neutralizing value of liming materials

Liming materials	Neutralizing value or CCE (%)
Calcium Oxide (CaO)	179
Calcium hydroxide [Ca $(OH)_2$]	136
Dolomite	108.7
Calcite ($CaCO_3$)	100
Basic Slag ($CaSiO_3$)	86

Nutrient content of organic manures (%)

Organic Manures	N	P_2O_5	K_2O
Farm yard manure	0.5	0.3	0.5
Compost(urban)	1.0	0.5	0.5
Compost(Rural)	0.5	0.5	1.0
Cow dung	1.5	0.5	0.5
Poultry litter	3.0	2.5	1.5
Night soil	5.0	3.0	2.0
Castor cake	5.5	2.0	1.5
Groundnut cake	7.0	1.5	1.5
Cotton seed cake	6.0	3.0	2.0
Safflower cake	8.0	2.0	2.0
Bone meal	3.0	20.0	-
Meat meal	10.5	2.5	0.5
Blood meal	10.0	1.5	1.0

Soil organic matter

Organic matter : It is the decomposed plant and animal residues. In addition to these materials soil organic matter contain living and dead microbial cell, microbially synthesized compounds and a number of derivatives produced as a result of microbial activity.

Soil Humus: After the active decomposition remaining residue and microbially synthesized product is called humus

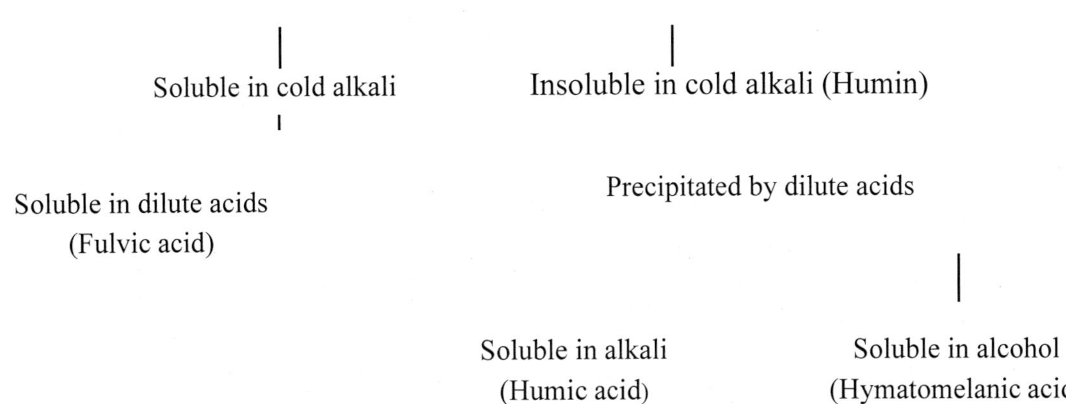

Nature of soil humus

- ❖ It is highly colloidal and amorphous nature
- ❖ Surface area and adsorptive capacity are much higher than that of clay colloids.
- ❖ CEC of humus ranges from 150 – 300 Cmol(p^+) kg ha^{-1}
- ❖ Humus has highest water adsorptive capacity
- ❖ Humic acid are weakly dissociated and determine the buffering capacity of soil.
- ❖ Humic substances are fulvic acid, humic acid and humin

Difference between fulvic and humic acid

Fulvic acid	**Humic acid**
Soluble in water acid & alkali	Only alkali soluble
Yellow to brownish yellow	Dark brown to black
Molecular weight is few 100 to few 1000	Molecular weight is 10000 – 200000
more acidic & hydrophilic groups	Rich in aromatic, hydrophobic groups

Humin:

- Insoluble fraction of humus is humin
- Not soluble in water, acid and alkali
- Very strongly bound to soil minerl matrix and difficult to extract

Chemical nature of humic substances

Humic substance	Elemental composition (%)				
	C	H	N	S	O
Humic acid Fulvic acid	54-59 41 - 51	3 –6 4 – 7	1 – 4 1 – 4	1 – 5 3 – 5	33 – 38 40 – 50
	Functional group (Cmol(p^+) kg ha $^{-1}$)				
Humic substance	Total acids	COOH	Phenolic	E_4/E_6	
Humic acid	500-900	150 – 600	200 – 600	3 –5	
Fulvic acid	700- 1400	500 – 1100	110 – 600	6 – 9	

Physical properties of humus

- ❖ It posses low plasticity and cohesion
- ❖ It imparts the colour to soil

pH	% of base saturation
5.0	25%
5.5	50%
6.0	75%

Colloidal properties

- ❖ They carry the negative charges
- ❖ It composed of C, H, O with trace amount of N, S, P & other element
- ❖ It adsorb Ca^{2+}, H^+, Mg^{2+}, K^+ and Na^+

Nutrient availability

Organic N, P, S ———— NH_4, NO_3, PO_4, SO_4
C: N <20:1 C: P < 200:1 C:S <200:1

Properties	Humus	OM
C:N	10-12:1	20-40 :1
Decomposition	Resistant	Easily

C:N ratio of different material

- ❖ Humus-10-12 :1
- ❖ Soil microorganism-8:1
- ❖ Soil organic matter-10:1
- ❖ Grain straw-80:1
- ❖ Corn/sorghum residues-60:1
- ❖ Microbes-4-9:1
- ❖ Legumes-25:1
- ❖ Rice straw-100:1
- ❖ Saw dust-400:1
- ❖ Undisturbed top soil-10-12:1

C:N & C:P Ratio

C:N & C:P Ratio	Process
C:N ratio <20:1	Mineralization (release of N)
C:N ratio 20-30:1	No release of N (Neither mineralization nor immobilization)
C:N ratio >30:1	immobilization
C:P ratio 200:1 or less	Mineralization
C:P ratio >200:1	Neither mineralization nor immobilization
C:P ratio 300:1 and above	immobilization

Priming action : The loss of native soil organic matter through the application fresh organic residue in the soil

- ☞ Organic matter is store house of nutrients in soil and acts as chelate.
- ☞ Tillage increases the oxidation of organic matter in soil.
- ☞ Accumulation of organic matter is high in moist soil than in dry soil.
- ☞ Soils having >20 % organic matter is called as organic soils.
- ☞ Organic matter content of Indian soil is 1 to 2 %.
- ☞ Van bemmelan suggested the colloidal character of humus
- ☞ Higher the exchange capacity of the soil the greater is the buffering capacity

Soil Testing : It is defined as a rapid chemical analysis to assess the available nutrient status and reaction of a soil then interpretation and evaluation of the soil values and fertilizer recommendation based on the result of the chemical analysis.

Soil testing laboratory

Type	India	Tamil Nadu
Stationary STL	396	19
Mobile STL	118	16
Total	514	35

Important soil testing methods

Analytical Parameter	Method	Author (s)
Available N	Alkaline $KMnO_4$ method	Subbiah and Asija (1956)
Available P in Neutral & alkaline soil	0.5M $NaHCO_3$ extractant	Olsen *et al.* (1954)
Available P in acid soil	Bray, 0.03M NH_4F in 0.25 M HCl	Bray and Kurtz, 1945
Available K and Na	Flame photometry	Stanford and English (1949)
Exchangeable Ca and Mg	Versenate titration	Jackson (1973)
Available S	Turbidimetry Barium chloride	Chesnin and Yien (1951)

Available iron, manganese, copper and zinc	DTPA Extraction (Atomic Absorption Spectrometer)	Lindsay and Norvell (1978)
Cation Exchange Capacity	Neutral Normal Ammonium Acetate method	Schollenberger and Dreibelbis (1930)
Organic carbon	Chromic acid wet digestion	Walkley and Black (1934)
Soil reaction (pH)	pH meter (soil water ratio 1:1 and 1:2.5; 1:2.5 Kcl)	Jackson (1973)
Electrical conductivity (EC)	Conductometry	Jackson (1973)
Mechanical analysis - Texture	International pipette method	Piper (1966)
Water holding capacity	Pressure plate method	Richards (1954)

Nutrient grades:

Nutrient	Low	Medium	High	Nutrient Index (Parker)	
Organic carbon	<0.5%	0.5 –0.75%	>0.75%	<1.5%	Low
Nitrogen	<280 kg ha^{-1}	280 -450 kg ha^{-1}	>450 kg ha^{-1}	1.5 – 2.5%	Medium
Phosphorus	<11 kg ha^{-1}	11 -22 kg ha^{-1}	>22 kg ha^{-1}	>2.5%	High
Potassium	<118 kg ha^{-1}	118 -280 kg ha^{-1}	>280 kg ha^{-1}		
Sulphur	<10 mg kg^{-1}	10-15 mg kg^{-1}	>20 mg kg^{-1}		

Micro-nutrients Critical Limits in Soil

Element	Critical Limit (ppm)	Critical Limit (ppm) for Tamil Nadu
Zinc (DTPA Extract, pH 7.3)	0.5 – 1.0	1.2
Copper (DTPA Extract, pH 7.3)	0.2 – 0.5	0.72
Iron (DTPA Extract, pH 7.3)	2.5 – 5.8	3.7-Non calcareous soil 6.3- Calcareous soil
Manganese (DTPA Extract, pH 7.3)	2 – 4	2
Boron (Hot Water Soluble)	0.5 – 1.0	-
Molybdenum (Ammonium Oxalate, pH 3.3)	0.05 – 0.2	-

Problem soil and management

Problem soil and their properties

Criteria	Saline Soil (Solonchak)	Alkali/Sodic (Solonetz)	Saline alkali	Degraded alkali
SAR	<13	>13	>13	<13(surface) and >13(subsurface)
ESP (%)	<15	>15	>15	>15
pH	<8.5	8.5 –10.0	<8.5	6 (surface) and 8.5(sub surface)
TSS (%)	>0.1	<0.1	>0.1	<0.1
EC(dSm^{-1})	>4	<4	>4	<4

Saline soil/ White alkali

- The soluble salts in these soils are predominantly of the chlorides and Sulphates of Na, Ca & Mg.
- Characterized by white encrustation of salts at surface.

Alkali soil/Black alkali : Carbonates and bicarbonates of sodium are the dominant salts.

Reclamation

- Alkaline soil - Gypsum ($CaSO_4 . 2H_2O$) application based on the soil pH
- Acid soil – Lime ($CaCO_3$) application based on the soil pH
- Saline soil - Leaching and sub surface drainage ideal for saline soils.

Salinity Grades		Sodicity grades	
Category	**EC (mSm^{-1})**	**Category**	**SAR**
Low salinity	0 –25	Low sodicity	0 –10
Medium salinity	25-75	Mediumsodicity	10 –18
High salinity	75- 225	High sodicity	18 –26
Very high salinity	225 – 500	Very high	>26

Resistant crops for saline and alkaline soil and problem soil

Soils	Tree sp.	Crops
Saline soils	*Casuarina , Prosopis juliflora, Acacia nilotica, Terminalia arjuna, Butea monosperma, Eucalyptus, Achres sapota andGlyricidia*	Paddy, Ragi, Sunflower, Sorghum, Green manure.Cotton, Sugatbeet, Maize, Barley, Safflower, Coconut
Alkaline soils	*Sesbania. Dalbergia, Prosopis, Albizzia, Neem, Pongamia, Glyricidia.*	Rice, Cotton, Millets, Medicgo, Berseem, Maize, Coconut, Sunflower and Sugar cane
Acid soils	*Eucalyptus, Teriticornis, Mango, Guava,and Anilotica sp.,*	Paddy, Wheat, Sorghum, Cotton, Vegetables and Sugarcane
Water Logged Soils	*Terminalia arjuna, jamun, bamboo*	Spontaneum, *Vetiver*
Calcareous soils	*Acacia Leucopholea, Swietenia sp. Tamarind,*	Paddy, Sorghum, Sugar cane and Castor
Gravel soils	*Acacia planifrons,A.lebbeck and Dalbergia sp.*	Sorghum, Castor andother millets

Integrated fertilizer recommendation based on nutrient requirements of major crops

Crop	Macronutrient fertilizer (kg ha^{-1})			Secondary Nutrient Fertilizer	Micronutrient	Organic Manures
	N	P	K			
Rice Wet seed Dry seeded Semi dry Rainfed	 120 100 100	 50 50 25	 50 50 25	-	$ZnSO_4$ 25 kg ha^{-1}	FYM 12.5 t ha^{-1} & Azospirillum 2 kg ha^{-1}
Sorghum Rainfed Irrigated	 40 90	 20 45	 20 45	-	MN mixture 12.5 kg or $ZnSO_4$ 25 kg, $FeSO_4$ 50 kg ha^{-1}	FYM 12.5 t ha^{-1} & Azospirillum 2 kg ha^{-1}
Cumbu Rainfed Irrigated	 40 70	 20 35	 20 35	-	MN Mixture 12.5 kg or $ZnSO_4$ 25 kg ha^{-1}	FYM/composted coir pith 12.5 t +2 kg ha^{-1} Azospirillum ha^{-1}
Maize Irrigated	135	62.5	50	-	MN Mixture 12.5 kg ha^{-1} TNAU-Maize maxim 3.0 kg ac^{-1} at tassel initiation and grain filling stage	FYM /composted coir pith 12.5 t ha^{-1} + 2 kg Azospirillum ha^{-1}
Redgram Rainfed Irrigated	 12.5 25	 25 50	 12.5 25	100 kg ha^{-1} gypsum as basal	MN Mixture 5kg ha^{-1},TNAU-Pulse wonder 2.25 kg ac^{-1} at 80 DAS	Seed treatment with Rhizobium and Phosphobacterium (0.6 kg ha^{-1}) or soil application (2 kg ha^{-1})
Blackgram and Green gram Rainfed Irrigated	 12.5 25	 25 50	 12.5 25	100 kgha^{-1} gypsum as basal	MN Mixture 5kg ha^{-1}, TNAU-Pulse wonder 2.25 kg ac^{-1} at 35 DAS	Seed treatment with Rhizobium and Phosphobacterium (0.6 kg ha^{-1}) or soil application (2 kg ha^{-1})
Groundnut Rainfed Irrigated	 10 25	 10 50	 45 75	400 kg ha^{-1} gypsum apply as basal or split at 35 -45 DAS	MN mixture 12.5 kg or $ZnSO_4$ 25 kg and 1% $FeSO_4$ spray TNAU-Groundnut rich 2.25 kg ac^{-1} at flowering and pegging stage	Chiseling + FYM / CCP 12.5 t ha^{-1} (for hard pan) Lime 2 t + FYM / CCP 12.5 t ha^{-1} (for surface crust) Rhizobium and Phosphobacterium seed treatment (0.6 kg ha^{-1}) or soil application (2 kg ha^{-1})
Sesame Rainfed	40	30	30	10 kg S (Gypsum 100kgha^{-1})	-	Rhizobium and Phosphobacterium seed treatment (0.6 kg ha^{-1}) or soil application (2 kg ha^{-1})

Sunflower	40	20	20	20 kg S Gypsum 100 kgha^{-1})	MN mixture 12.5 kg or $ZnSO_4$ 25 kg ha^{-1}	Rhizobium and Phosphobacterium seed treatment (0.6 kg ha^{-1}) or soil application (2 kg ha^{-1})
Sugarcane	275	63	115		MN mixture 25 kg ha^{-1} and TNAU sugarcane booster @ 2 , 3 and 4 kg ac^{-1} at 45, 60 and 75 days after planting	Soil application Azospirillum or Azotobacter (2 kg ha^{-1})
Cotton	40	20	-	$MgSO_4$ 10 kg ha^{-1}	-	
Castor	45	15	15	Gypsum 200kg ha^{-1}	-	Soil application (2 kg ha^{-1})

2.4. SOIL MICROBIOLOGY

The sub-discipline of soil science which study the microorganisms present in the soil, their functions and the effect of their activities on the character of the soil and the growth and health of plant life.

Population of micro organism in soil

Soil and pH	Bacteria	Fungi	Actinomycetes
Rhizosphere soil	6778×10^6	918×10^3	83×10^4
Non rhizosphere soil	25×10^6	68×10^3	1×10^4
pH requirement	6.5 - 8.0	4.5 - 6.5	Neutral

Heat Requirements

- Psychrophiles $<10°$ C
- Mesophiles 20-40°C
- Thermophiles $>45°$ C

Microbial function for plant nutrient availability

- N_2 fixation
- PO_4 solubilization
- Organic matter decomposition and synthesis of humus
- Nitrification and denitrification
- Protein decomposition and ammonification

Different type of microbes in soil

Symbiotic Nitrogen fixing bacteria: Rhizobium

Associative symbiosis: Azospirillum

Free living Nitrogen fixing bacteria: Azotobacter, Beijerinkia, Clostridium

Ammonifying bacteria: Bacillus, Pseudomonas

Nitrifying bacteria: Nitrosomonas, Nitrobacter

Denitrifying bacteria: Pseudomonas, Bacillus, Paracoccus

Cellulose decomposing bacteria: Cytophaga, Sporocytophaga, Angiococcus, Polyangium,Clostridium

Cellulose & hemi cellulose decomposing fungi: Penicillium, Trichoderma

Lignin decomposing fungi: Basidiomycetes

Protein decomposing fungi: Fusarium, Aspergillus, Mucor

Humus forming fungi: Alternaria, Aspergillus

Mycorrhizae: Ecto Mycorrhizae-Amanita, Boletus; Endo Mycorrhizae- Phoma, Rhizoctonia

Phosphate solubilising organisms: Pseudomonas, Bacillus bacteria and Penicillium, Aspergillus fungi

Gum producing bacteria: Azotobacter, Beijerinkia, Rhizobium

Biofertilizer: Biofertilizer are ready to use live formulates of such beneficial microorganisms which on application to seed, root or soil mobilize the availability of nutrients by their biological activity in particular and help build up the micro-flora and in turn the improve the soil health.

Types of biofertilizer

Nitrogen

- Rhizobium (50-100 kg N ha^{-1}) for legume crops.
- Azotobacter and Azospirillum (20-40 kg N ha^{-1}) for cereal crops.
- Azotobacter Acetobacter for sugarcane only.
- Blue –Green Algae (BGA) and Azolla for low land paddy.

Phosphorous

- Phosphobacteria recommended for all crops to be applied with rhizobium, azotobacter, azospirillum, acetobacter and AM Fungi

Mode of Application

Seed treatment : Depending on the seed rate, the required quantity of rice cruel is boiled in water and cooled. 600 g (3pockets) of nitrogenous biofertilizers such as azospirillum(Cerelas), rhizobium (Legumes), Acetobacter (Sugarcane) and 600 g (3pockets) of Phosphobacteria inoculum recommended seeds per hectare is mixed in the rice cruel solution and sprinkled over the seeds. Then the seeds are thoroughly mixed to spread the inoculums over the entire surface of the seeds. Seed are dried under shade and have to be sown as soon as possible.

Seedling root dip: For rice crop, one day before transplanting 1 kg azospirillum and phosphobacteria are mixed in the nursery and the roots of seedlings are kept for at list minimum 15-30 minutes.

Soil treatment: Azosprillum for cereals, rhizobium for pulses, acetobacter for sugarcane and phosphobacteria for all the crops @ 2 kg / ha are mixed in 200 kg of compost and kept overnight. This mixture is incorporated in the soil at the time of sowing or planting.

Algal bio -fertilizer (Blue-green algae): These are mainly used for low land rice cultivation. Among algal bio - fertilizers,major strains are Tolypothrix, Nostoc, Anabaena, Aulosira and Cylindrospermum. The amount of nitrogen fixed by Blue Green Algae (BGA) ranges from 15 -45 N kg/ha.Standing water of 2 to 10 cm in the field is a pre - requisite for the growth of blue -green algae. BGA inoculum 10 kg/ha is applied after transplantation of rice crop in the main field.

Azolla: Among Azolla bio -fertilizers the important species are *Azolla caroliniana, A. nilotica, A. filiculoides, A. mexicana, A. microphyllaan d A. Pinnata*, out of which *Azolla pinnata* is most commonly used strain in India. It fixes nitrogen due to Anabaena species of BGA present in the lobes of azolla leaves and it supplies 30 -40 Kg nitrogen /ha.

P-solubilizing biofertilizer: Soil bacteria like *Pseudomonas* and *Bacillus* and fungi like *Penicillium* and *Aspergillus* are capable to convert the insoluble phosphate present in soil which are not available to plants in to soluble forms by secreting organic acids like formic, acetic, propionic and lactic acid etc. The acids lower down the pH of soil and bring the dissolution of bounded forms of phosphate. Among phophatic biofertilizers, *Bacillus megatherium* termed as *Phosphobacteria* are prepared and used commercially.

Vesicular-Arbuscular Mycorrhizae (VAM): It is the symbiotic association of fungi with roots of vascular plants. The main advantage of mycorhizae lies in extension of the penetration zone of root -fungus system in the soil and facilitating an increased efficiency of water and nutrient uptake by the host plant. VAM has important effects on plant phosphorus uptake and availability of other elements like zinc, copper, potassium, sulphur, manganese, magnesium and iron.

Benefits of bio-fertilizers in agriculture

- Biofertilizer can add 20-45 kg N/ha year *(eg. Rhizobium sp* 50-100 kg N/ha year; *Azospirillum, Azotobacter:* 20-40 kg N/ha /yr; Azolla: 40-80 kg N/ha; BGA: 20-30 kg N/ha) under optimum soil conditions.
- Biofertilizers such as *rhizobium, BGA and azotobacter spp* stimulate production of growth promoting substance like vitamin-B complex, Indole acetic acid (IAA) and Gibberellic acids etc.
- Arbuscular mycorrhiza (AM fungi) enhance the uptake of P, Zn, S and water, leading to uniform crop growth and increased yield
- *Azolla-Anabaena* grows profusely as a floating plant in the flooded rice fields and can fix 100-150 kg N/ha /year in approximately 40-60 tones of biomass produced

Vermicompost : It is the process of converting waste into valuable manure by using earthworms.

- *Eisenia fetida* (European worm)
- *Eudrillus eugeniae(* African worm) - more efficient for vermicomposting
- *Perionyx excavatus*(Indian worm) - more tolerent to high temperature
- Culturing of earth warm is ***wormini***

Constituents of biogas	
CH_4 50 –60 %	H_2 5-10 %
CO_2 30-40 %	N_2 4-6%

Interaction of soil microbes and their role in agriculture

Neutralism : Each organism behaves independently and this kind as interaction probably doesn't exist in the soil communities

Symbiosis: Both organisms where both the partners derive mutual benefit. This association has been noticed in biological nitrogen fixation (BNF)

Proto-cooperation: Both are mutually benefited & it is not obligatory for their existence or performance of a particular activity

a. Synergism between AM-legume and rhizobium (P - Soil fertility - N Fixation)
b. Synergism between PO_4 solubilizer and rhizobium

Commensalism: It is the association a partner derives the benefit of the company of the associate without affecting it eg. Cellulose decomposer having enzyme cellulase C_1 act on un-degraded cellulose. Later the partially decomposed cellulose can be metabolished by soil microbes.

Harmful effect

Ammensalism: A partner suppress the growth of the other partner by producing toxin like antibiotics harmful gaseous compound like NH_3, CO_2, ethylene and sulphur

Competition: It is the universal phenomenon of coexistence in a community where species are is suppressed as the two species struggle for the limited supply of nutrients.

Parasitism: It is an association where one partner lives in or on the body of the other and feeds on the body fluid of the host.

Predation: It is the direct feeding of one organism by other.

2.5. QUALITY OF IRRIGATION WATER

Criteria for evaluation of Irrigation water

- Salinity hazard
- Salt index
- Boron concentration
- Soluble Sodium Percentage
- Nitrate concentration
- Sodium hazard
- Bicarbonate hazard
- Chloride concentration
- Magnesium Hazard
- Lithium

Salinity classes (Wilcox 1948)

- C_1 and C_2 – Suitable for irrigation purpose
- C_3 and C_4 – Not suitable for irrigation purpose

Water class	EC mS/m	Salt concentration	Remarks
C_{1-} Low	0 – 25	< 0.16	Safely used
C_{2-} Medium	25 – 75	0.16 – 0.50	Moderate leaching
C_{3-} High	75 – 225	0.50 – 1.50	Not used
C_{4-} Very high	225 – 500	1.50 – 3.0	Not used

Sodium Hazard

$$\text{Sodium Absorption Ratio (SAR)} = \frac{Na^+}{\sqrt{(Ca^{2+} + Mg^{2+})/2}}$$

Adj SAR= SAR (1+ (8.4 – pH_c) - Safe limit is <13

Water classes	SAR value	Remarks
S_1- low S_2 – medium S_3 – high S_4 – very high	0 – 10 10 – 18 18 - 26 > 26	No hazard Appreciable management Unsatisfactory Unsatisfactory

Salt index = (Total Na – 24.5) - (Total Ca-Ca in $CaCO_3$) x 4.85
(Safe limit is between -24.5 to 0)

Bicarbonate Hazard

Residual Sodium carbonate RSC = $(CO_3^{2-} + HCO_3^{-}) - (Ca^{2+} + Mg^{2+})$

<1.25	-	Water used safely
1.26 – 2.5	-	Certain management needed
>2.5	-	Unsuitable

Boron concentration

Water classes	Boron concentration (ppm)			Remarks
	Sensitive	Semi tolerant	Tolerant crop	
Very low	< 0.33	< 0.67	< 1.00	Used safely
Low	0.33 – 0.67	0.67 – 1.33	1 – 2	Certain management needed
Medium	0.67 – 1.00	1.33 – 2.00	2 – 3	Unsuitable for irrigation
High	1.00 – 1.25	2.00 – 2.50	3 – 3.75	
Very high	>1.25	> 2.50	> 3.75	

Chloride concentration

$$\text{Chlorine Index} = \frac{Cl + NO_3}{CO_3^{2-} + HCO_3^{-} + SO_4^{2-} + Cl^{-} + NO_3}$$

Water classes	Chloride (meq/lit)	Water classes	Chloride (meq/lit)
Excellent	<4	Slightly Usable	7- 12
Good	4 – 7	Not suitable for irrigation	12 – 20

Soluble Sodium percentage

$$\text{Soluble Sodium percentage (SSP)} = \frac{Na^{+}}{Ca^{2+} + Mg^{2+} + Na^{+}} \times 100 \quad \text{(Safe limit is <60)}$$

Other water quality indices

- Magnesium absorption ratio $= \frac{Mg^{2+}}{Mg^{2+} + Ca^{2+}} x100$
- Potential salinity $= Cl + ½\ SO_4$ (Safe limit is <5)
- Permeability index $= \frac{(Na^{+} + HCO_3^{-})}{Ca^{2+} + Mg^{2+} + Na^{+}} x100$
- Salt concentration $= \frac{EC}{100}$
- Total Dissolved Salt (TDS) = EC (dSm^{-1}) x 640 or EC (msm^{-1}) = 10^{-2} x 640
- Exchangeable Sodium Percentage (ESP) = Adj SAR
- ESR = 0.015 SAR
- Osmotic pressure (OP) = 0.36 x EC
- Millimhos/cm x 1 = Desi siemen/m
- Micromhos/cm x 0.1 = Millisiemen/m

Irrigation Water Quality

Particulars	EC (dSm^{-1})	pH	SSP	Cl (mel^{-1})	SAR	RSC (mel^{-1})
Excellent	< 0.5	6.5-7.5	< 30	> 2.5	> 1	< 1.0
Good	0.5-1.5	7.5-8.0	30-60	2.5-5.0	1-2	1.0-1.25
Fair	1.5-3.0	8.0-8.5	60-75	5.0-7.5	2-4	1.25-2.0
Poor	3.0-5.0	8.5-9.0	75-80	7.5-10	4-8	2.0-2.5
Very poor	5.0-6.0	9.0-10	80-90	10-12.5	8-15	2.5-3.0
Unsuitable	> 6	> 10	> 90	> 12.5	> 15	> 3.0

2.6. SOIL EROSION AND CONSERVATION

Soil erosion : It is defined as detachment of soil particles their transportation from one place to another and deposition elsewhere through water, wind, coastal waves, snow, gravity and other forces.

Process of soil erosion

- Detachment of soil particles
- Transportation by splashing, rolling and dragging
- Deposition at another place

Type of soil erosion

- Geological erosion
- Accelerated soil erosion

Geological erosion : it refers to natural or normal erosion. It is the regular of the action of water, wind, gravity and glaciers

Acceleration : Soil erosion exceeds the normal rate and becomes unusually destructive and unproductive called accelerated erosion. It occupies 50% area in India.

- National average soil loss is 16.35t / ha/yr
- Permissible limit is between 4.5 – 11.2 t /ha / yr
- 5334 Mt of soil being eroded
- Loss of soil nutrients though water is 5.4 to 8.4 Mt

Factors affecting soil erosion

Erosion = f(cl, v , t, s, h)

- Climate: Intensity, kinetic energy, amount, duration and frequency of rainfall and wind velocity
  - Velocity of 1mm diameter raindrop = 4.5 m/sec.
  - Velocity of 5mm diameter raindrop= 9 m/sec
  - Velocity of water flow is doubled its erosive power increases four times and carrying power 64 times
- Vegetation: Crops, cropping system and accompanying vegetation (Thickness of canopy and root system).
- Topography: Land slope, degree and length
- Soil : Physical, chemical and biological properties of soil
- Biotic interferences

Water erosion : It is the removal of soil from the land surface by water including runoff from melted snow and ice

Different type

- Splash erosion
- Sheet erosion
- Rill erosion
- Gully erosion

Splash erosion : The falling rain drop is capable of generating a force of almost 14 times its weight

Causes

- Beating action breakdown the soil aggregates
- Detachment of soil particles
- Displacement of soil particles

Sheet erosion : It is the uniform removal of thin layers of soil from the surface of the earth caused by sheet flow

Causes

- ❖ Surface creeping of soil and deposition or washes away with drainage water
- ❖ Saltations
- ❖ Suspension

Rill : It is the removal of soil by water form small channels which develop due to concentration of surface flow along the slope.

Gully : Rills with more than 30cm depth are generally called gullies.

Classification of gullies

Classes	Depth (m)	Length(m)
Very small	<3 m deep	< 18 m
Small	3 m deep	> 18 m
Medium	3-9 m	> 18 m
Deep and narrow gullies	9 m deep	< 18 m

Wind erosion : Detachment, transportation and deposition of loose top soil or sand by the action of wind. The different types of wind erosion are

- ☞ Saltation
- ☞ Suspension
- ☞ Surface creeping

Saltation : It is the movement of soil by short series of bounces along the surface of lands at height not more than 1 meter and size of the particles is 0.1 to 0.5 mm

Suspension : Movement of soil particles through suspension and size of the particles is less than 0.1 mm

Surface creeps : Movement of coarse grained particles varying between 0.5 to 1.0 mm is called creep.

Methods of estimating erosion

Soil and water erosion	Wind erosion
Universal soil loss equation (Wischmeir and Smith 1978) A = R x K x Lx S x C x P Where A – Computes soil loss / unit area R – Rainfall factor, K – Soil erodability factor L – Slope length factor, S – Slope gradient C – Crop management, P – Erosion control practice	Chepil & Wood ruff (1963) E = f (I x k x c x l x v) I-Soil erodability index (0.84 mm) C-Climatic factor K-Soil surface roughness L-Unsheltered field V-Vegetative cover
Sedimentation yield : Total runoff can be calculated by using Coshocton Runoff = Run off in litres / Area of plot m^2	

Type of degradation	Area M ha
Water erosion	148.9
Wind erosion	13.5
Chemical degradation	13.8
Loss of nutrients	3.7
Salinization	10.1
Physical degradation	11.6
Total degraded land	187.8

Conservation measure for wind erosion

- Creation of permanent grass strips across the prevailing wind direction
- Roughening of the soil surface
- Stubble mulching
- Tillage provide cloddy surface
- Strip cropping
- Wind breaks

Soil conservation measures for water erosion

- Biological measures
- Mechanical measures
- Other methods

Biological measures :

- Contour cultivation: Growing of crops along the contour
- Strip cropping: Growing of crops across the contour
  - Contour strip cropping
  - Field strip cropping
  - Buffer strip cropping
  - Wind strip cropping
- Cropping system
- Crop rotation
- Mixed cropping
- Tillage practices
- Mulch tillage / Stubbles mulching
- Cover cropping

Mechanical methods :

- Contour bunding is suitable for 2-6% slopping area having <600 mm rainfall and light textured red soil.
- Graded bunds is suitable for 2-6% slopping area having >600 mm rainfall and heavy textured black soil
- Broad base terrace suitable for 6 –10% slopping area
- Contour stone wall suitable for 10 –16 % slopping area
- Bench terracing suitable for 16 –33% slopping area
- Trenches suitable for > 33% slopping area

2.7. REMOTE SENSING AND GIS

Remote Sensing: It is the acquisition of information about an object or phenomenon, without making physical contact with the object.

- In remote sensing optical wave length from 0.3 to 15 ìm are used
- Human eye responds to wave length between 0.4 to 0.7 ìm and which is also photo synthetically active radiation range
- Near infra red wave length ranges between 0.7 to 1.0 ìm and middle infra red ranges between 1.3 to 3.0 ìm
- Velocity of light in air is 3×10^8 m/s
- Simple Vegetation Index = reflectance in infra red band / reflectance in red band

$$\text{Normal Vegetation Index} = \frac{\text{Reflectance in IR band} - \text{Reflectance in Red Band}}{\text{Reflectance in IR band} + \text{Reflectance in Red Band}}$$

- Vegetation index is used in identifying the crop stress that can be used in crop management
- LISS stands for Linear Image Self Scanner, WiFS stands for Wide Field Scanner and PAN stands for Panchromatic Scanner

Characteristics of Sun-synchronous Indian Satellites

Satellites	Height (Km)	Repetivity	Ground Resolution (m)	Application
IRS-1A / 1B	904	22 days	72.5 (LISS-I) 36.5 (LISS-II)	Crop acreage and production estimation, drought monitoring and assessment,flood risk zone mapping and flood damage assessment, irrigation command area status monitoring, land use and land cover mapping, forest survey, wetland mapping and coastal studies.
IRS-1C	817	24 days	6.25 (PAN)188 (WIFS)	
IRS-1D	817	25 days	5.2 (PAN)169 (WIFS)	

Radar Satellite-1 (RISAT-1)	536	25 days	Synthetic Aperture Radar (SAR) management of natural like flood and cyclone.	Agriculture, particularly paddy monitoring in kharif season and disasters
RESOURCESAT-1	817	24 days	5.8 (LISS-4) 23.5 (LISS-3) 56 (AWiFS)	Crop monitoring and condition Assessment Crop canopy water stress
RESOURCESAT-2	822	24 days	5.8 (LISS-4) 23.5 (LISS-3) 56 (AWIFS) multispectral	Crop yield estimates Damage assessment
CARTOSAT – 1	618	5 days	2.5 (PAN)	Cartographic applications like mapping, urban and rural infrastructure development and management, as well as application in Land Information (LIS) and Geographical nformation System (GIS).
CARTOSAT – 2A	635	5 days	1.0 (PAN)	
CARTOSAT - 2B	637	4-5 days	1.0 (PAN)	

Application of remote sensing in agriculture

Soil resource management

- ❖ Soil resource inventory
- ❖ Assessment of soil related constraints
- ❖ Development of management strategies

Crop health monitoring

- ❖ Assessment of drought
- ❖ Monitoring crop condition and fore warming of short falls

Crop inventory

- ❖ Production forecasting
- ❖ Changes in crop pattern
- ❖ Cropping system analysis

Monitoring in season agricultural operation

- ❖ Progress of sowing and total crop area estimation
- ❖ Progress of harvest

Command area management

- ❖ Crop water requirement
- ❖ Water logging
- ❖ Crop violation
- ❖ Canal water seepage

Watershed management

- ❖ Characterization
- ❖ Development
- ❖ Prioritization
- ❖ Monitoring

Land use planning

- ❖ Assessment of land use and land covering
- ❖ Monitoring the land use changes
- ❖ Land evaluation or evaluation of potential land use option

Water availability

- ❖ Monitoring surface water bodies
- ❖ Ground water exploration

Land degradation

- ❖ Extent of waste land
- ❖ Problem soil (water logging and salt affected soil)

Damage assessment

- ❖ Damage due to cyclones and floods
- ❖ Large scale incident of pest and disease

Geographic Information System (GIS)

A geographic information system, commonly referred to as a GIS, is an integrated set of hardware and software tools used for the manipulation and management of digital spatial (geographic) and related attribute data for decision making, planning and management of land use, natural resources, environment, transportation, urban facilities and others

Functions of GIS

Functions	Sub functions
Data acquisition and preprocessing	Digitizing, Editing, Topology building, Projection transformation, Format conversion, Attribute assignment.
Data base management and retrieval	Data archival, Hierarchical modeling, Network modeling, Relational modeling, attribute query, Object oriented attribute
Spatial measurement and analysis	Measurement operations, buffering, Overlay operations, Connectivity operations
Graphic output and visualization	Scale transformation, Generalization, Topographic map, Statistical map

Software used in Geographical Information System

- ArcGIS
- GEOMETICA
- IDRISI
- ILWIS (Integrated Land and Water Information System)
- MapWindow GIS
- ERDAS IMAGINE
- MapINFO
- gvSIG
- GRASS
- SAGA GIS
- Quantum GIS

GLOBAL POSITIONING SYSTEM (GPS)

Global Positioning System (GPS) is a satellite-based positioning system, which by virtue of its special characteristics, is revolutionizing the tasks of navigation and surveying. GPS uses a system of coordinates called World Geodetic System 1984(WGS 84).

GPS applications

- Surveying and mapping, on land, at sea and from the air. Includes geophysical and resource surveys, GIS data capture surveys, etc.
- Land, sea and air navigation including enroute as well as precision navigation, cargo monitoring, vehicle tracking, etc.
- Search and rescue operations, including collision avoidance and rendezvous functions.
- Spacecraft operations.
- Defense applications.

Application of GPS in agriculture

1. **Soil Sampling** : Georeferenced soil samples can be collected using GPS and it will regularly monitored over the decades

Sampling Methods

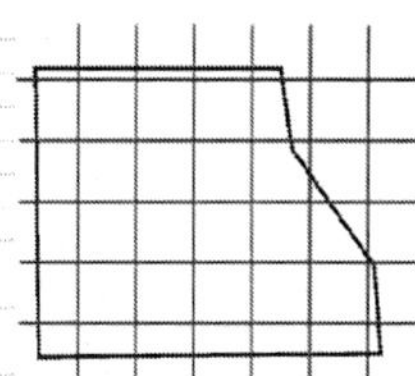

Grid sampling: Intensive sampling of entire field

- Data collected for each cell or point
- Multiple samples combined into each cell or point sample

Directed sampling: Intensive sampling of particular target areas

- Sampling zones established based on knowledge of field
- GPS used to locate sample points.
- Areas of interest intensely sampled, others lightly sampled

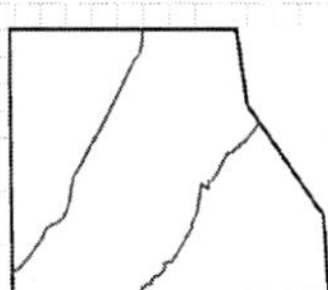

2. Field mapping with GPS

GPS technology is used to locate and map regions of fields such as high weed, disease and pest infestations. Rocks, potholes, power lines, tree rows, broken drain tile, poorly drained regions and other landmarks can also be recorded for future reference. GPS is used to locate and map soil-sampling locations, allowing growers to develop contour maps showing fertility variations throughout fields. The various datasets are added as map layers in geographic information system (GIS) computer programs. GIS programs are used to analyze and correlate information between GIS layers.

3. Variable Rate Technology (VRT)

It is a technology that allows variable rates of fertiliser application, seeding, chemical application and tillage throughout a single paddock. The rate is changed according to a preset map or through information gathered "on the go" by sensors. VRT is most commonly used in conjunction with mapping information such as yield maps or satellite maps, but benefits can still be gained from VRT without such practices.

Benefits of VRT

- Save inputs without wasting (upto 80%)
- Save cost and time
- Reduce pollution
- Site specific management
- VRT integrates well with precision farming

Two approaches to VRT

1) **Map based VRT** involves creating application maps that describe the varying amounts of input needed throughout the field. Application maps are produced from yield, topography, soil, nutrient or weed maps which have been 'ground-truthed' to give specific details of inputs required throughout a paddock. The application maps are interpreted by small computers called controllers that increase or decrease the amount of input according to the map.It involve collecting and processing certain amounts of data, greater amounts of data collected over longer periods of time can create more accurate maps.

2) **Sensor based VRT** utilises sensors to collect data, such as soil properties or crop characteristics, "on the go". This information is processed and used to vary the amount of input applied. This technology does not require detailed maps or extensive decision making prior to application.

Yield Monitoring Systems

Yield monitoring systems typically utilize a mass flow sensor for continuous measuring of the harvested weight of the crop. The sensor is normally located at the top of the clean grain elevator. As the grain is conveyed into the grain tank, it strikes the sensor and the amount of force applied to the sensor represents the recorded yield. While this is happening, the grain is being tested for moisture to adjust the yield value accordingly. At the same time, a sensor is detecting header position to determine whether or not yield data should be recorded. Header width is normally entered manually into the monitor and a GPS, radar or a wheel rotation sensor is used to determine travel speed . The data is displayed on a monitor located in the combine cab and stored on a computer card for transfer to an office computer for analysis. Yield monitors require regular calibration to account for varying conditions, crops and test weights.

GPS system consists of three segments:

Space segment : The space segment (SS) is composed of the orbiting GPS satellites or Space Vehicles (SV) in GPS parlance. The GPS design originally called for 24 SVs, eight each in three approximately circular orbits.

Control system: The control segment is composed of a master control station (MCS), an alternate master control station, four dedicated ground antennas and six dedicated monitor stations.

User segment: This component consists of the GPS receivers and the user community. GPS receivers convert SV signal into position, velocity and time estimates. This process requires four satellites to compute the four dimension of X, Y, Z (position) and time. With this ability, GPS has three main functions; navigation, precise positioning and time and frequency dissemination.

Scientist and their contribution

Scientist	Contribution
Russel & Martin	Y index
Walkley & Black	Wet combustion method
Thompson & Later way	Ion Exchange
Wiegner	Reversibility &Equivalence of exchange Reaction
Adams & Holmes	Synthetic resin
Krishnamoorthy, Davis, Overstreet	Nearest Neighbour Model
Jenny	Oscillation volume
Kappan	Total acidity
Sokonov	Exchange acidity
Boussinghault	Legume can utilise the atmospheric Nitrogen
Beijerink	Isolate Rhizobium from root nodules
Kellner & Dyer	Fixation of potassium
E.J. Russel	Theory of bonding of clay- clay particle
Chepil	Dry sieving method
Tiulin & yoder	Wet sieving method
Kostiakov	Equation on infiltration
O'neal	Permeability class
Schofield	PF scale & ratio law
Ficks law	Gaseous exchange between soil & atmosphere
Atterberg	Soil consistency limit
Arnon & stout	Criteria of essentiality of nutrients
Nicholas	Functional or Metabolism nutrients
Sprengel	Calcium & magnesium
Sachs & Knobs	Sulphur
Warrington	Boron
Gris	Iron
Ma Hargue	Manganese
Sommer & Lipman	Zinc
Sommer	Copper
Arnon & stout	Molybdenum
Broyer	Chlorine
Bray & Kurtz	Phosphate estimation in Acid soil
Olsen	Phosphate estimation in slightly acid & Neutral soil
Das	Phosphate estimation in calcareous soil
Von Liebig	Law of Restitution & Law of minimum & law of limiting nutrient
Mitcherlich	Law of diminishing return & Pot culture Technique
Mehlich	*Aspergillus niger* method
Sackett & Stewart	Soil plague method
Winogradsky	*Azotobacter* method

Schoonover	Gypsum requirement
Shoemaker	Lime requirement
Cate & Nelson	Critical soil test level approach
Dockuchaev	Factors of soil formation
Jenny	Classified the factors of soil formation as active & passive factors
Milne	Soil catena
Dockuchaev	Concept of zonality
Jenny	Expressed soil age in terms of time
Mohr & Vanbaren	Five stages of soil formation
Buchanan	Coined the term Laterite
Vilenski	Climatic classification of soil
Virtanen	Part of Nitrogen fixed is excreted as free Amino acid
Dockuchaev & Sibirtzev	Law of adaptability
Chepil & Wood	Wind erosion equation
Weinmaeir & Smith	Universal soil equation
Puri	Salt index
Baldwin, Kellog & Thorp	Elaborated marbut's classification
Leather	First imperial agricultural chemist & Father of Soil Science & Agrl. Chemistry in India
Wadia, Krishnan & Muderjee	Prepared soil map of India in 1935 based on geological formation
Schokalsky	Prepared soil map of India in 1932 showing 4 soil groups
Viswanath & Ukil	Prepared soil map of India in 1943 showing 14 classes
Reychaudhuri	Soils of India , their classification , occurrence, and properties
Murhty	Benchmark soils of India
Dokuchaev	Genetic system of classification
Marbut	Morphogenetic system of classification
Dachaufour	Coined the term Lessivage
Dalton	Atomic hypothesis
Neils bohr	Atomic structure for hydrogen
De broglie	Quantum mechanical model of atom
Doeberiner	Law of triads
Newlands	Law of octaves
Kossel & Lewis	Octet rule
Heisenberg	Ortho & Para hydrogen
Pierre & Marrie curie	Discovered polonium & radium
Schmidt & Curie	Thorium
Debierne	Actinium
Rutherford	Effect of electric & magnetic field on radioactive rays
Libby	Carbon dating
Thomson	Discovered electron
Rutherford	Discovered proton & spinthariscope
Chadwick	Coined the term neutron

Anderson	Positron
Yukawa	Pi Meson
John Dalton	All matter was composed of atoms
Blackman	Law of limiting factor
Blackman	Electro chemical theory of nutrient uptake &theory of relativity
Nymith	Electro ultrafiltration theory
Ramamoorhty	Target yield concept
Ramamoorthy & Subramanium	Equilibrium Phosphate potential
Beckett	Phosphate equilibrium potential
Russel	E – value
Lorentz	Coined the term electron
Goldstein	Discovered positive rays
Bothe & Becker	Found neutron
Lawrence	Cyclotron
Henry Becquerel	Radioactivity
Curie & Joliot	Artificial radioactivity
Pauli	Neutrino & anti neutrino
J.J. Thompson	Found e/m value
R.A. Milikan	Found e value(electronic charge)
Rutherford	Alpha particle scattering experiment
Mosley	Determined atomic number or proton No.
Ottoman & Stassmann	Nuclear fission
Cameroon	Soil is an inexhaustible source of nutrients
Baule	Percent yield concept & nutrient sufficiency concept
Bray	Nutrient mobility concept & percent sufficiency concept
Nicholas	Ultra micronutrient concept
Marshall	CEC is next to photosynthesis & pH is physical phenomenon
Epstein	Carrier theory
S.V.Desai & S.C.Biswas	Biogas
La cassageue	Auto-radiogaphy
Waksman	Lignin theory
Van Helment	Water essential for plant
Woodward	Water –transpiration
Lewis & Randol	Ionic strength
Hevesy & Hoffer	Isotopic dilution
Reinganum	Auto radio activities
Jenny & Overstreet	Root exchange
Devaux	Root CEC
Sorensen	pH concept
Silen	pE concept
Troug	Prescription approach & father of soil testing

Colwell	Integrated soil test approach
Beaufils	DRIS concept
Libby	Radio carbon dating
Aslyng & Latter Schofield	Lime potential
Beckett	Q / I (Quantity / Intensity)
Stevenson	Polyphenol theory of organic matter
Parker	Fertility or nutrient index
Lindsay & Norvell	DTPA extractable micronutrients
Fried & Dean	A value
Larsen	Y or L value
Mitscherlich	Mathematical relation indicating relationship between plant growth response and growth factor
Chapman	Dignostic technique for soil and plants
Holmes	Total micronutrient
Lindsay & Norvell	Available micronutrients
Wollvy	Father of soil physics
Dokuchaev	Normal and extranormal soil
Fallov	Residual and transported soil
Hilgard	Black alkali and white alkali soil
G.D. Smith	Seventh apprximation
Kononova	Work on organic matter
Ramser	Rational formula for watershed
Andres	Liquefication of gas
J.B.Lawes	Super phosphate
Nernst	Third law of thermodyanmics
Bragg	X ray diffraction
Murthy	Benchmark soil
Hole	Horizonaion
Kellog	Latosols
Buchanan	Laterite
Robinson	International pipette method
Atterberg	Plasticity limits

Points to remember

- Soil development study is Pedology
- Crop husbandry study is Edapology
- ‘O’ Horizons dominated by organic material normally occur in forest
- Soil material less than 2 mm constitutes soil sample
- Soil particle density varies within 2.5-2.7 and bulk density between 1.4 and 1.8 Mg m^{-3}
- Gravitational water is above field capacity (-0.1 to -0.6 bar)

- Capillary water is held between -0.1 and 31 bar and mostly available for plants.
- Capillary water held within 15 bar is relatively easily available to the plants.
- Hygroscopic water is held tightly lower than - 31 bars
- Feldspar is an anhydrous alumino silicates of potassium, sodium and calcium
- 1:1 Type clay minerals - Kaolinite
- 2:1 Type clay minerals – Montmorillonite, Illite and Vermiculite
- 2:2 Type clay minerals - Chlorite
- Total Land Capability classes are I to VIII and I to IV are arable
- Irrigability classes are I to V
- Hydraulic conductivity: Ca –soil > K –soil >Na – soil
- Alluvial soil is a largest and most important soil group of India
- Saline soil - ECe >4 dS/m, ESP <15% and pH <8.5
- Salinity above 4 d S/m is harmful to crops
- Alkali soil - ECe <4dS/m, ESP > 15% and pH>8.5
- Saltation - 0.1 to 0.5 mm size
- Suspension - Less than 0.1mm
- Surface creep - Larger than 0.5 mm smaller than 1.0mm
- Mechanical analysis (Texture) -Stocks Law (1851)
- Dr. J.W. Leather – Father of Soil Science and Agricultural Chemistry in India
- DESSIFER - Software used for fertilizer recommendation
- Acid soil: Al^{3+}, Fe^{2+} and Mn^{2+} are found to be toxicity in nature
- N, P, K, Ca and Mg are available more in the pH 5.5 – 6.5
- Zn, Mn and Fe are less available in the pH 5.5 – 6.5
- pH increases from 5.5 to 7.5 the available phosphorus changes from $H_2PO_4^-$ to HPO_4^{2-}
- Plant nutrients are extracted using di-acid 5:2 (H_2SO_4 and $HClO_4$) and tri-acid 9:3:1(HNO_3, $HClO_4$ and H_2SO_4)

CROP IMPROVEMENT

3

M. Govindaraj and P.Arunachalam

3.1. INTRODUCTION TO PLANT BREEDING

What is Plant Breeding?: Plant breeding is an art and science which deals with changing genotypes of the plants, to adapt them to the service of human being. (or) The application of genetic and cytogenetics principles in crop improvement are called plant breeding.

Specific objectives of plant breeding: The main objectives of the plant breeding is to improve the characteristics of crop plants which are agronomically and economically desirable

- ✓ To increase the yield of the existing crop plant by developing "superior strains"
- ✓ To improve the quality of seed i.e., size, colour, flavour, nutritional quality
- ✓ To impart tolerance to abiotic stress like drought, frost, diseases, salinity and biotic stress like pest and diseases
- ✓ To develop early maturity and non-lodging varieties
- ✓ To modify agronomic characters such as plant height, tillering, branching, etc.,
- ✓ To evolve better strains, that can be respond well to fertilizers and minerals
- ✓ To evolve the varieties that can be cultivated under varied seasons, environmental conditions and geographical regions
- ✓ To produce better strains that can be handled easily during harvest
- ✓ To evolve varieties to prevent the premature drop of leaves, fruits and seeds

Three specific strategies were employed during the Green Revolution

1. **Plant improvement:** The green revolution centered on the breeding of high-yielding, disease resistant, and environmentally responsive (adapted, responsive to fertilizer, irrigation, etc.) cultivars.
2. **Complementary agronomic package:** Improved cultivars are as good as their environment. To realize the full potential of the newly created genotype, a certain production package was developed to complement the improved genotype. This agronomic package included tillage, fertilization, irrigation, and pest control.
3. **Favorable returns on investment in technology:** A favorable ratio between the cost of fertilizer and other inputs and the price the farmer received for using this product was an incentive for farmers to adopt the production package.

- Norman E. Borlaug (Mexico) is the Father of green revolution (Wheat)
- Dr.M.S. Swaminathan, the Father of green revolution in India is a pioneer in mutation breeding

History of Plant Breeding

- Around 700 BC Babylonians and Assyrians pollinated date palm artificially
- Thomas Fairchild produced the first artificial plant hybrid (*Fairchilds' mule*) by crossing cornation with sweet William in 1717.

- Mather recognized the process of natural selection in maize.
- German Scientist Joseph Koelrouter carried out *systematic* studies on plant hybridization and produced many hybrids in tobacco.
- Thomas Andrew knight (English Scientist) was perhaps the first to use hybridization for practical plant improvement and produced new varieties of fruits trees and ornamental plants.

Achievements and milestones in plant breeding

Year	Selected milestones in plant breeding
9000 BC	First evidence of plant domestication in the hills above the Tigris river
3000 BC	Domestication of all important food crops in the Old World completed
1000 BC	Domestication of all important food crops in the New World completed
700 BC	Assyrians and Babylonians hand pollinate date palms
1694	Camerarius of Germany first to demonstrate sex in plants and suggested crossing as a method to obtain new plant type
1716	Mather of USA observed natural crossing in maize
1717	Fairchild created first artificial hybrid (carnation × sweet william)
1727	Vilmorin Company of France introduced the pedigree method of breeding
1753	Linnaeus published Species plantarium. Binomial nomenclature born
1762-66	Koelreuter of Germany made extensive crossing in Tobacco
1847	Reid's Yellow Dent" maize developed
1857	Hallet in England practical single plant selection in wheat, oats and barley.
1866	Mendel published his discoveries in Experiments in plant hybridization, cumulating in the formulation of laws of inheritance and discovery of unit factors (genes)
1899	Hopkins described the ear-to-row selection method of breeding in maize
1900	Mendel's laws of heredity rediscovered independently by Correns of Germany, de Vries of Holland, and von Tschermak of Austria
1903	The pure-line theory of selection developed by Johanson
1904-1905	Nilsson-Ehle proposed the multiple factor explanation for inheritance of color in wheat pericarp
1908-1909	Hardy of England and Weinberg of Germany developed the law of equilibrium of populations
1908-1910	East published his work on inbreeding
1909	Shull conducted extensive research to develop inbreds to produce hybrids
1917	Jones developed first commercial hybrid maize
1926	Pioneer Hi-bred Corn Company established as first seed company
1934	Dustin discovered colchicines
1935	Vavilov published The scientific basis of plant breeding
1940	Harlan used the bulk breeding selection method in breeding
1944	Avery, MacLeod, and McCarty discovered DNA is hereditary material

1945	Hull proposed recurrent selection method of breeding
1950	McClintock discovered the Ac-Ds system of transposable elements
1953	Watson, Crick, and Wilkins proposed a model for DNA structure
1970	Borlaug received Nobel Prize for the Green Revolution
1970	Berg,Cohen, and Boyer introduced the recombinant DNA technology
1994	FlavrSavr" tomato developed as first genetically modified food produced for the market
1995	Bt corn or cotton developed
1996	Roundup Ready® soybean introduced
2004	Roundup Ready® wheat developed

List of various Biometrical Techniques used in Crop Improvement

Applications/Uses	Biometrical Techniques Used
1. Assessment of Variation	1. Measures of Dispersion 2. Components of Genetic variances 3. Metroglyph Analysis 4. D^2 Statistics
2. Selection of Elite Genotypes	1. Correlation Analysis 2. Path Coefficient Analysis 3. Discriminant Function Analysis
3. Choice of Parents and Breeding Procedures (a) Analysis of Several Single Crosses (b) Analysis of several Three way Crosses (c) Analysis of Individual Cross	1. Diallel Cross Analysis 2. Partial Diallel Analysis 3. Line x tester Analysis 4. Triallel Analysis 5. Generation mean analysis 6. Biparental cross analysis
4. Assessment of Varietal Adaptability	1. Stability Analysis Models. 2. AMMI analysis

Variation: variation in morphology, physiology, and behavior exist among individuals in a natural population.

Heredity: offspring resemble their parents more than they resemble unrelated individuals.

Selection: some individuals in a group are more capable of surviving and reproducing than others. (i.e., more fit).

Pre-breeding: Pre-breeding includes the introduction, adaptation, evaluation, and improvement of germplasm resources for use in breeding programs. Pre-breeding usually does not provide directly new cultivars for the growers. It rather develops germplasm resources that are either directly or indirectly used to develop new cultivars. Pre-breeding is not a recent concept and has been an important component in the development of present-day single-cross hybrids. Nowadays pre-breeding includes inbred development and hybrid parent breeding and its disseminations.

Comparison of pre-breeding and traditional breeding

Applications/Uses	Biometrical Tech
1. It is also known as genetic enhancement	It is also known as sustainable plant breeding
2. It leads to genetic enhancement of germplasm	It leads to development of productive cultivars/ hybrids
3. It leads to value addition	It does not lead to value addition
4. It leads to broadening the genetic base of population	In improved cultivars the genetic base becomes narrow
5. The chief breeding method is backcross methods	All breeding methods such as introduction, selection, hybridization and mutation are used
6. The end products are improved germplasm lines	The end product is cultivar or hybrid
7. The end product is used as parent or developing improved cultivar, hybrid	The end product is used for commercial cultivation
8. It involves adapted and non-adapted genotypes in crossing programme	It includes only adapted genotypes
9. It is a long term breeding programme	It is a short or medium term breeding programme
10.It is taken up by public sector plant breeding organizations	It is taken up both public and private sector organizations

Technology / Tools in Plant Breeding

Technology / tool	Common use of the technology/tool
Classic / traditional tools	
Emasculation	Immature anthers from hermaphrodite flower to make famale flower
Hybridization	Crossing un-identical plants to transfer genes or achieve recombination
Wide crossing	Crossing of distantly related plants
Selection	Primary tool for discriminating among variability
Chromosome counting	Determination of ploidy characteristics
Chromosome doubling	For manipulating ploidy for fertility
Male sterility	To eliminate need for emasculation in hybridization
Triploidy	To achieve seedlessness
Linkage analysis	For determining association between genes
Relatively advanced tools	
Mutagenesis	To induce mutations to create new variability
Tissue culture	For manipulating plants at the cellular or tissue level
Haploidy	Used for creating extremely homozygous diploid (dihaploid)
Isozyme markers	To facilitate the selection process
In situ hybridization	To assist in successful inter-specific crossing
More sophisticated tools	
DNA markers	

RFLP	More effective than protein markers (isozymes)
RAPD	PCR-based molecular marker
Advanced technology	
Molecular markers	SSR, SNPs, etc.
Marker-assisted selection	To facilitate the selection process at allelic/genetic level
DNA sequencing	To generate ultimate physical map of an organism
Plant genomic analysis	Studying the totality of the genes of an organism
Bioinformatics	Computer-based technology for prediction of biological function from DNA computational biology especially using genetic information
Microarray analysis	To understand gene expression and for sequence identification

Various kinds of varieties

Types of Variety	Brief description
1. Variety of common knowledge	A variety whose existence is a matter of common knowledge at the time of filing the application.
2. Extant variety	All released and notified varieties which have not been protected.
3. Candidate variety	A variety that has to be registered under PVPA.
4. Reference Variety	All released and notified extant varieties that are in seed production chain.
5. Example Variety	A variety that is used for comparison of a particular character.
6. Farmer's Variety	A variety that has been developed by a farmers and used for commercial cultivation for several years.

3.2. PLANT REPRODUCTION

Plants reproduce in two ways by sexual and asexual reproduction.

i. Asexual / vegetative / reproduction / Agamogenesis: It occurs without the fusion of gametes (reproductive) cells. Asexual reproduction can be brought about artificially by means of leaf cuttings, stem cuttings, root cuttings, etc. Plants that originate from asexual reproduction are usually identical to the parent plant.

ii. Sexual reproduction (**gamogenesis**) involves the union of a male and a female gametes cell **(Amphimixis)**. From this union a seed and ultimately a new plant is produced. The plants originating from sexual reproduction are often quite different from their parents and from each other. Because of this possibility for variation, sexual reproduction of plants is the method used by plant breeders in developing new strains and varieties.

Fertilization : It is in the ovule that the male and female germ cells unite to produce diploid zygote is called **fertilization**. Once fertilization has occurred development of the ovule begins. The result is a seed. Fusion of remaining sperm with secondary nucleus to form triploid primary endosperm nucleus is known as **triple fusion**. The parent flower that furnishes the pollen is called the pollen parent; the one that bears the seed is called the seed parent. **Autogamy** is the transfer of pollen grains from the anther to the stigma of the same flower. **Allogamy** is the transfer of pollen grains from the anther to the stigma of another flower of the same plant (Geitanogamy) or flower of another plant (Xenogamy).

Complete flower is having calyx (sepal), corolla (petal), stamens and pistil. Stamen contains anthers and filament. Pistil has stigma, style and ovary.

- Microsporogenesis: pollen mother cell (pmc) undergoes meiosis cell division to produce four haploid cells called microspores
- Microgametogenesis: development of male gametophyte (vegetative and generative nucleus) from functional microspore
- Megasporogenesis: megaspore mother cell (mmc) undergoes meiosis to produce four haploid cells. Three cells degenerates and remaining one functional megaspore
- Megagametogenesis: development of female gametophyte (eight nucleated structure i.e., egg cell, antipodals, synergids and polar nuclei) from functional megaspore

During fertilization one sperm nucleus unites with egg to form diploid (2n) zygote (first fertilization) and tube nucleus unites with the two polar nuclei to form triploid (3n) endosperm (second fertilization). These processes collectively called as *double fertilization.*

- The ovule is attached with ovary through *funiculus.*
- At maturity stage funiculus leaves a scar on seed coat known as *hilum.*
- A pore in the ovule wall of the integument through which the pollen tube enters to fertilize the egg called as *micropyl*
- The region of integument origin and directly beneath the funicular attachment is known as *chalaza.*
- The area between micropyl and chalaza called as *raphe.*

Self pollinated crops: 1-5% cross pollination occur eg. rice, wheat, blackgram, greengram, groundnut, tomato

Often cross pollinated crops: 5-50% cross pollination occur eg. sorghum, redgram, sesame, cotton, brinjal, bhendi

Cross pollinated crop: more than 50% cross pollination occur eg. maize, pearl millet, sunflower, castor, cabbage, watermelon

Mechanism Promoting Self-Pollination :

i. Cleistogamy: Flower do not open – Wheat, Oats, Barley

ii. Chasmogamy: Flowers open after pollination - Rice

iii. Stigma closely surrounding anthers - Tomato, Brinjal

iv. Stamens and Stigma hidden by floral organs – Pulses

v. Stigma elongate through staminal column

Mechanism Promoting Cross-Pollination :

i. Dicliny/ Unisexuality:

a) Monoecy : Presence of staminate and pistillate flowers in the same plant and in same inflorescence (castor, mango, banana, coconut) and in separate inflorescence – Maize

b) Dioecy – Male and female flowers in different plants – Papaya, datepalm

ii. Dichogamy:

a) Protantry - Pistils mature before stamens – Bajra

b) Protandry - Stamens matures before pistils – Maize, Sugarbeets

iii. Herkogamy: Spatial separation anther and stigma - Stigma are covers with a waxy film - Lucerne or alfalfa

iv. Combinations of more than one mechanism eg. Maize with monoecy and protandry

v. Self – incompatibility: Failure of pollen from a flower to fertilize same flower or other flower in the same plant – Primula , *Nicotiana sanderae*

vi. Male sterility: Absence of functional pollen grains

Apomixis is an asexual type of reproduction in which the plant embryos grow from egg cells without being fertilized by pollen or the male part of the plant.

Classification of Apomixis

Basis of classification	Types of Apomixis	Brief description
(i) Site of origin	Parthenogenesis	Embryo develops from egg cell.
	Apogamy	Embryo develops from either synergids or antipodal cells.
	Apospory	Embryo develops from the embryosac which has developed from the cell of archisporeum/nucellus or integument.
	Adventive Embryony	Embryo develops directly from the diploid cell of wither nucellus or integument.
(ii) Occurrence	Recurrent Apomixis	Embryosachs diploid cells and embryo develops from the diploid cells.
	Non-Recurrent Apomixis	Embryosac consists of haploid cells and embryo develops from haploid cell.
(iii) Frequency	Obligate Apomixis	The reproduction occurs by apomictic means only.
	Facultative Apomixis	The reproduction occurs by both sexual and apomictic means.

Difference between Apomixis and Amphimixis

Particulars	Apomixis	Amphimixis
1. Term first used or coined by	Winkler 1906	Weismann 1891
2. Sexual Fusion or union of male and female gametes	Does not involve	Involved
3. Gene combination or gene flow	Not possible	Possible
4. Segregation	Does not occur	Occurs
5. Rapid development of inbred lines	Possible	Not possible
6. Conservation of heterosis	Possible	Not possible
7. Type of production	Asexual	Sexual
8. Types	Parthenogenesis, Apogamy, Apospory, Adventive	Autogamy, Allogamy
9. Frequency in plants	Low	High

3.3. METHODS IN PLANT BREEDING

A. PLANT INTRODUCTION

- Dr. Harbajan Singh called as Indian Vavilo
- Plant introduction is a process of introducing plant with specific characters from their area of origin into a new area.
- If introduced from foreign country, it is called exotic collection (EC) and from same country, it is known as indigenous collection (IC).
- Plant introduction is *easiest and quickest* method of plant breeding.
- Introduced plants may be used directly for cultivation or may be used after subjecting to selection, field trials and hybridization with indigenous strains (secondary introduction).
- New plants are introduced in the form of cuttings or seeds or stubbles.
- Generally the adaptation and the performance of the introduced variety are poor because of different climatic conditions.
- The introduced variety can be improved by rising several generations until it is well adapted to new environmental conditions.
- The adjustment of newly introduced plant to new or changed climatic conditions is called *acclimatization.* The extent of acclimatization is determined by the mode of pollination, range of genetic variability, environmental conditions and duration of the crop.
- Great care must be taken while importing a plant material because it may carry pathogens along with it; hence, introduced plant material is thoroughly subjected to "plant protection" and "plant quarantine regulations".
- The introduced plants are maize, groundnut, tea, tobacco, potato, apple, cashewnut, pine apple, grapes, cherries, guava, chillies, sweet potato, eucalyptus, sapota, custard apple, bougainvillea, sorghum, etc.,

Types of genepool and their description

Type of gene pool (Designation)	Definition/brief description
1. Primary gene pool (GP1):	Intermating is easy and leads to production of fertile hybrids. It refers to genotypes of the same species. It is of prime breeding importance.
2. Secondary gene pool (GP2):	It leads to partial fertility on crossing with primary genepool. It includes related species.
3. Tertiary gene pool (GP3):	It leads to production of sterile hybrids on crossing with primary genepool. It includes unrelated species.
4. Quaternary gene pool (GP4):	It refers to crop cultivars that have been developed through biotechnological approaches. It includes transgenic varieties and other value added genotypes.

- National Bureau of Plant Genetic Resources (NBPPR), New Delhi was established in 1976 to helps in plant introduction in India

The other organizations which are also involved in plant introduction are

- Forest Research Institute (FRI), Dehradun
- ICAR - National Rice Research Institute (NRRI), Cuttack formerly Central Rice Research Institute (CRRI)

- International Crops Research Institute for Semi-Arid Tropics, ICRISAT (near Hyderabad)-
- Central Potato Research Institute (CPRI), Shimla
- Botanical Survey of India, Kolkata - Medicinal and Aromatic plants
- Central Plantation Crop Research Institute, Kasergode
- Sugarcane Breeding Institute, Coimbatore

Merits

- The easiest, quickest and economical method of crop improvement
- Introduced plants serve as a good source of parental material (genetic variation) in hybridization experiments
- Introduced plants can be subjected to selection to get better results
- The introduced varieties can be released directly for cultivation (primary introduction)
- The strains developed in one part of a country can be cultivated in another part

Demerits

- Plant introduction is associated with the introduction of new plant diseases, pest and weeds. They may multiply in the new climate and cause serious damage to the introduced variety and local variety.
- The pathogens invaded from other countries aare (Fungal and Bacterial)
  - *Phytophthora infestans* – Late blight of potato (from Europe, 1883)
  - *Hamiliea vastatrix* – Coffee rust (from Ceylon, 1876)
  - *Urocystis tritici* – Flag smut of Wheat (from Australia)
  - Mycoplasma – Bunchy top of banana(from Ceylon)
- Some of the dangerous weeds introduced into our country are: *parthenium hysterophorus* (along with wheat imports from USA), *Argemone mexicana*, *Lantana camera*, *Eicchornia* etc.

Achievements of plant introduction

Some of the introduced plant varieties which are performing well in India are

Crop	Important Variety	Introduced from
Rice	IR-8, IR-21, IR-28	Taiwan
Maize	Dlxle-11, 12 Texas-21	USA
Wheat	Ridley, Sonara 63	USA
Oat	Kent	Australia
Tomato	Sloux	USA
Tobacco	Varginia Gold	USA
Sweet potato	RA-17	China
Soya bean	Lee, Hill Davis	USA
Onion	Texas grano early	USA
Grape	Beauty seedless	USA
Pea	Rimpus	Germany
Watermelon	Asshyamato	Japan
Cinchona	Varieties	S.Africa

B. SELECTION

- Selection is the *oldest* breeding method.
- The process of selection of individual plants or a group of plants which show superior characters is called plant selection.
- Selection process is the basis for all crop improvement methods since the origin of domestication of crop plants.
- Selection is of two types: Natural and Man-made (Artificial).

a. Natural Selection

- Natural selection eliminates all but favorable gene combinations (superior characters) from amongst the numerous random variants
- The natural selection is a dynamic guiding influence
- Many differences between species, sub-species and varieties have arison as a result of such a selection pressure
- Naturally existing population are *not homogenous*
- Even in the case of plants which are fertilized over a number of generations, heterogeneity can still be demonstrated
- Mendelian recombination of genes provides inexhaustible variability to breeder
- The ecotypes and the cultivated crop plants are developed through natural selection (cultivated wheat is a natural allohexaploid)
- Ecotypes provide the basis for artificial selection and hybridization

b. Artificial selection

- The artificial selection of plants with desirable characters and superior gene combinations from mixed population is described as artificial selection
- Man is the major agent which influences the plant selection continuously under domestication
- The many domesticated cultivated plants have arisen from wild ancestors by selection consciously or unconsciously applied on them by man considerably
- Selection and mutation are powerful tools in the hands of the breeder to change the characters of plants to his advantage

The different artificial selection methods is narrated below

i. Pureline selection

- Pureline selection involves selection of progeny of singly, self-pollinated, homozygous desirable plant from a mixed population
- The basis for the pureline selection is phenotype
- W.L.Johanssen (Denmark) gave the concept of pureline selection and called single hereditary lines as purelines. Pureline selection is also called head to row selection or progeny selection or inbred selection
- Johanssen defined a pureline as "the descendants of single homozygous, self fertilized organism having undergone no germinal change"
- Pureline selection is practiced in the mixed population of a self pollinated crop
- The variety developed by pureline selection is uniform in phenotype and genotype

- Differences in the phenotypic characters of the progeny of a pureline are due to the effect of environmental conditions

Some varieties developed through pureline selection are Rice (MTU-3), Wheat (NP-4, NP-12), Cotton MCU, Nandyala14), Tobacco (NP-28, NP-63, T-59), Groundnut (TMV-3) and Castor (HC-1, HC-6)

Merits

- Easy method of crop improvement
- To improve local self-pollinated plants
- These varieties are impressed by cultivators as the population of pureline variety is uniform in phenotype and genotype
- This method is also applicable to cross fertilized crops to produce purelines

Demerits

- Pureline selection is lengthy, laborious and expensive as it involves 7-8 generations of selfing
- Heterosis is not associates with pureline selection
- New characters cannot be introduced into the improved variety
- It is not possible to improve a variety beyond a certain level of homozygosity
- The adaptation of the variety to varied environments is limited as it is not associated with genetic variations

ii. Mass Selection (purification of existing varieties)

- In mass selection plants with desirable superior characters are selected on the basis of phenotype from the mixed cross-pollinating population
- Mass selection is applicable to cross-pollinated plants
- Through mass selection, the genetic pool and level of the population is improved through selection and bulking the superior genotypes already existing
- Mass selection does not bring out any changes in the genotype
- It helps in isolating different phenotypes from a mixed heterogenous population
- It has a limited use for self pollinated plants

Important strains developed through mass selection are Sorghum (T-22), Potato (K-122), Groundnut (TMV-1, TMV-2), Cotton (Dharwar American, Cambodia)

Merits

- Purification of existing varieties
- Easiest and quickest less expensive method of crop improvement
- Does not need much scientific knowledge, skill and control of pollination
- Newly produced variety need not be tested
- Exclusive method of selection to improve wild or local varieties
- It is more an art than a scientific method
- Developed strains are well adapted to different environmental conditions because they are selected from a large number of plants (selected from heterogeneity)
- Retains considerable genetic variations in improved varieties

Demerits

- This selection method utilizes already existing genetic variability and it cannot create genetic variability
- Strains developed are not uniform but exhibit great variation
- The results of mass selection are temporary
- The heterozygosity becomes more complex if there is no control over pollination
- The concentration in the development of new strain is only on female parent
- Not possible to determine that a superior character is homo- or heterozygous
- Mass selection in generally not practiced in self pollinated crops
- It is not a good method to increase the yield because the yield is subjected to environmental pressures and variation
- Because of genetic variability desirable qualities may diminish continuously

iii. Clonal Selection (vegetatively propagated plants)

- The progeny of a single plant obtained by vegetative propagation is called **clone**
- Vegetative propagation is through underground stems, tuberous roots, bulbils, stem cuttings, sub aerial stems etc.,
- The plants produced from a clone or the variations between different clones are observed and the plants with desirable characters are selected is called clonal selection
- All the plants of this progeny are similar in phenotype and genotype because there is no involvement of crossing over, recombination and sexual fusion
- Mutations can only change the characters
- Minor variations in the progeny are due to environment
- The main aim is to isolate the best genotype already present in the population of vegetatively propagated crop on the basis of phenotype only
- This selection method is practiced in vegetatively propagated plants like potato, banana, turnip, sugarcane, orange, grape, garlic, sugarbeet, mango, etc,.

Improved strains of vegetatively propagated plants through clonal selection

Potato	Kufri red (Darjeeling), Kufri Safed (Phulwa), Red Round
Mango	Mundapa poddaneelam
Banana	Bombay green (Developed from Dwarf Cavendish banana), High Gate, Pede Monthan
Sweet oranges	Yuvraj Blood Red

Merits

- The varieties developed by clonal selection are stable in phenotype and genotype even after several generations as there are no segregation of characters
- Clonal selection is suitable for the improvement of vegetatively propagated plants
- Heterozygosity and hybrid vigour can be maintained and preserved for several generations
- Hybrid vogour is also maintained through tissue culture and apomixes if they also can reproduce sexually

Demerits

- It is not applicable to sexually reproducing seed plants
- Plants with new phenotypes and genotypes cannot be produced by clonal selection in exclusively vegetatively propagated plants (except through induction of mutations)
- Clonal selection is useful only to isolate the superior genotype from a mixed population

C. HYBRIDIZATION

- The crossing between two dissimilar plant genotypes to get offsprings with new desirable combinations of characters as a result of genetic recombination is called hybridization
- It creates new genetic combinations and genetic variations
- German Botanist Joseph Koelreuter was the first person obtained hybrids in tobacco
- Desirable characters present in a strain can be transferred to another strain by hybridization

Hybridization involves the following general procedure

a) Selection of parents

- It is the *first step* in hybridization
- The plants possessing desirable characters are selected as parents from local collections
- Of the two parents selected, at least one should be a well adapted proven variety to local environmental conditions

b) Selfing of parents

- In this step the plants selected are self-pollinated to attain homozygous parental characters
- In cross pollinated crops homozygous plants are produced by a process called *inbreeding* and the progeny obtained is known as *inbred line*
- An inbred line is homozygous for some traits

c) Emasculation

- The removal of stamens or anthers or killing the pollen grains of a bisexual flower without affecting the female reproductive organ is described as *emasculation*
- The purpose of emasculation is to prevent self pollination and self fertilization in female parent
- In monoecious plants (castor) the male flowers or the male inflorescences (maize) are removed during emasculation
- Removal of stamens is done by the foreceps (wheat, oat, barley) or by hot water treatment at 40-45^{0}C for ten minutes (rice, sorghum) or by producing male sterile lines

d) Bagging

- Immediately after emasculation, the flowers or inflorescences of the female parent is covered with a butter paper or polythene bag to prevent random cross pollination
- In monoecious species like maize or where pollen is used from male flowers, these too are bagged to maintain the purity of pollen grains
- The bag is labeled exhibiting the date of emasculation, the date of artificial pollination and the names of male and female parents

e) Artificial Pollination

- Mature, fertile and viable pollen grains are placed on a receptive stigma
- The procedure consists of collecting pollen from freshly dehisced anthers and dusting them on the stigmas of emasculated flowers
- The collection and dusting pollen is carried out with brushes
- Immediately after the artificial cross pollination, the flower or the inflorescence of the female parent is covered with the bag

f) Raising F_1 Generation

- Mature seeds from female parent are sown separately to raise F_1 generation/hybrid plants
- F_1 plants are selfed and seeds are collected to raise F_2 generation

g) Raising F_2 Generation

- The recombinants in F_2 are carefully noted to select plants with desirable characters
- In **self pollinated crops** different selection methods of selection are applied to isolate the plants with desirable traits from the segregating progeny are
  - **i. pedigree method**: Progenies of each selected plant in F_2 generation (25-30) are selected in succeeding generations until genetic purity is reached
  - **ii. Bulk population method**: Selection is delayed until a later generation usually the F_5 or F_6 after hybridization at which time segregation will virtually have ceased
  - **iii. Back Cross method**: this method is a form of recurrent hybridization by which a superior characteristic may be added (disease resistant) to an otherwise desirable variety or parent (donor parent), for example Genes for stem rust resistant were added to a widely adapted variety of (Recurrent parent)
- In **cross pollinated crops** the following five methods applied to produce hybrids and to isolate the plants with desirable traits from the segregating progeny
  - **i. Single cross** involving two inbreds (A x B)-Single cross hybrids are common in sorghum, bajra
  - **ii. Double cross** involving four inbreds (A x B) x (C x D)-Double cross hybrids are common in maize and they retain hybrid vigour for a few generations
  - **iii. Three way cross** involving many inbreds (A x B x C)
  - **iv. Top crossing** involving inbred and open pollinated variety
  - **v. Synthetic cross** involving many inbreds that combine well with each other
- Synthetic or composite varieties are used for the improvement of forage crops, sugar beets, maize, bajra and other cross pollinated crops
- Hybrid variety development by any of the methods is tested for adaptation to the environment, diseases, pests and for yield at different localities
- If the variety is found to be stable and desirable, the seed is multiplies and distributed to cultivators
- Based on the taxonomic relationships of the two parents, 4 types of hybridization are recognized

Multiple crossing: A complex system in which 8 to 16 varieties are systematically crossed in pairs. These multiple crosses are produced by crossing pairs of parents, and then crossing pairs of F_1's until all parents enter into a common progeny as given below

	A x B	C x D		E x F	G x H
F_1's	AB	x CD		EF	x GH
F_2's	ABCD		x	EFGH	
			ABCDEFGH		

This system of crossing has the advantage of bringing together quickly combinations of genes from several varieties

Based on the taxonomic relationships of the two parents, 4 types of hybridization are recognized

i) Intra-varietal : Intra-varietal hybridization is carried out between two different plants of the same variety. It is useful in improving the self-pollinated crops. The two plants with different genotypes of the same variety are selected to produce a new plant with desirable characters

ii) Inter-varietal hybridization or intraspecific: Inter-varietal hybridization is carried out between two different compatible species. It is useful in the production of superior varieties in self pollinated crops and in some cross pollinated crops.

iii) Interspecific or intrageneric

- Inter-specific hybridization is carried out between two different compatible species belongs to the same germs to transfer one or two desirable characters from the wild to cultivated species
- This type of cross is mainly used to produce disease, drought and pest resistant varieties
- The resistance to tobacco mosaic virus is transferred from *Nicotiana glutinosa* to *N. tabacum* by interspecific hybridization and mutation breeding
- ***Devitej*** cotton interspecific hybrid is obtained between *Gossypium hirsutum x G. herbaceum* and ***Deviraj*** from *G. hirsutum x G. arboreum*

iv) Intrageneric

- Intrageneric hybridization is carried out between two different genera of the same family. Hence this type is also called wide/distant cross
- Muntzing (1979) obtained a new genus of manmade cereal Triticale between *Triticum aestivum* (wheat) and *Secale cereale* (rye) which is being cultivated in China and European countries
- Karpenchenko, a Russian geneticist synthesized a new genus, *Raphano brassica* (Rabbage) from *Raphanus sativa* (radish) and *Brassica oleracea* var. capitata (Cabbage)

Merits

- Generates new genetic combination/variations
- Superior characters like high yield, promising seed etc., brings into a single variety
- Production of more vigorous and better adapted strains (heterosis)
- Creation of new genotypes, species and new genera

Demerits

- This technique is handled by experienced plant breeder as it involves technical and scientific skill
- It is tedious, time consuming and expensive
- Production of sterile hybrids

- It brings unusual and undesirable characters also into the new hybrid

Improved strains of crop plants through hybridization

Rice	:	Jaya, Padma, Krishna, Sabarmati, Deccan-101
Wheat	:	NP-165, NP-710 (= NP52 x NP165)
Maize	:	Ganga-3, Ganga-7, Ganga safed
Cotton	:	Deviraj (*Gossypium hirsutum* x *G arboreum*), MCU, Lakshmi (Gadagi x CO2)
Potato	:	Kufrikuber (*Solanum eurtilobatum x S. Andigenum x S. tuberosum*)
Tomato	:	Pusa red plum (*Lycopersicum esculentum* x *L. Pimpinellifolium*), Pusa Red, Pusa Early (Dwarf), Pusa Ruby (Sioux x Maruti)

Hybrid vigour or heterosis

- Superiority of F_1 hybrid of two genetically dissimilar parents is called *heterosis*
- F_1 hybrids are heterozygous for one or more characters
- Heterosis is measured in terms of size, growth rate, yield, disease resistance etc.,
- In 18th and 19th centuries the hybrid vigour was noticed by Koelreuter, Beele etc.,
- The term heterosis was coined by G.H. Shull (1914) who worked on maize
- Balanced heterosis will be obtained by crossing and mutational heterosis is by mutation breeding
- Cross pollinated and asexually reproducing species are highly heterozygous. These plants usually exhibit severe reduction in fertility and vigour on inbreeding called **inbreeding depression**. Again heterosis is recovered by crossing the inbred lines (single or double cross)
- Hybridization between unrelated strains generally leads to increased vigour
- Hybrid vigour is best maintained in vegetatively reproducing crops as there is no genetic recombination (Apple, Guava, Mango, Dhalia, Chrysanthemam). Hybrid vigour is also maintained by tissue culture and by introducing apomictic genes/development of apomictic lines.

Genetical basis of heterosis is explained by

i. Dominance hypothesis (Davenport); heterosis is due to the presence of favourable dominant genes in hybrids (F_1 hybrid Aa Bb Cc Dd Ee)

ii. Over dominance hypothesis (Shull and East): heterosis is due to superior heterozygous gene combination

Genetic constitution, genetic diversity, genetic base and adaptation in breeding population developed by various breeding methods

Breeding methods	Outcome	Genetic constitution (heritable variation)	Genetic diversity	Genetic base	Adaptation
A. Self pollinated crops					
1. Pureline selection	Variety	Homozygous & homogeneous	Nil	Narrow	Narrow
2. Mass selection	Variety	Homozygous & homogeneous	Moderate	Broad	Wide
3. Backcross / pedigree / bulk and single seeds descent methods	Variety	Homozygous & homogeneous	Narrow	Narrow	Nil
4. Heterosis breeding	Hybrid	Hetrozygous & homogeneous	Nil	Narrow	Narrow
5. Multilines	Variety	Homozygous & heterogeneous	High	Broad	Wide
B. Cross pollinated crops					
6. Mass selections	Variety	Heterozygous & heterogeneous	High	Broad	Wide
7. Synthetic breeding	Variety	Heterozygous & heterogeneous	High	Broad	Wide
8. Composite breeding	Variety	Heterozygous & heterogeneous	High	Broad	Wide
9. Heterosis breeding	Variety	Heterozygous & heterogeneous	Nil	Narrow	Narrow
10. Land Races	Primitive Cultivars	Heterozygous & heterogeneous	High	Broad	Wide
11. Clonal Selection	Clonal Variety	Heterozygous & heterogeneous	Nil	Narrow	Narrow

Similarities and dissimilarities between purelines and inbred lines

Particulars	Purelines	Inbred Lines
A. Similarities		
1. Genetic constitution	Homozygous and homogeneous	Almost homozygous and homogeneous
2. Genetic base	Narrow	Narrow
3. Adaptation	Narrow	Narrow
4. Heritable variation	Absent	Absent
5. Buffering capacity	Poor	Poor
6. Reproduction	Sexual	Sexual
7. Selection	Ineffective	Ineffective
8. Produce	Highly uniform	Highly uniform
B. Dissimilarities		
1. Relevant to	Self pollinated species	Cross pollinated species
2. Mode of pollination	Self pollination	Cross pollination
3. Maintained by	Natural self pollination	Inbreeding or artificial self pollination
4. Developed	A single homozygous plant by selfing	A heterozygous base population
5. Used as	A variety	In developing hybrids, synthetics and composites

D. POLYPLOID BREEDING

- An organism having more than two sets (diploid) of chromosomes is called polyploidy.
- The production and utilization of polyploids for improvement of crops is called polyploidy breeding
- Polyploid plays an important role in the evoluation of new species
- Polyploids are the change in chromosomal number (heteroploid)
- Polyploids could be euploids (with multiples of chromosome sets) or aneuploids (with changed chromosome numbers with the same set)
- Euploids could be either autopolyploids or allopolyploids (amphidiploids)
- **Autopolyploid** : A polyploid with more than two copies of a same genome (A_1 A_2 A_3 A_4 etc)
    i. Based on the number of sets of chromosomes the autopolyploids are classified into autotriploids (3 sets), autotetraploids (4 sets), auto-pentaploids (5 sets), auto-hexaploids (6 sets) etc.,
    ii. Sterile autotriploids are useful to produce seedless fruits in watermelon and sugar-beet
    iii. Autotetraploids are employed and exploited in barley, maize, tobacco and banana
    iv. Though autopolyploids are vigorous they are highly sterile. Hence, they are highly employed in vegetatively propagating plants and where the seeds are important
- **Allopolyploid :** A polyploid containing two or more different genomes. They are developed from two or more species of same genus or two distantly related genera belonging to the same family
    i. The F_1 hybrids are highly sterile because it contains different (heterologous) sets of chromosomes (AB). The F_1 hybrid is treated with colchicines to double the heterologous sets of chromosomes of each species

ii. Hence, thus hybrid (ampidiploid) behaves as fertile diploid during meiosis though it is sterile allotetraploid

iii. Allopolyploids are more vigourous than diploids, better adapted to wide range of environmental conditions, robust growth, thick green large leaves, fruits and seeds and highly tolerable to gene loss and mutations

iv. Allopolyploids is utilized in producing new species (Triticale, Rabbage, Giant napier grass, etc)

v. Allpolyploids is a source of new genetic variations in the existing species

Colchicine is an alkaloid drug extracted from the seeds and bulbs of *Colchicum autumnale* plant. Colchicine inhibits the formation of spindle fibre or their divergent orientation during metaphase, leading to the doubling of chromosome number.

Polyploid / Heteroploid	
A. Aneuploid (2n + few)	**B. Euploid**
i. Hyperploid (Addition of chromosomes) Tetrasomic (2n+2) Trisomic (2n + 1) Double trisomic (2n + 1 +1)	**i. Autopolyploids** Autotriploid (3x) Autotetraploid (4x) Autopentaploid (5x) Autohexaploid (6x) Autooctaploid (8x)
ii. Hypoploid (Deletion of chromosomes) Monosmic (2n - 1) Nullisomic (2n-2) Double monosomic (2n-1-1)	**ii. Alloploids** (Amphidiploid) Allotetraploid ($2x_1$+$2x_2$) Allohexaploid ($2x_1$+$2x_2$+$2x_3$) Allooctaploid ($2x_1$+$2x_2$+$2x_3$+$2x_4$)
x – basic set of chromosome (genome number), 2n – somatic chromosome number, n – gametic chromosome number	

E. Mutation breeding

- Mutations are sudden heritable changes in the genotype of the organisms
- Short legged sheep mutants were discovered by an English farmer Hugo de Vries coined the term mutation in 1900.
- Mutations are either natural or induced
- The rate of induced or artificial mutation is very high than natural mutations
- Individual carrying mutations are called *mutants*
- Mutations which are produced artificially by mutagens are described as induced *mutations*
- The agents which induced mutations are known as *mutagens* or mutagenic agents
- Mutations generates gene or point mutations and chromosomal (structural or number) mutations
- The mutagens may be physical or chemical agents
  - The physical mutagens are X-rays, Gamma rays, Beta rays, Neutrons, UV radiations (ionized or non-ionized)
  - The chemical mutagens are mustard gas, nitrous acid, ethyl methanesulfonate (EMS), nitroso methyl urea (NMU), Methyl ethanesulfonate (MES), methyl methanesulfonate (MMS), Diethyl sulphate (DES), Malic hydrazide (MH), acridine dyes, sodium, azide, etc.,

- Most of the mutants produced are recessive
- Mutations can create genetic variability
- For the first time **Muller** (1927) induced mutations in Drosophila with X-rays
- **Stadler** (1928) induced mutations in barley with X-rays
- Sharbati Sonora-64 an amber coloured variety of wheat was obtained by irradiation of Sonara-64 with X-rays **(Dr. M. S. Swaminathan)**

Strains developed by mutation breeding

Rice	Jagannath (γ-rays induced mutant of T 141), IIT 60, K 84, HM 95
Wheat	NP 836
Cotton	Indore 2, Pusa Ageti, MCU-7
Castor	Aruna
Jute	JRO 514, JRO 412
French Bean	Pusa parvati
Tomato	Pusa parvati
Groundnut	TG 1, TG 38, TG 51
Sugarcane	CO 8162 (γ-irradiation of CO 527)

- Plant material is used to irradiate are seeds or seedling. In the vegetative propagated plants underground stems and buds are subjected to mutagens
- Irradiation is commonly used for inducing mutations, CO^{60} or C^{14} are employed for producing gamma rays
- First gamma garden at Long Island, New York, USA, First gamma garden of India is established in Bose Research Intitute (1959), Calcutta presently these are also available at IARI, New Delhi, Bhabha Atomic Research Centre, Trombay

Metits

- To create new variations among the organisms
- To improve specific characters of a well adapted high yielding variety
- Mutation breeding is an alternative method when all natural variability is used

Demerits

- Mutations are mostly recessive and cannot be detected easily
- Most induced mutations are lethal
- To select desirable mutants, plant breeder has to screen large populations
- Desirable and promising mutations are commonly associated with undesirable characters
- Large number of induced mutations is undesirable
- Undesirable characters are eliminated by laborious time consuming and expensive back cross method
- Most of the mutations are not stable and generally get reverted

F. SOMATIC CELL or PARASEXUAL HYBRIDIZATION or PROTOPLAST FUSION TECHNOLOGY

- Harris and Watkins (1965) discovered and demonstrated protoplast fusion from different animal species (mouse and man) easily because there is no cellulose cell wall. But, the hybrid cells formed were not developed into animal because the animal cells fail to differentiate *in vitro* conditions

- The process of obtaining hybrids by methods other than sexual methods is described as *parasexual hybridization*. Parasexual hybrids are obtained by *in vitro* fusion of protoplasts (protoplast fusion or somatic cell hybridization)
- Protoplasts of plant cells are obtained by removing cell wall through enzymatic digestion using cellulases and pectinases
- Effective fusion of protoplasts is facilitated by PEG (polyethylene glyccol) or sodium nitrate
- The fusion product (heterokaryotic cell) is cultured in a nutrient medium and allowed to grow into a hybrid plant
- This method is useful in cases where crosses between unrelated and distantly related species and genera cannot be obtained by conventional methods
- Many interspecific and intergeneric hybrids have so far been produced as follows
  - Raphanobrassica (from *Raphanus sativus* x *Brassica oleracea var caritata*)
  - Erucobrassica (from *Nicotiana tabacum* x *Pentunia inflorata*)
  - Solanopersicon (from *Solanum tuberosum* x *Lycopersicon esculentum*)
  - Daturotropa (from *Daturia innoxia* x *Atropa beliadone*)
- Hybrids from protoplast fusion are very important for introducing some useful and desirable genes such as those of disease resistance, male sterility and herbicidal resistance.

3.4. GENETICS

Genetics is a branch of biology deals with the heredity and the variation of inherited characteristics. Gregor Johann Mendel, who is known as the "father of modern *genetics*". He conducted his research on pea plant (*Pisum sativum*). Mendel presented his paper on "*Experiments on Plant Hybridization*", at the Natural History Society of Brünn in Moravia in 1865 and published in a German natural science journal 1966.

Rediscovery of Mendel's Work (*1900)*: Three scientists namely Dutch biologist *Hugo de Vries*, German plant geneticist *Carl Correns*, and Austrian plant breeder *Erich von Tschermak-Seysenegg* 'rediscovered' the Mendel's work and later so called Mendel's laws.

Reasons for Mendel's success:

✓ Lucky to chose simple, easily-identified characteristics (traits) which has two distinct forms, eg., Plant height (tall or dwarf), flower colour (purple or white), seed colour (green or yellow)

✓ Peas plants are easy to grow, have a short life cycle and are normally self-pollinated

✓ He developed 'pure-breeding' parent plants and controlled the pollination process carefully

✓ Kept accurate records of large number of plants and offspring and analyzed his results mathematically

Mendel's Laws

1. Law of segregation
2. Law of independent assortment

Law of segregation

Hybrids or heterozygotes of F1 generated from the parents have two contrasting characters or allelomorphs of dominant and recessive nature. These alleles remain together for long time but do not contaminate or mix with each other and separate or segregate at the time of gametogenesis, so that each gamete receive only one allele of a character either dominant or recessive.

As gametes formed are always pure (carry either dominant or recessive allele not both), this law is also referred principle of segregation or principle of purity of gametes.

Example

Parents Red flower (RR) × White flower (rr)

F_1 Pink (Rr)

F_2 3Red : 1White - Phenotypic ratio 1RR (Red) : 2Rr(Red) : 1rr(White) – genotypic ratio

Law of independent assortment

When the parents differ from each other in two or more pairs of contrasting characters or factors then the inheritance of one pair of factor (gene) is independent to that of the other pair of factors.

Example

Pea plant, seed characters shape and color considered.

Parents Yellow Round × Green Wrinkled

F_1 Yellow Round

F_2 Phenotypic ratio = 9Yellow Round : 3Yellow Wrinkled : 3Green Round : 1Green Wrinkled Genotypic ratio = 1:2:1:2:4:2:1:2:1

n = no. of genes segregating	Monohybrid	Dihybrid	Trihybrid
No. of gametes & phenotype produced (2^n)	2	4	8
No. of genotype produced (3n)	3	9	27
Minimum no. of plants required (4n)	4	16	64

Extensions or Deviations of Mendelian concepts are incomplete dominance, co-dominance, linkage, multiple alleles, gene interactions, lethal genes, polygene, pleiotropic gene effect, cytoplasmic or maternal effect and environmental effects.

GENE INTERACTIONS

Di-genic interactions	F_2 Phenotypic ratio	F_2 Genotypic ratio
Supplementary recessive epistasis	9:3:4	1:2:1:2:4:2:1:2:1
Dominant epistasis	12:3:1	
Additive epistasis / Duplicate genes with cumulative effect	9:6:1	
Complementory / Duplicate recessive	9:7	
Inhibitory epistasis	13:3	
Duplicate dominance	15:1	

Differences between Dominance and Epistasis

Features	Dominance	Epistasis
1. Gene	It involves interaction between two alleles of the same gene	It involves interaction between alleles of two or more different genes
2. Expressed in	Heterozygotes only	Both in homozygotes and heterozygotes
3. Recessive gene	Recessive allele can express only in hemizygous /homozygous condition.	Recessive allele can exhibit masking effect

Difference between Mendelian and Cytoplasmic inheritance Patterns

Particulars	Mendelian inheritance	Cytoplasmic inheritance
1. Governed by	Nuclear genes	Plasmagenes
2. Segregation	Present	Absent
3. Segregation occurs during	Meiosis	Mitosis
4. Reciprocal differences	Not observed	Regularly observed
5. Transmission	Biparental	Uniparental; often biparental
6. Mapping	Genes can be easily mapped on Chromosomes	Mapping is feasible in selected cases
7. Location of genes	In nuclear chromosomes	In chloroplast DNA or mitochondrial DNA
8. Linkage map	Linear	Circular

Differences between Sex-influenced and Sex-limited Traits

S.No.	Sex-influenced traits	Sex-limited traits
1	Genes controlling these traits are located in autosomes	Genes controlling these traits are located in X chromosmes or autosomes.
2	The dominance relationship depends on the sex of the individual.	Dominance relationship is not affected by sex.
3	Expressed in both the sexes.	Expression limited to one sex.
4	Expression of such characters appears to be influenced by sex hormones. E.g., pattern baldness in man, horns in sheep, whore forelock in human.	Expression of such traits depends on a particular hormone present in the concerned sex, e.g., premature baldness in man, milk production in cattle, breast development in women, beard development in man.

Differences between Sex-linked and Sex-limited Traits

Sex-linked traits	Sex-limited traits
Genes controlling these traits are located in X chromosome.	Genes controlling these traits are located in X chromosome or autosomes.
They can express in both the sexes	Their expression is limited to only one sex.
They include such traits, which are not concerned, with sex, e.g., haemophilia, colour blindness, etc. in humans.	They include secondary sex characters, e.g., milk production in cattle, breast development in women, beard development in men, etc.

CHROMOSOMAL ABERRATIONS : Any change in 2n number of chromosomes or structure of chromosomes

a. Numerical aberrations

Term	Description	Symbol
Heteroploid	Change from $2n$ state	
A. Aneuploid	One or few chromosomes extra or missing from $2n$	$2n\pm$few
1. Hypoploid	Chromosome number less than $2n$	$2n$-few
I. Monosomic	One chromosome missing	$2n-1$
II. Double monosomic	Two non-homologous (each from a different pair) chromosomes missing	$2n-1$ -1
III. Nullisomic	One chromosome pair missing	$2n-2$
2. Hyperploid	Chromosome number more than $2n$	$2n$ + few
I. Trisomic	One extra chromosome	$2n+1$
II. Double trisomic	Two nonhomologous (each from a different pair) chromosomes extra	$2n+1+1$
III. Tetrasomic	One chromosome pair extra	$2n+2$
B. Euploid	Number of genomes different from two (2x)	
1. Haploid	Gametic chromosome number	n
I. Monoploid	Only one genome present	x
II. Polyhaploid	Haploid of a polyploid	
2. Diploid	Two copies of a single genome present	$2n = 2x$
3. Polyploid	More than two copies of a single genome or two or More genomes present in (usually) two copies each	
I. Autopolyploid	More than two copies of the same genome present	
(a) Autotriploid	Three copies of the same genome	$2n = 3x$
(b) Autotetraploid	Four copies of the same genome	$2n = 4x$
(c) Autopentaploid	Five copies of the same genome	$2n = 5x$
(d) Autohexaploid	Six copies of a single genome	$2n = 6x$
II. Allopolyploid	Two or more distinct genomes; usually, two copies of each genome are present	$2n = 2x_1 + 2x_2$, etc.
(a) Allotetraploid	Two copies each of two genomes	$2n = 2x_1 + 2x_2$
(b) Allohexaploid	Two copies each of three genome	$2n = 2x_1 + 2x_2 + 2x_3$

(c) Allooctaploid	Two copies each of four genomes	$2n = 2x_1 + 2x_2 + 2x_3 + 2x_4$
III. Segmental-polyploid	Two copies each of two or more largely similar (but not the same) genomes	$2n = 2x' + 2x$
IV. Auto allopolyploid	Two or more distinct genomes; at least one genome has more than two copies	$2n = 4x_1 + 2x_2$

Note: x_1, x_2, x_3, etc., are distinct genomes, while x and x' are largely similar (but not the same) genomes

b. Structural aberrations

i. Deletions: A portion of the chromosome is missing or deleted.

ii. Duplications: Addition of chromosome segment, resulting in extra genetic material.

iii. Inversions: A portion of the chromosome has broken off, turned upside down and reattached.

v. Translocations: A portion of one chromosome is transferred to non-homologous chromosome.

Findings in Genetics and cytogenetic

Scientist	Year	Findings
Pythagoras	600 BC	Vapour theory of inheritance
Empeddes	485-425 BC	Fluid theory of inheritance
Robert hook	1666	Discovered the cell and term cell
Leeuwen hock	1674	Father of microbiology
Redi and Spallanzani	1626-1698, 1729-1799	Disproved the spontaneous theory
Swammerdam and Punnet		Preformation theory
Camerarius	1694	Sexes in plants
Maupertius	1698-1759	Particulate theory of inheritance
Spallanzani	1785	Artificial insemination (dogs)
C.F.wolff and K.E. VonBaer	1733-1794, 1792-1876	Theory of epigenesist
Louis Pasteur	1822-1895	Father of modern microbiology/ Fermentation
Linnaeus	1702-1778	Taxonomy
Koelreuter	1733-1850	Hybridization in plants/Systematic breeding
Lamark	1744-1829	Theory of acquired characters
Charles Darwin	1838	Theory of pangenesis
August Weismann	1834-1914	Forerunner of modern genetics, germplasm theory of inheritance
Robert brown	1831	discovered nucleus
M.J. Schleiden	1838	Discovered nucleoli
M.J. Schleiden and T. Schwann	1838, 1839	Cell theory
R. Virchow	1858	Cell linkage theory
Darwin and Wallace	1858	Theory of evolution through natural selection
Darwin	1859	Origin of species

Gregor Johann Mendel independence	1865	Law of segregation, law of assortment, particulate inheritance
O.Hertwig and Fol	1865-1877	Discovered fertilization in animals
Huxley	1868	Book "The physical basis of life"
E.Strasburger	1877-1884	Discovered fertilization in plants, cell division, cytoplasm, nucleoplasm and chromosome
Freterick Meischer	1869	Discovered nucleic acid
W. Flemming	1882	Discovered mitosis, lambrush chromosome, chromatin
A.F.W. Schimper	1883	Discovered plastid
Weismann, Von Kolliker, Strasburger	1885	Chromatin theory of inheritance
W.Waldeyer	1888	Term- chromosome, discovered Centromere
Henking	1891	First observed sex chromosome
R.Altmann	1894	Discovered mitochondria and named as bioplast, introduce the term nucleic acid
C.Benda	1897	Coined the term mitochondria
Camilo Golgi	1898	Discovered golgi complex
De Veries, Carl Correns and Von Tschermak	1900	Rediscovery of Mendel laws
Hugo de Veries	1900	Term- mutation
Karl Landsteiner	1900	Discovered A B O blood group in man
Mc Clung	1901	X- chromosome
Sutton and boveri	1903	Chromosomal theory of inheritance
William Bateson	1906	Term – genetics, allelomorph, homozygote, heterozygote,F1,F2, epistatic genes
Johannsen	1909	Term -Gene

Nobel Prize Winner in Genetics and Cytogenetics

Scientist	Year	Area
T. H. Morgan	1933	Theory of the genes
H. J. Muller	1946	Radiation genetics
Beadle and Tatum	1958	Gene expressions
Sanger	1958	Primary structure of the protein
Kornberg	1959	Isolation of DNA polymerase
Watson and Crick	1962	3D structure of DNA
Jacob and Monod	1956	Operon concept

Paltimore and Demin	1975	Reverse transcriptase
Niranberg and Korana	1978	Genetic code
Arber Nathans and Smith	1972	Restriction endonuclease
Berg Gilberg and Sanger	1980	Recombinant DNA technology
Mc Clintock	1983	Transposable element
Kary Mullis	1993	PCR technique
Robertsand smith	1993	Split genes and RNA property
Leland H. Hartwell, Tim Hunt and Sir Paul M. Nurse	2001	Discoveries of key regulators of the cell cycle
Sydney Brenner, H. Robert Horvitz and John E. Sulston	2002	Genetic regulation of organ development and programmed cell death
Roger D. Kornberg	2006	Molecular basis of eukaryotic transcription
Venkatraman Ramakrishnan, Thomas A. Steitz and Ada E. Yonath	2009	Structure and function of the ribosome

Differences between Qualitative and Quantitative Traits

Particulars	**Qualitative Traits**	**Quantitative Traits**
Definition	Traits those are governed by one or few genes are called as qualitative characters. *e.g. Seed shape, seed color, seed form*	Traits those are governed by many or several genes are called as quantitative traits. *e.g. number of seeds, height, grain yield*
1. Gene involved	One or few	Several
2. Effect of individual gene	Large and detectable	Small and undetectable
3. Variation	Discontinuous	Continuous
4. Gene action	mostly non-additive type i.e. epistastic or dominance gene interaction	Gene action mostly of additive
5. Grouping into distinct classes	Possible	Not possible
6. Effect of environment	Little	High
7. Metric Measurement	Not possible	Possible
8. Analysis/studied by	Based on frequencies and ratios (Mendelian genetics or population genetics)	Based on means, variances and covariances (Biometrical genetics)
9. Stability	High	Low to medium
10. Transgressive segregation	not possible	Possible

3.5. CELL BIOLOGY

What is Cell? : Cell is a basic, structural and functional unit of life of plants and animals.

The word cell comes from the Latin word cellula, meaning, a small room. The term was coined by Robert Hooke (1665) when he compared the cork cells he saw through his microscope to the small rooms monks lived in.

Any cellular organism may contain only one type of cell either prokaryotic cell or eukaryotic cell.

- Cell- word coined by Robert Hooke
- Nucleus term coined by Robert Brown
- Virchow-new living cell arise only by reproduction of pre-existing living cells
- Protoplasts (Cytoplasm + Nucleus) coined by Purkinje

A Brief Description and Function of Different Cellular Components

Cellular Component	Description	Function
A. Cell Wall	Plant cells are surrounded by a non-living and rigid cellulosic coat, called cell wall; a differentiated cell has three distinct parts, viz., middle lamella, primary and secondary cell walls.	Provides definite shape, mechanical support and strength to tissues and organs.
1. Middle lamella	Cementing layer between cells mainly consists of pectin.	Gives integrity to tissues.
2. Primary cell wall	Present between middle lamella and secondary cell wall, mainly consists of cellulose.	Definite shale and mechanical support.
3. Secondary cell wall	Present between primary cell wall and plasma membrane/cell membrane	Additional strength.
B. Protoplast	A plant cell devoid of cell wall.	
1. Cell membrane	Present around the cytoplasm, consist of protein, cholesterol and phospholipids.	Maintains chemical integrity of cells; controls inflow and outflow of materials; site of many important reactions.
2. Cytosol	Fluid part of cytoplasm, enclosed within cell membrane, surrounds the nucleus.	Most of the organelles are suspended in cytosol; site of bulk of cellular activity.
3. Cytoplasm	Semi-fluid substance, surrounded by cell membrane; consists of cytosol, organelles and other structures.	Houses organelles; vesicles and inclusions; site of all cellular activities, except those related to chromosomes.
4. Endoplasmic reticulum (ER)	An extensive network of membrane -enclosed spaces, extended throughout the cytoplasm from the plasma membrane to the nuclear envelope; consists of vesicles, tubules and cisterns.	Rough ER, serves as site for attachment of ribosomes; smooth ER makes steroids; provides site of protein modification and trafficking.

(i) Rough E.R.	Ribosomes are attached to the outer surface of endoplasmic reticulum membrane.	Protein synthesis and modification; also trafficking.
(ii) Smooth E.R.	Endoplasmic reticulum not associated with ribosomes.	
5. Ribosome	Particles of about 200 Å diameter composed of RNA and protein, generally attached to the outer surface of ER membrane and some are also freely present in cytosol.	Serves as the site for protein synthesis.
6. Golgi apparatus	Stacks of disk like membranous structures; consists of 2-7 flat cisternae stacked close to each other	Involved in formation of cell wall, lysosomes, vacuoles; packaging plant of cells.
7. Mitochondria	Organelle surrounded by double concentric membranes, inner one being infolded; an average cell may have 200-800 mitochondria; consists of proteins, lipids, RNA and DNA	Site of Kreb's cycle and electron transport system; converts energystored in food into the form of ATP.
8. Plastids	Two membrane-bound disccoid shaped organelles, originate from proplastids; contain DNA.	
(i) Leucoplasts (a) Amyloplasts	Colourless plastids. Contain starch grains.	 Starch storage
(b) Proteinoplasts (c) Elaioplasts (ii) Chloroplasts (iii) Chromoplast	Contain proteins. Contain fats. Green colour plastids, many contain starch grains.Coloured plastids	Protein storage Lipid storage Light energy trapped, transformed and used to convert CO_2 and water into carbohydrates.Provide colour to fruits etc.
9. Centriole	Present in animal cells.	Formation of spindle fibres.
10. Microbodies	Glyoxysomes, peroxisomes and spherosomes are collectively called microbodies; diameter 1-2nm; these are proteinacious in nature and are bounded by a single membrane.	
(i) Glyoxysome	Usually seen in the cells of oilseeds, e.g., peanut, castor; contain several enzymes of glyoxylate cycle.	During germination, these bodies function in the glyoxylate cycle where fat is converted into carbohydrates.
(ii) Peroxisomes	In C_3 plants, photorespiration takes place in peroxysomes; absent in C_4 plants.	During photorespiration in these organelles, glycolic acid is oxidised to glyoxlic acid and finally converted into amino acid, glycine.
(iii) Spherosomes	Present in the cytoplasm; rich in hydrolysing enzymes, such as esterases, proteases, phosphatases and ribonucleases.	It is believed that these organelles are involved in the storage and translocation of lipids.

11. Cytoskeleton	The microfilaments, microtubules and intermediate filaments constitute cyto-skeleton; these elements consist of proteins.	Plays an important role in maintaining the structure and function of cells.
(i) Microfilaments	Rod-like structures containing the protein actin	Provides structural support and assist in cell movement.
(ii) Microtubules	Hollow cylindrical structures.	Assist in movement of cilia, flagella and chromosomes; provide transport system.
12. Lysosome	Only present in animal cells, membrane bound spheres, contains hydrolytic enzymes.	Digestion of food material and phagocytosis of germs.
13. Vacuoles	Membrane surrounded, often large spaces in the cytoplasm, loaded with mineral, water, anthocynins.	Maintain turgidity of cell.
14. Nucleus	Spherical structure surrounded by nuclear envelope; contains nucleolus, nuclear sap and chromatin fibres.	
(i) Chromatin fibres	Thread like structures; consist of DNA and proteins (histone and non-histone).	Control cellular activities and determine biological characteristics of organisms.
(ii) Nucleolus	Rounded mass present in nucleus; associated with sat-chromosomes.	The place of ribosomes synthesis.

Difference between Prokaryotic and Eukaryotic cell

Criteria	Prokaryotic cell	Eukaryotic cell
Multicellular forms	Rare	Common
Nucleus	Nucleoid region (No real nucleus)	Real nucleus with double membrane.
Cytoplasmic membrane	Only unit membrane system of majority of prokaryotes. Internal compartmentalization by unit membrane system is absent.	Presence of a multiplicity of unit membrane systems. Many of internal unit membranes are structurally different from cytoplasmic membranes.
	Selective barrier between cell and external environment than does membrane of eukaryote cell.	Less selective barrier than prokaryotic cytoplasmic membrane.
Exocytosis and Endocytosis	Absent.Hence it lacks biological properties that are dependent on the capacity for endocytosis.	Observed
Respiration	Among aerobic bacteria, respiratory electron transport system is incorporated into cell membrane.	Machinery of respiration is incorporated into the inner membrane of mitochondrion
Cell membrane	Sterols are not found in significant amount in the cell membrane of prokaryote.	Cell membrane contains sterols.

Cell wall	Except for mycoplasma, thick cell wall is always present, which contains amino sugars and muramic acid.	Not always present, if present, contains simple substances like cellulose, mannose, xylans.
Cell division	Binary fission (simple division)	Mitosis and Meiosis
DNA	Double stranded, circular or linear	Linear, double stranded.
Histone	DNA never associated with histone proteins.	DNA associated with histone
Ribosomes	70S type (50S+30S)	80s type (60S+40S)
Protein synthesis	Starting amino acid in protein synthesis - methionine	Starting amino acid in protein synthesis - N-formylmethionine.
Cell movement	Flagella	Flagella and cilia containing microtubules; lamellipodia and filopodia containing actin
Flagella	Simple in structure.	Complicated in structure, consisting of 2 central and nine peripheral double fibrils.
Means of reproduction	Asexual means in the haploid state (diploid state may occur rarely)	Diploidy is the character of many groups of eukaryotes

Difference between mitosis and meiosis

Mitosis	Meiosis
By Flemming Equational division 5 stages	By farmer and more Reduction division 2 stages (Meiosis I & Meiosis II)
Prophase (Longest Phase) Chromatin condensation chromosome visible nucleolus & nuclear envelope disappear	**Meiosis I** **Prophase : Leptotone :** Crown looks like a loose ball of knitting wool, RNA synthesis occurs **Zygotene :** Pairing of homologous chromosomes, Synaptenemal complex is seen (synapsis) **Pachytene :** Complete bivalent, Tetrad, Crossing over. **Diplotene :** Chiasmata seen as a result of separation of homologous, Chiasma terminalization. **Diakinesis:** Chromosome condensation nucleoli disappears, chiasma terminalization
Metaphase Chromosome arranged in a equatiorial plane	**Metaphase l** Bivalents orient at the middle of the equatorial plate
Anaphase (Shortest Phase) Division of centromere chromosome moves to opposite pole **Telophase :** Nucleolus reappears, nuclear membrane reorganized	**Anaphase I :** the centromere do not divide. Homologous move to the opposite pole, this reduces the number of the chromosomes **Telophase I** : Chromosome do not uncoil completely

Cytokinesis	Telophase II
Division of cytoplasm, Two identical daughter cells, similar to mother cell	Four haploids are produced and generates in to spores/gametes.

Inter phase: S Phase-DNA Synthesis, G1 and G2 Phase- RNA and protein synthesis

Types of chromosome : Chromosome can be categorized on the basis of two criteria such as number of centromeres and position of centromeres.

1. Types of chromosome on the basis of number of Centromeres

Type	Number of centromeres
Acentric	Nil
Monocentric	One
Dicentric	Two
Tricentric	Three
Polycentric	Many
Holocentric	Diffused (Here, microtubules are attached along the length of chromosome)

2. Types of chromosome on the basis of position of centromere

Type	Position of centromere	Shape	Details	Examples
Metacentric	Centre (median)	v-shaped	equal arms	tradescantia, amphibians
Sub-metacentric	Near centre (sub-median)	J or L shaped	Two unequal arms	human being
Acrocentric	Near at one end (sub-terminal)	One arm very long	short and another	grasshopper
Telocentric	Terminal	Rod like		drosophila

Chromatin and its types

Chromatin is the basic material of chromosome. Chromatin can be of two types depending on staining properties. They are euchromatin and heterochromatin

Euchromatin

- Euchromatin is portions of chromosome with lightly packed DNA, partially condensed and stain lightly. DNA is packed in 3 to 8nm fibers.
- Genes situated on this portion are structural genes, these genes undergo replication and participates in the active transcription.
- Most of this chromatin disperses after completion of mitosis.
- Euchromatin comprises the genetically most active portion of the genome, have role in phenotypic expression of genes.

Heterochromatin

- Heterochromatin is the portions of chromosome with tightly packed DNA, always remains in the condensed state and stain dark. DNA is tightly packed in the 30nm fibers
- Definition by Heitz: those regions of the chromosome that remains condensed during interphase and early prophase and form so called chromocentre
- Late replication, limited transcription
- Have high content of repetitive DNA sequence
- Contain very few structural genes

Chromosome contents (DNA) : DNA is the chemical and mostly inside the nucleus of the cell that carries biological instructions generation after generation. Biological instruction means hereditary or genetic information used in the day to day functioning of all living organisms (including some of the viruses) and development of new young ones.DNA is a short form of Deoxyribose Nucleic Acid. What these three words mean is given below.

Deoxyribose :	A pentose sugar (with 5 carbons), does not have oxygen atom at the 2' carbon
Nucleic :	Named because these molecules were first identified in the nucleus of the cell
Acid(phosphoric acid) :	Out of three reactive hydroxyl groups of the phosphoric acid, two are used for DNA backbone formation or DNA chain formation. The third one gives the DNA an acidic property

Components of DNA : DNA is a polymer and monomers are deoxyribonucleotides (we will call nucleotides). Components of the DNA are Nitrogen bases, Sugar and Phosphate

Nucleotides and Nucleosides

	Nucleotides	**Nucleosides**
Description	Each nucleotide is made up of three parts namely a sugar, a phosphate and a nitrogen base. [phosphate covalently bonded to 3' or/and 5' hydroxyl group(s)] DNA contains four types of nucleotides. Type of nucleotide depends on nitrogen base.	Each nucleoside is made up of two parts - a sugar and a nitrogen base. [No phosphate]
Nitrogen base	**Name of Nucleotide (Base + S + P)**	**Name of Nucleoside (Base + Sugar)**
Adenosine (A)	Deoxyadenylate	Deoxyadenosine
Thymine (T)	Deoxyguanylate	Deoxyguanosine
Guanine (G)	Deoxycytidylate	Deoxycytidine
Cytosine (C)	Deoxythymidylate	Deoxythymidine

Nitrogen bases

- There are 4 different nitrogen bases in a DNA molecule, two purines and two pyrimidines.
  - o *Purines* : adenine and guanine are derivative of pyrimidines, two ringed structure and bigger than pyrimidines
  - o *Pyrimidines* : cytosine and thymine are one ringed structure
  - o Adenine pairs Thymine with two hydrogen bonds
  - o Cytosine pairs Guanine with three hydrogen bonds
- Sequence of nitrogen bases determine what biological information is written in DNA strand which is used to for build and maintain an organism.

How DNA could differ from RNA?

DNA (Deoxyribose Nucleic Acid)	RNA (Ribonucleic acid)
Usual genetic material	Genetic material of some viruses
Usually double stranded	Mostly single stranded (sometimes may be double stranded)
Pentose sugar is deoxyribose	Pentose sugar is ribose
The nitrogen bases present are adenine, guanine, cytosine and thymine	Thymine is replaced by Uracil
Adenine pairs with thymine and guanine pairs with cytosine	Adenine pairs with Uracil and Guanine with cytosine
Pairing of bases is throughout the length of molecule	Pairing of bases is only in the helical region
Fewer uncommon bases	More uncommon bases
Replicate and transcribe RNA synthesize RNA chain)	Do not replicate and transcribe (in certain viruses,
Large number of nucleotides (4.3 million)	Fewer nucleotides (up to 12000)
Function - To carry genetic message coded on DNA	Function - To synthesize proteins using information

Comparison among different forms of DNAs

S.No.	Characteristics	A-DNA	B-DNA	C-DNA	Z-DNA
1.	Coiling	Right-handed	Right-handed	Right-handed	Left-handed
2.	Pitch	Ca. 28Å	Ca. 34Å	Ca. 31 Å	Ca. 69Å
3.	Base pair per turn	11	10 (10-10.6)	9.33	12
4.	Diameter	23 Å	19 Å	19 Å	18 Å
5.	Vertical rise per base pair	2.56 Å	3.38 Å	3.32 Å	5.71 Å
6.	Sugar-phosphate-backbone	Regular	Regular	Regular	Zig-zag

The Gene (alternate forms of alleles): Johanssen (1909) described Mendel's 'Factor' as 'Gene'. Seymour Benzer (1962) recognized three units of gene cistron, recon and muton. Nowadays, many times cistron itself is considered as gene.

List of Important Models, Theories and Hypothesis

S.No	Theory/hypothesis/model	Proposed by
1.	Cell lingage theory	R.Virchow (1855)
2.	Cell theory	M.J. Schleiden and T. Schwann (1938-39)
3.	Central dogma of life	F.H.C.Crick (1958)
4.	Chiasmatype or one-plan theory	F.A. Janssen (1909,1924)
5.	Chromosomal basis of sex determination	E. Wilson (1909)
6.	Chromosomal theory of inheritance	W.Sutton and T.Bovery (1902)
7.	Classical or two-plane theory	L.W. Sharp (1934)
8.	Copy choice theory of recombination	J.Belling (1933), J. Lederbergh (1955)
9.	Dosage compensation	H.J. Muller (1948)
10.	Double helix model of DNA	J.D. Watson and F.H.C. Crick (1953)
11.	Epigenesis theory	K.W.Wolff (1738-1794)
12.	Evolution through natural selection	C. Drawin and A.R. Wallace (1858)
13.	First genetic map (in Drosophila)	A.H. Sturtevent (1913)
14.	Fluid mosaic model of cell membranes	S.J.Singer and G.L.Nicholson (1972)
15.	Folded-fiber model of cell membranes	E.J. DuPraw (1965)
16.	Founder principle of speciation	A.R. Templeton (1980)
17.	Genic balance theory of sex determination	C.B. Bridges (1925)
18.	Germplasm theory	A.Weismann (1833)
19.	Hardy-Weinberg law	G.H. Hardy and W. Weinberg (1908)
20.	Hybrid DNA models of recombination	H.L.K. Whitehouse (1963), R. Holliday (1964)
21.	Inheritance of acquired characters	J.B. deMonet-Lamarck (1809)
22.	Law of homologous series in variation or Law of parallel variation in plants	N.I. Vavilov (1926, 1951)
23.	Law of independent assortment	G.J.Mendel (1865)
24.	Law of segregation	G.J.Mendel (1865)
25.	Lyon hypothesis	M.F.Lyon (1969)
26.	Model for prokaryotic promoter	D. Pribnow (1975)
27.	Multiple factor hypothesis	H. Nilsson-Ehle (1909)
28.	Mutation theory	Hugo de Vries (1901)
29.	Neutral gene theory of molecular evolution	M.Kimura(1968)
30.	Nucleosome-solenoid model of chromatin fibres	R.D. Kornberg and Thomas (1974)
31.	One-gene-one-enzyme hypothesis	G.W.Beadle and E.L.Tatum (1941)
32.	One-gene-one-polypeptide hypothesis	V.M. Ingram (1957)
33.	Operon concept of gene regulation	F. Jacob and J. Monod (1961)
34.	Origin of new genes by duplication followed by mutation accumulation	S. Ohno (1970)

35.	Origin of species through natural selection	C. Darwin (1859)
36.	Pangenesis	C. Darwin (1809-1882)
37.	Precosity theory of synapsis	C.D. Darlington
38.	Preformation	J. Swammerdam and Bonnet (1720-1793)
39.	Protoplasm theory	O.Hertwig
40.	Pureline theory	W.L.Johannsen (1903)
41.	Replicon model	J. Monod and S. Brenner (1963)
42.	Rolling circle model of DNA replication	W. Gilbert and D. Dressler (1968)
43.	Semiconservative replication of chromosomes	J.H.Taylor (1957)
44.	Semiconservative replication of DNA	M. Messelson and F.W. Stahl (1963)
45.	Synaptonemal complex	Moses (1956)
46.	Theory of organic evolution	C. Darwin
47.	Theory of polyploidy as basis of evolution	O.Winge
48.	Theory of sudden and rapid changes leading to macro-evolution (or punctuational model of evolution)	R. Goldschimds
49.	Triplet codon	G.Gammow (1954)
50.	Wobble hypothesis of codon-anticodon base pairing	F.H.C.Crick (1966)

Sobriquets (Nick-names)

Sobriquent	**Used for**
Drosophila of plant kingdom	*Arbidopsis thaliana*
Fairchild's mule	Hybrid between sweet William and carnation
Genetic bit	Nucleotide
Jumping genes	Transposons
Lamarckism	Theory of inheritance of acquired characters
Man-made cereal	Triticale
Molecular glue	Ligase
Molecular scissor	Restriction enzymes
Natural genetic engineer	*Agrobacterium*
People's choice reaction	Polymerase chain reaction
Queen of genetics	*Drosophila melanogaster*
Secret of life	DNA
Powerhouse of cell	Mitochondria
Physical basis of life	Protoplasm
Suicidal bag	Lysosome
Teminisim	Reverse transcription

3.6. CROP BIOTECHNOLOGY

Biotechnology term was assigned by Karl Ereky (1919). It is a new tool in crop improvement and the major applications of biotechnology are

- Widening the genetic variability through plant tissue culture & genetic engineering
- Speeding up the process of conventional plant breeding through molecular breeding
- Evolving novel genotypes through rDNA Technology

PLANT TISSUE CULTURE

- Haberlandt is known as Father of Plant Tissue Cullture (PTC)
- Totipotency is a ability of single cell (isolated) to develop into a whole (adult) organism
- 1902 – First attempt in PTC by Haberlandt
- 1962 – Production of first haploid plants from pollen grains of Datura by Guha and Maheswari (Indians)
- 1963 – Development of MS medium – Murashige & Skoog
- ICGEB- International Centre for Genetic Engineering and Biotechnology, Italy Headquarters, India, South Africa.

Major MS Media Components and their functions

- Sucrose serve as a food source or carbon source
- Auxin induce formation of roots
- Cytokinin induce formations of shoots
- Abscissic acid induce embryogenesis
- GA_3: Elongation of internodes and growth of meristem or buds
- Agar: Solidifying agent can act as substrate

Media pH is normally 5 to 6. Alchocol (70%) (or) Hg Cl_2 (or) Sodium hypochloride used as sterilization agents. Clonal propagation in vitro first used for orchids

Applications of plant tissue culture

1. Meristem culture (Morel & Martin): This method is useful for mass multiplication of *virus free* plants. Less than 0.5 mm size apical meristems are used. The plants produce by this method are *virus free* because

i. Multiplication rate of meristem cells faster than virus cell multiplication
ii. No vascular connections in meristem cells
iii. Production of growth hormones and other metabolites are high

2. Anther culture is developed through pollen culture (Androgenesis) / ovule culture (Gynogenesis), the resultant are *haploids.* Colchichine treatment for doubling of chromosomes in haploids to produce *double haploids* (Homologous) plants.

3. Embryo culture / embryo rescue: This technique is useful to

- *In vitro* pollination to overcome pre-fertilization barrier

- Embryo rescue to overcome post-fertilization barrier
- Useful in Grapes and Banana (produce seedless fruits) propagation

4. Somatic hybridization/ protoplast fusion helps to overcome all barriers in hybridization (wide) including failure of flowering or meiosis etc. In this technique the enzymes macerozyme and cellulase are used to degrade cell wall. Protoplasts are united by the fusinogens like providing polyethylene glycol, high pH and high Ca. Used to transfer of cytoplasmic genomes as well as nuclear genome.

5. Production of secondary metabolites: Secondary metabolites are organic compounds that are not directly involved in the normal growth, development, or reproduction of an organism. Secondary metabolites are produced by limited number of species. Secondary metabolites often play an important role in plant defence. Humans use secondary metabolites as medicines, flavorings, and recreational drugs. The secondary metabolites can be produced by tissue/cell culture methods. Example Shikonin from Lithospermum sp, Taxol from taxus plant etc.

6. Conservation of germplasm: Especially recalcitrant seeds, vegetative propagated species, elite and rare germplasm *etc.,* are stored by this method. For long term storage, stored-196° C in liquid nitrogen

MARKERS

1. Morphological markers are based visual traits like stem pigmentation
2. Biochemical marker are based on proteins like iso-enzymes/isozyme
3. Molecular markers / DNA marker

Molecular Marker

- ✓ Land mark of chromosome at DNA level
- ✓ It is a DNA sequence; whole inheritance can easily be detected and monitored
- ✓ There are two classes of molecular markers; PCR based and Non- PCR based markers (RFLP)

PCR Based RAPD markers are dominant nature, non-reproducible, mainly used to study variability between genotypes.

Hybridization based markers are Restriction Fragment Length Polymorphism (RFLP), Amplified Fragment Length Polymorphism (AFLP) and Simple Sequence Repeat (SSR) or microsatellites, they are co-dominant and reproducible in nature.

Polymerase Chain Reaction, machine devised by Dr. Kery Mullis. The main steps employed in **PCR cycle** are

1. Denaturation of double stranded DNA at 94 ⁰ C
2. Annealing temperature at 55-65 ⁰ C /36 ⁰ C (depends on primer sequence)
3. Primer extension / synthesis of template DNA by Taq polymerase at 72⁰ C

- *Taq* (*Thermas aquaticus*) polymerase
- *Tth* (*Thermus thermophilus*) polymerases

RELP	-	Restriction fragment length polymorphism
AFLD	-	Amplified fragment length polymorphism
SSR	-	Simple sequence repeats
STMS	-	Sequence Tagged Microsatellite sites
SNP	-	Single Nucleotide Polymorphism
STS	-	Sequence Tagged sites

Uses of Molecular Markers

1. Constructions of genome map
2. Gene mapping and tagging
3. Marker assisted selection (MAS)
4. Characterization of germplasm

- First bacterial genome sequenced is *Haemphilus influenza.*
- First Multicellular genome sequenced is *Carnorhabditis elegans* and eukaryotic genome
- First plant sequenced is *Arabidopsis thaliana*
- Rice genome sequenced in 2005 and it has up to 24,000 genes. Rice genome sequencing (done in India for long arm of 11th chromosome in the variety Nipponbare of Japan)

Transgenic is the plant contains foreign genes by transfer of genetic material from across the biological system (Between the species, between the genus etc.,)

- Two major method employed in gene transfer are
  i. Vector mediated gene transfer through *Agrobacterium tumifacians*. Mainly for dictoyledons and some monocots. The vector used is Ti- plasmid is used.
  ii. Direct transfer by particle bombardment or biolistics method, chemical methods, electroporation, lipofection and micro injection methods

Classification of Genetic Polymorphism

Basis of Classification	Types of Polymorphism	Brief description
1.Nature of polymorphism	(i) Balanced polymorphism	It is due to superiority of heterozygote over homozygotes.
	(ii) Transient polymorphism	It is limited to a particular period.
	(iii) Regional polymorphism	It is found in different regions of the habitat.
	(iv) Neutral polymorphism	It is neutral in the surviving genotypes.
	(v) Unisexual polymorphism	It is confined to one sex only.
	(vi) Cryptic polymorphism	It refers to chromosomal polymorphism.
2.Detection of polymorphism	(i) Phenotypic polymorphism	It is detected on the basis of morphological markers.
	(ii) Biochemical polymorphism	It is detected on basis of protein markers.
	(iii) Molecular polymorphism	It is detected on the basis of variation in the DNA markers.

- 1976 : First Genetic Engineering company is **Genetech**
- 1984 : First transgenic plant developed in Tobacco
- 1994 : First transgenic variety cultivated in USA was *Flavr Savr* of Tomato for delayed ripening (now not in cultivation)

- 2002 : First transgenic crop approved by GEAC for release in India in cotton (MECH12, 162& 184)
- 1994 : First transgenic variety by cal gene using anti sense RNA technology
- Golden rice - β carotene rich, developed by Ingopotrykus and Bayer
- Barnase-Baster system in *Brasica napus* is called as Transgenic Male sterility system
- *Flavrsavr* and *Endless Summer* varieties of tomato developed based on anti-sense RNA technology
- *Roundup Ready* is the variety of soybean develop for glyphosate herbicide resistance

Bacillus thuringiensis : *Bt* named by Berliner, Identified in thuringien region of Germany. Bt toxins are crystal proteins called as δ-*endotoxin.* Bt gene type are

- ✓ Toxins produced by *Cry 1A(a),Cry 1A(ba),Cry 1A(c), Cry 1B,Cry1C,Cry1D* and *Cry 1E* genes are effective against lepidopteran pest
- ✓ *Cry IIA* toxins effective against leplidopteran and dipteron
- ✓ *Cry IIIA,Cry IIIB* and *CryIIIC* toxins effective against coleopteran pest
- ✓ *CryIVA,CryIVB,CryIVC* and *CryIVD* toxins effective against dipteran pest
- ✓ *CryV* and *Cry VI* against nematodes

Transgenic facts

- ☞ Highest area covered by soyabean among transgenics
- ☞ Highest area in transgenic crop is cultivated in USA
- ☞ Maximum trangenics available in rapeseed and mustard
- ☞ Maximum transgenics is developed for herbicide resistance trait
- ☞ India occupies 4th position in area

Institutions regulates the research and commercial release of transgenic in India are

- ✓ Institutional of biosafety committee (IBSC)
- ✓ Review committee on genetic manipulation(RCGM)
- ✓ Monitoring and evaluation committee (MEC)
- ✓ Genetic engineering approval committee (GEAC)

Genome sequencing in some organisms

Species	Year	Genome size (Mb)	No. of genes identified
Prokaryotes			
Haemophillus influenza	1995	1.83	1740
Mycoplasma genitalium	1995	0.58	500
Methanococcusjannaschi	1996	1.66	1750
E.coli	1997	4.64	4,400
M. tuberculosis	1998	4.41	4,000

Eukaryotes			
Yeast (*S. cerevisiae*)	1996	12.00	5,800
Nematode (*C.elegans*)	1998	97.00	19,000
Fruitfly (*D. melanogaster*)	2000	180.00	13,600
Arabidopsis thaliana	2000	125.00	25,500
Human (*Homo sapiens*)	2001	3200.00	40,000
Rice (*Oryza sativa*)	2002	430.00	56,000

Estimated genome size in some crop plants

Name of Crop	Botanical Name	Genome Size (Mb)
Field Crops		
Rice	*Oriza sativa*	400
Sorghum	*Sorghum bicolour*	1000
Maize	*Zea mays*	2500
Barley	*Hordeum vulgare*	8000
Bread wheat	*Triticum, aestivum*	16000
Chick pea	*Cicerarietinum*	1000
Pigeonpea	*Cajanus cajan*	1500
Field pea	*Pisum sativum*	4800
Soybean	*Glycine max*	2000
Tomato	*Lycopersicon esculentum*	950
Fruit Crops		
Mango	*Mangifera indica*	350
Citrus	*Citrus sp.*	385
Banana	*Musa sapientum*	873

Main characteristics of major types of molecular markers

Characteristic	RFLP	SSR	AFLP	RAPD	SNP
Types of visualization	Single locus	Single locus	AFLP Multi-loci	RAPD Multi-loci	SNP Single locus
Allelism	Codominant	Codominant	Dominant	Dominant	Codominant
Type of polymorphism	Sequence	No. of repeats	Sequence	Sequence	Sequence
Level of polymorphism	Good	Excellent	Excellent	Good	Excellent
Polymorphism at the locus alleles	2 to 5 alleles	Multiple alleles	Presence/ absence	Presence/ absence	2 alleles
Quantity of DNA needed	Large	Small	Small	Small	Small
Quality of DNA needed	Good	No restrictions	Good	Good	Good

Reproducibility	Good	Good	Good	Low	Good
Time	Long	Fast, once markers are developed	Fast	Fast	Fast, once markers are developed
Cost	Expensive	Average	Average	Average	Expensive
Technical difficulty	High	Low	Medium	Medium	High

Some achievements of transgenic technology in crop plants

Crop	Character improved	Gene transferred from
1. Cotton	Resistance to Helicoverpa Herbicide resistance	*Bacillus thuringiensis* Microbial gene
2. Maize	Resistance to European corn borer Herbicide resistance	*Bacillus thuringiensis* *Arabidiopsis thaliana*
3. Wheat	Herbicide resistance	Streptomyces
4. Tobacco	Insect resistance Freezing resistance Cold resistance Herbicide resistance	Tripsin Inhibitor gene from cowpea Winter flounder fish *Arabidopsis thaliana* *Acaligeneeutrophus*
5. Potato	Protein quality High protein content Resistance to X and Y virus Starch content Herbicide resistance	Serum albumin gene from human Amaranthus Potato Virus X and Y *Escherichia coli* Streptomyces
6. Tomato	Insect resistance Herbicide resistance Resistance to tomato mosaic virus Freezing resistance	*Bacillus thuringiensis* Streptomyces Tomato mosaic virus Fish (winter flounder)
7. Rape seed	Male sterility system Herbicide resistance	*Bacillus amyloliquefaciens* *Arabiodopsis thaliana*
8. Soybean	Increased methionine Herbicide resistance	Bertholletia excels microbial gene Microbial gene
9. Linseed	Herbicide resistance	Microbial gene
10. Alfalfa	Protein quality	Ovalbumin gene of chicken
11. Petunia	Orange flower colour	Corn (*Zea mays*)

Various plant characters improved by transgenic technology in field crops

Character Improved	Example/Related to	Crop
Insect Resistance	*Helicoverpa armigera*	Cotton
	European corn borer	Maize
	Budworm	Tobacco
	Fruit borer	Tomato
Disease Resistance	Brown spot	Tobacco
	Seedling blight	Tobacco
	Potato leaf roll	Potato
	Virus X and Y	Patato
	Tomato mosaic virus	Tomato
	Seedling blight	Rape seed
Herbicide Resistance	Glyphosate	Cotton, flax, soyabean
	Gluphosinate	Wheat, potato, tomato and rapeseed
	Bromoxynil	Cotton
	2-4D	Cotton, tobacco
	Sulphonyl urea	Rapeseed and Flax
Quality traits	Protein quality	Potato, Alfalfa and tobacco
	Starch content	Potato
	Increased methionine	Soybean and rapeseed
Male sterility	Bar and barnase system	Rapeseed
Resistance to abiotic stress	Freezing resistance	Tomato, tobacco
	Cold resistance	Tobacco

CROP PROTECTION

4

L. Allwin, M. Rajesh and P. Anuradha

4.1. PESTS OF FIELD CROPS

Insect pests are the major constraints limiting agricultural productivity growth. It is estimated that herbivorous insects eat about 26 percent of the potential food production. Emerging problems of insecticide and disease resistance, secondary pest outbreak and resurgence further add to the cost of plant protection. Annual crop losses due to insect pests and diseases in India are estimated to be 18 percent of the agricultural output. Main causes of emerging pest and disease are favourable prevailing weather condition, absence of natural enemies, abundant food supply, use of pesticides, use of fertilizers and irrigation.

PESTS OF RICE

Borers and Defoliators

Sl.No	Pest	Scientific name	Symptoms	Nature of damage
1	Paddy stem borer	*Scirpophaga incertulas*	In the vegetative stage, dead heart seen in the affected tillers and in the reproductive stage, white ear may be seen	
2	Paddy gall midge	*Orseolia oryzae*	The central shoot instead of producing leaf produces a long tubular structure.	onion shoot, silver shoot or anaikomban. The feeding by the maggot and the larval secretion, which contains an active substance called cecidogen, is responsible for cell proliferation of the meristematic cells and gall formation. Parasitiod Platygaster oryzae naturally attack on rice gall midge.
3	Swarming caterpillar	*Spodoptera mauritia*	Nurseries found completely eaten away by the caterpillars overnight	Under severe infestation crop gives the appearance of grazed plants.
4	Rice case worm	*Nymphula depunctalis*	Plants stunted, caterpillars hanging on the leaf edges in a tubular case	ladder like appearance of the removed leaf tissue

Sl.No	Pest	Scientific name	Symptoms	Nature of damage
5	Rice skipper	*Pelopidas mathias*	Leaves folded longitudinally and scrapped patches in such places	
6	Leaf folder/ roller	*Cnaphalocrocis* medinalis	Leaves folded longitudinally silk and or transversely with scrapped patches in such places.	
7	Grass hoppers	*Hieroglyphus banian*	The nymphs and adults nibble leaves and also earheads in the early stages	

PESTS OF RICE- SUCKING PESTS

Sl.No	Pest	Scientific name	Symptoms	Nature of damage
1	Thrips	*Stenchaetothrips biformis*	Affected nurseries present a pale yellow colour with brown tips	Silvery streaks are seen on the leaves. Later, the leaves curl longitudinally and begin to dry from the tip downwards. In severe cases, the entire nursery may dry up and fail to produce seedling.
2	Green leafhopper	*Nephotettix virescens*	Affected plants become pale yellow in colour and get stunted in growth. If the plants are tapped large number of leafhoppers may be seen jumping to water	Plants are predisposed to fungal and bacterial infection through feeding and ovipositional punctures. Nymphs and adults exude sticky, whitish honeydew, which attracts sooty mould (that reduces the photosynthetic rate). It also transmits plant diseases such as dwarf, transitory yellowing, yellow dwarf and rice tungro virus (Tungro is transmitted during short feeding period).

Sl.No	Pest	Scientific name	Symptoms	Nature of damage
3	Brown planthopper/ Fulgorid	*Nilaparvata lugens*	hopper burn, in circular patches.	During sustained feeding, it excretes a large amount of honeydew. It also acts as vector of the virus diseases like grassy stunt, wilted stunt and ragged stunt. (Transmission of persistant ragged stunt and grassy stunt virus require more time. Sheath blight and stem rot incidence was high in BPH infested plants.)
4	Whitebacked planthopper	*Sogatella furcifera*	Damage in the form of **hopperburn** appears uniformly in a rice field, whereas it appears as circular patches in the case of BPH	Feeding puncture and lacerations caused by ovipositor predispose the plants to pathogenic organisms and honeydew excretion encourages the growth of sooty mould. It is not a vector of any viral disease.
5	Mealy bug	*Brevennia rehi*	Stunted, circular patches may be seen in the fields.	The affected tillers remain stunted with yellowish curled leaves. When the attack is severe, it inhibits panicle emergence. This type of disease is called as Soorai disease in Tamil Nadu.
6	Earhead bug/ Gundhi bug	*Leptocorisa acuta*	Leaves turn yellow and later rusted from tip downwards. Appearance of numerous brownish spots at the feeding sites / shrivelling of grains.	Sucking of the milky sap causes ill-filled/ partial filled and chaffy grains. The straw gives off-flavour that is unattractive to cattle.

PESTS OF SORGHUM (BORERS, LEAF FEEDERS AND SAP FEEDERS)

Sl.No	Pest	Scientific name	Symptoms	Nature of damage
1	Shoot fly	*Atherigona soccata*	Dead heats or drying of central shoots or production of profuse side tillers in main plants.	
2	Stem borer	*Chilo partellus*	Presence of circular holes on the unfolded leaves and dead hearts in the early stages	
3	Pink Stem borer	*Sesamia inferens*	Presence of dead heart.	The young larvae after hatching, congregate inside the leaf whorls and feed on folded central leaves causing typical 'pin hole' symptoms
4	Slug caterpillar	*Thosea apierens*	Defoliation	They feed on the leaves and defoliate. Apart from this they cause lot of irritation on the people who work in the field. Harvest is made difficult and it is reported that even cattle do not relish the fodder.
5	Leaf roller	*Marasmia trapezalis*	Rolled up leaves in which the larvae are found feeding and longitudinal patches on leaves whose tips dry are the clear symptoms	
6	Earhead bug	*Calocoris angustatus*	No external symptom will be visible. The earheads should be tapped either on the palm or a piece of cardboard. A number of brownish or greenish nymphs and adults can be seen. On the developing grains small brownish spots will be visible. In severe infestation, the grains get shriveled without maturing and the earheads appear uneven	
7	Sorghum midge	*Contarinia sorghicola*	The flowers appear damaged; earheads are devoid of grains and during severe attack, may appear blighted.	

PESTS OF FINGER MILLET

Sl.No	Pest	Scientific name	Symptoms	Nature of damage
1	Pink borer	*Sesamia inferens*	Deadheart.	
2	White borer	*Saluria inficita*	Deadheart.	
3	Flea beetle	*Chaetocnema pusaensis*	Shot holes in the leaves of young plants.	
4	Ragi Root Aphid	*Tetraneura nigriabdominolis*	wilting	

PESTS OF MAIZE

Sl.No	Pest	Scientific name	Symptoms	Nature of damage
1	Stem fly	*Atherigona orientalis*	Dead heart.	
2	Corn worm/ Earworm	*Helicoverpa armigera*	Damage on grains and presence of broken grains in the earhead	
3	Cutworm	*Agrotis ipsilon*	Cutting of tender stem and defoliation	
4	Phadka grasshopper	*Hieroglyphus nigrorepletus*	Leaves defoliated from the margin; plants often bared.	

PEARL MILLET AND ITALIAN MILLET

Sl.No	Pest	Scientific name	Symptoms
1	Shoot fly	*Atherigona approximata*	Dead hearts in young plants; corkscrew or chaffy grains in the upper portion and well-developed grains in the lower portion of earhead in mature crop
2	Earhead midge	*Geromyia penniseti*	Grainless glumes with white pupal case attached to the tip of the spikelet.
3	Leaf beetle	*Lema downsei*	Whitening and drying of leaves leading to burnt up appearance
4	Stem borer	*Anadastus parvulus*	Scrapped leaves, plants wither and in severe cases.
5	Stem fly	*Atherigona destructor*	Dead hearts in young plants.

PESTS OF PULSES- REDGRAM AND CHICKPEA

Borers

Sl.No	Pest	Scientific name	Symptoms	Nature of damage
1	Gram pod borer	*Helicoverpa armigera*	In the early stages, plants seen defoliated. Boreholes seen on the pods and affected pods have no seeds.	Young larva feeds on tender leaves, buds, flowers, and subsequently it bores into the pods

Sl.No	Pest	Scientific name	Symptoms	Nature of damage
				and feeds on the seeds with its head and part of the body only thrust inside, the rest remaining outside. A single larva may destroy 30-40 pods before maturity
2	Blue butterfly	*Lampides boeticus*	Bore holes on buds, flowers, green pods and matured pods	The larva bores into the buds, flowers and green pods just within couple of hours after hatching and feeds inside the developing grains
3	Gram blue butterfly	*Euchrysops (Catochrysops) cnejus*	Presence of regular, big, circular borehole on the flowers and pods. Presence of flat slug like green coloured larva on affected flowers or young pods. Black ants hovers around the plants	Larva bores into the buds, flowers and green pods and feeds inside the developing grains.
4	Plume moth	*Exelastis atomosa*	Pods are scrapped in the early stages, later boreholes seen on the pods and seeds eaten away.	Young larvae bore into the unopened flower buds for consuming the developing anthers. Grown up larvae first scrape the surface of the pods and then bore into pods. The larvae never enter the pod completely.
5	Spotted pod borer	*Maruca testulalis*	Presence of semi-solid excreta at the junction of the borehole. Young shoot with dried tip, large scale dropping of flowers. Larva present inside the webbing of leaves, flowers and young pods, faecal material accumulates outside the borehole	It feeds on the seeds by boring into the pods.
6	Spiny pod borer	*Etiella zincknella*	Entrance hole in the green pod disappears and leaves little evidence that the pod is infested. In pods, the larva devours many seeds. The pod	Young larva bores into floral parts, making rough and irregular incision

Sl.No	Pest	Scientific name	Symptoms	Nature of damage
			always contains a mass of frass and held together by a loosely spun web.	
7	Field bean pod borer	*Adisura atkinsoni*	Affected pods and flowers have irregular bore holes	Young larva bores into floral parts and pods and feeds on it.
8	Redgram podfly/ 'Tur' podfly	*Melanagromyza obtusa*	Shriveled pods and seeds. Damaged seeds become unfit for consumption and also do not germinate. However, the attack of the fly remains unnoticed by the farmers due to the concealed mode of life of this insect within the pods	Young maggot attaches itself on the immature seed inside the pod. In the beginning it feeds on the surface and thereafter mines into seeds and makes galleries just under the seeds epidermis, causing a ring like track. (One seed is enough for the development of a maggot. It never leaves the pod in which it enters once and completes its maggot stage)
9	Stem fly	*Ophiomyia phaseoli*	Drooping of the tender leaves and yellowing characterize serious damage of young plants. The sites where maggot and pupae are present become swollen and start rotting. Older plants show stunting but are not usually killed	Maggot is the damaging stage. It mines sub-epidermally through the leaves. Plants are most seriously affected at the seedling stage, where stem is tunneled

FLOWER FEEDERS

Sl.No	Pest	Scientific name	Symptoms	Nature of damage
1	Blister beetle	*Mylabris pustulata*	Presence of blister beetles on the flowers. The adult beetles feed on flowers, leaves and tender panicles, thus preventing grain formation	These beetles are diurnal and general feeders
2	Flower webber	*Eublemma hemirrhoda* *Eublemma silicula*	Webbing of flowers	Larva webs the flowers on the inflorescence in greengram, blackgram

PESTS OF PULSES- BLACKGRAM, GREENGRAM, LABLAB AND COWPEA

SAP FEEDERS

Sl.No	Pest	Scientific name	Symptoms	Nature of damage
1	Bean aphid	*Aphis craccivora*	Presence of aphid colonies on the leaves, terminal shoots and pods. In severe cases, the plants wither and vigour reduced	
2	Thrips	*Ayyaria chaetophora*	Leaves mottled with characteristic silvering on lablab, green gram, black gram and cowpea.	
3	Whitefly	*Bemisia tabaci*	Yellowing of leaves, plant vitality reduced, development of sooty mould, plant dies in case of severe attack	Nymphs and adults suck the plant sap and also transmits yellow mosaic virus (YMV).
4	Green leafhopper	*Empoasca kerri* *Empoasca binotata*	Affected leaves turn pale and then rust-red. They curl downwards; in severe cases, show 'hopper burn' symptom, dry and fall to the ground	
5	Pod bug	*Riptortus pedestris* *Clavigrella gibbosa*	Nymphs and adults cause substantial damage to pods and also to stem, leaves and flower buds. Attacked pods show pale-yellow patches. When the attack is heavy, the pods shrivel up. The grains in the attacked pods remain shriveled and extremely small	
6	Lablab bug/ Stink bug	*Coptosoma cribraria*	Bugs cluster on the plant parts	

LEAF FEEDER

Sl.No	Pest	Scientific name	Symptoms
Sl.No	Pest	*Scientific name*	Symptoms
1	Leaf webber	*Eucosma critica*	During vegetative stage of the crop, the caterpillar damages leaves by webbing, while at the floral stages of the crop they enter the buds, flowers and pods and feed on the immature seeds.
2	Leaf folder	*Anticarsia irrotata*	Leaves folded together
3	Lab-lab leaf miner	*Cyphosticha coerula*	Leaves with large irregular papery mines on the dorsal surface

PESTS OF OILSEEDS – MUSTARD

Sl.No	Pest	Scientific name	Symptoms
1	Mustard aphid	*Lipaphis erysimi*	Suck the sap from under surface of leaves
2	Mustard sawfly	*Athalia lugens proxima*	Grub nibbles the margins of tender leaves and pod

GROUNDNUT

LEAF FEEDERS

Sl.No	Pest	Scientific name	Symptoms	Nature of damage
1	Red hairy caterpillar	*Amsacta albistriga*	Defoliation	All the leaves are eaten away by the larvae leaving the main stem alone. Larvae are voracious feeders. During severe attack, the caterpillars move in groups destroying field after field.
2	Leaf miner/ Surul poochi/ soybean leaf miner	*Aproaerema modicella*	Brown blotches in midrib; terminal leaflets with white patches; webbed together, distorted and finally get dried up	Minute larvae wander briefly and they mine into the leaves and eat the green matter of the leaves. Later they web several leaves together and feed inside the enclosed space thus formed.
4	Tobacco caterpillar	*Spodoptera litura*	Skeletonization, defoliation	
5	Gram pod borer	*Helicoverpa armigera*	Larvae feed on the leaves irregularly	

SUBTERRANEAN PESTS

Sl.No	Pest	Scientific name	Symptoms	Nature of damage
1	Stem borer, Ver poochi	*Sphenoptera perotetti*	Branches may droop and the plants show wilting and may die when the grubs reach the root.	Grubs tunnel the stem just below the soil surface and main roots, tunnel containing elongate flat-headed grubs
2	White grub	*Holotrichia spp*	Wilting of plants in patches. Wilted plants do not have taproot and rootlets	Both the grubs and adults do the damage. The grubs feed on the root and rootlets. Adult beetles feed on

Sl.No	Pest	Scientific name	Symptoms	Nature of damage
				the foliage and are capable of defoliating plants and even trees like neem, banyan. Feeding is carried out during the night
3	Termite	*Odontotermes obesus*	Wilting of plants in patches. Taproots of wilted plants penetrated; bore holes in pods. Damage to soft tissue on pod (scarification) leaving thicker portion intact. Termites found hovering in and around plants.	
4	Pod borer (Earwig)	*Euboriella (Anisolabis) stali*	Young pods showing bore holes plugged with excreta, sand particles or discoloured pulps; pods without kernels	
5	Pod bug	*Elasmolomus sordidus*	Freshly harvested pods having shrivelled kernels	
6	Millipede	*Spirostreptus sp*	Empty pods with hole	Young one and adult bore into soil attack developing pods and away internal kernel.

SAP FEEDERS

Sl.No	Pest	Scientific name	Symptoms
1	Aphids	*Aphis craccivora*	Wilting of tender shoots during hot weather. Leaves mottled with chlorotic or dark green spots and plants stunted
2	Leafhopper	*Empoasca kerri*	Tip yellowing in leaves in the form of 'V' shape later turn brown, dry and brittle, called 'hopper burn'. Also cause distortion and whitening of veins, leaf curling and plants stunted
3	Thrips	*Caliothrips indicus*	Older/lower leaves showing white spots/marks or streaks intermingled with black excreta on the upper surfac
4	Thrips	*Scirtothrips dorsalis*	Tender leaves showing yellowish green patches on the upper surface and brown necrotic areas and silvery sheen on the lower surface
5	Thrips	*Frankliniella schultzei*	Young/ terminal leaves showing white scars, transmit bud necrosis

PESTS OF GINGELLY

Sl.No	Pest	Scientific name	Symptoms	Nature of damage
1	Leaf webber Leaf roller, Sesame webworm, Til leaf roller, Simsim webworm, pod caterpillar	*Antigastra catalaunalis*	Top leaves rolled and webbed together and damaged in vegetative phase; flowers and young capsules bored at reproductive phase	
2	Gall fly	*Asphondylia sesami*	Gall formation	Maggots feed on floral parts and cause malformation like galls (tumors) on the buds, which fail to develop
3	Sphinx, Death's head moth, hawk moth	*Acherontia styx*		Massive larva is voracious feeder on leaves defoliating the whole plant
4	Leaf hopper	*Orosius albicinctus*	Phyllody	Vector of sesamum phyllody

PESTS OF CASTOR

Sl.No	Pest	Scientific name	Symptoms	Nature of damage
1	Shoot and capsule borer	*Conogethes (=Dichocrocis) punctiferalis*	Capsules with bore holes, damaged capsules webbed together, peduncle and capsules show galleries made of silk and frass	In early stage, larvae are present at junction of the petiole feed on petiole and young leaves. They also bore into capsule and peduncle
2	Castor semilooper	*Achaea janata*		First instar larva nibbles on the leaves while the second instar makes holes and late instars eating the leaves completely. Caterpillars destroy leaves, bark and capsules. Moths suck citrus fruit, so called fruit sucking moth, doing great damage to citrus
3	Semilooper	*Parallelia algira*		Larva feeding on castor leaves

Sl.No	Pest	Scientific name	Symptoms	Nature of damage
4	Castor slug caterpillar	*Latoia (=Parasa) lepida*	The slug-caterpillars feed on the leaves. When they are young they remain in clusters and feed by scrapping the green matter of the leaves on the underside. Affected leaves may dry	
5	Tussock caterpillar	*Notolopus (=Orygyia) postica*	Caterpillars of the moth feed on the leaves	

PESTS OF COTTON- SUCKING PESTS

SUCKING PEST

Sl.No	Pest	Scientific name	Symptoms	Nature of damage
1	Leafhopper / Jassid	*Amrasca devastans*	Yellowing of leaves, crinkling, backward curls of leaves, bronzing and hopperburn, plants become stunted	Nymphs and adults remain between the veins on the undersurface of the leaves and suck the sap.
2	Cotton aphid	*Aphis gossypii*	Young plants become weak, leaf curl up and wither	Nymphs and adults suck the plant sap and lower the plant vitality. They produce honeydew, which results in development of black sooty mould. The damage is more severe in young plants.
3	Thrips	*Thrips tabaci, Scirtothrips dorsalis*	Silvery sheen on the undersurface of the leaves. When infestation in heavy, leaves shrivel and have ragged edges. Attacks leaf and flowers	Nymphs and adults lacerate the tissue on the under surface of the leaves and suck up the sap.
4	Whitefly	*Bemisia tabaci*	White chlorotic spots appear on leaves, which coalesce forming reddish yellow area, which extends from veins to the outer edges. Leaves fall prematurely. Number and quality of flowers squares and bolls gets reduced. Nymphs and adults secrete honeydew, which lead to sooty mould formation.	Nymphs and adults remain in cluster and desap. It also acts as vector of cotton leaf curl virus, vein-clearing disease in bhendi, tobacco leaf curl and cassava mosaic

Sl.No	Pest	Scientific name	Symptoms	Nature of damage
5	Mealy bug	*Ferrisia virgata*	Presence of mealybugs on tender shoots, leaves, stem and bolls. Yellowing of leaves initially and later the plants die. Leaves and bolls fall prematurely	Nymphs and adults remain in-group and suck the sap. Generally the damage is noticed towards the end and in places where ratooning is practiced

PESTS OF COTTON

BOLLWORMS

Sl.No	Pest	Scientific name	Symptoms	Nature of damage
1	Spotted bollworm	*Earias vittella*	Terminal shoots wither and droop; shedding of buds, flowers and bolls, flaring of bracts, bore holes on bolls plugged with excreta, bolls eaten and lint discoloured	Larva bores into the tender top shoot during vegetative stage and during reproductive stage they bore into the buds, flowers and bolls
	Spiny bollworm	*Earias insulana*		
2	Pink bollworm	*Pectinophora gossypiella*	Premature shedding of the buds and bolls, infested flowers are spun together to form rosette shaped bloom and boll fails to open, premature opening of the bolls, seeds destroyed, lint quality declines seed germination reduced, presence of interlocular burrow in the opened bolls	Larvae feed inside the buds, flowers and bores into bolls. They feed on the developing anthers and style and occasionally ovary. The external injury caused by them heals up.
3	American bollworm	*Helicoverpa armigera*	Regular circular boreholes on bolls, presence of granular faecal pellets outside the bore hole.	Caterpillar cuts a hole on the boll and feed on the boll by thrusting their heads alone inside. The affected bolls may rot due to fungus attack.
4	Red bollworm	*Rabila frontalis*	Bolls showing irregular bore holes the bolls and feed	Larvae remain inside

COTTON FOLIAR FEEDERS

Sl.No	Pest	Scientific name	Symptoms	Nature of damage
1	Leaf roller	*Sylepta derogata*	Leaves rolled in the form of trumpets and fastened by silken threads, defoliation	Larva remains within the fold and eat away the leaves. A single larva can damage several leaves.
2	Ash weevil	*Myllocerus spp*	Holes in the leaves, leaf margins notched, roots eaten up, plants wilt and they come off easily when pulled	Grubs remain underground and feed on the root system. Adults they damage the above ground parts

STEM WEEVIL, SURFACE AND LEAF WEEVILS

Sl.No	Pest	Scientific name	Symptoms	Nature of damage
1	Stem weevil	*Pempherulus affinis*	Swellings of stem at the point of attack (collar region) and the plant break off from this point, when heavy winds blow	Adults feed on the bark while the young grub cuts through medullary rays, tunnels round the stem along the cambium and feed on the soft portion
2	Shoot weevil	*Alcidodes affaber*	Galls in the stem and petiole, defoliation	Grubs bore into stem and petiole causing gall like swelling. Adults feed on leaves, buds and tender terminal shoots
3	Surface weevil	*Atactogaster finitimus*	Severing of seedlings	Adults cut the stem of seedlings at the time of germination under rainfed condition in the black soil tract of Tirunelveli district
4	Stem borer	*Sphenoptera gossypii*	Stem tunneled, stem swollen at the base and dries up.	The grub burrows into the stem feeding on the internal tissues

PESTS OF SUGARCANE

Sl.No	Pest	Scientific name	Symptoms	Nature of damage
1	Early shoot borer	*Chilo infuscatellus*	Deadheart in 1-3 month old crop, which can be easily pulled out, rotten portion of the straw coloured dead-heart emits an offensive odour. A number of bore holes at the base of the shoot just above the ground level	Caterpillar bores into the central shoot and feeds on the internal tissue. This cause interference in the translocation of plant sap and damage the sugar storing tissues
2	Internode borer	*Chilo sacchariphagus indicus*	Internodes constricted and shortened, with a number of boreholes and fresh excreta in the nodal region. Affected tissues reddened	Caterpillars attack sugarcane plants after 3 months of planting. They bore into the canes near the nodes; entry holes are plugged with excreta. Entry is generally confined to the first five internodes.
3	Top borer	*Scirpophaga excerptalis*	Deadheart in grownup canes, which cannot be easily pulled; deadheart reddish brown in colour; parallel row of shot holes in the emerging leaves and red tunnels in the midribs of leaves; bunchy tip due to the growth of side shoots	Caterpillar are mainly found in the apical portion of the canes, boring through the growing point and down the upper joints until it reaches the sappy portion of the stem, there it feeds on the tissues and destroys the cane. They also bore into the unfolded leaves preferably into the midrib, mining its way to the base

SUCKING PESTS

Sl.No	Pest	Scientific name	Symptoms	Nature of damage
1	Whiteflies	*Aleurolobus barodensis*	Yellowing of leaves, turn pinkish and later gradually dry. Infested leaves look white, while those below are black due to the growth of sooty mould on the honeydew	
2	Whiteflies	*Neomaskellia bergii, Neomaskellia andropogonis*	Black, grey or white dot like pupae on the undersurface of leaves	
3	Leaf hopper	*Pyrilla perpusilla*	Yellow leaves, covered with black sooty mould; top leaves gets dried up and lateral buds germinate.	Nymphs and adults remain in groups on the lower surface of leaves and suck the plant sap. They secrete honeydew, which attract sooty mould development. Severe attack results in loss of sugar content in the canes, to the extent of 34% and growth is suppressed
4	Mealybug	*Saccharicoccus sacchari*	Pinkish oval insects beneath leaf sheath on the nodes, with whitish mealy coating, main cane stunted also attack roots. Sooty mould develops on the plant	
5	Black winged bug	*Proutista moesta*		Nymphs and adults suck the plant sap

SUBTERRANEAN PESTS

Sl.No	Pest	Scientific name	Symptoms	Nature of damage
1	Termites	*Odontotermes obesus*	Poor germination of setts (after planting), characteristic semi-circular feeding marks on the margin of the leaves in the standing crop. Entire shoot dries up and can be pulled out. Setts hollow inside and may be filled with soil. Cane collapses if disturbed; rind filled with mud.	
2	Root borer	*Emmalocera depresella*	Young shoots with deadhearts which cannot be pulled out easily and do not emit any smell. Grown up canes will be with yellow leaves	Caterpillars bore at the base of a stem, which is very close to the root. Although they do

Sl.No	Pest	Scientific name	Symptoms	Nature of damage
			not actually bore into the roots, but since they do so near the soil surface they are called root borers. It differs from the shoot borer damage in the way that the dead-heart when pulled out the whole plant comes out and it does not possess bad smell. Besides central leaf whorl, some side leaves also dry	
3	Whitegrub	*Holotrichia consanguinea*	Drying of crown, preceded by yellowing and wilting of leaves. Affected canes come off easily when pulled. Cause extensive damage to roots	
4	Root aphid	*Tetraneura javensis*	Root colonized with aphids; movement of black ants	

PESTS OF GREEN MANURE CROPS AND FORAGE CROPS

SESBANIA

Sl.No	Pest	Scientific name	Symptoms	Nature of damage
1	Sesbania stem borer	*Azygophleps scalaris*	Drying of sesbania plants bores into the stem the faecal	Stoutly built larva and feeds, matter getting accumulated at the entrance hole
2	Stem weevil	*Alcidodes bubo*	Drying of the terminal branches or whole plant, lodging of the plants during monsoon seasons; affected stem with numerous bore holes and swelling symptoms. Presence of apodous grub inside the stem.	

LEAF FEEDERS

Sl.No	Pest	Scientific name	Symptoms	Nature of damage
1	Green semilooper	*Pericyma glaucinans*	Defoliation of plants preceding with leaflets damage	
2	Green looper	*Semiothisa pervelgata, Semiothisa emersaria*	Defoliation of plants preceding with leaflets damage	
3	Brown looper	*Hyposidra succesaria*	Defoliation of plants preceding with leaflets damage	
4	Leaf webber	*Striglina scitaria*	Webbing of terminal leaflets into conical structures	In the early stage the larva twists the

Sl.No	Pest	Scientific name	Symptoms	Nature of damage
				terminal portion of the leaflet in the form of a small cone and lives inside scraping the green matter. As it grows, it webs together the leaflets and lives inside the tunnel of the web and ultimately pupates inside it
5	Mottled emigrant	*Catopsilia pyranthe*	Defoliation of the leaflets	

SUNHEMP

Sl.No	Pest	Scientific name	Symptoms	Nature of damage
1	Hairy caterpillar	*Utethesia pulchella*	Defoliation of the plant	Larvae feeds on the leaves
2	Flea beetle	*Longitarsus belgamensis*	Adult makes minute holes on leaves, grub feeds on the roots by mining	
3	Sunhemp mirid	*Ragmus importunitas*	Minute chlorotic spots, later coalescing to cause yellowing of leaves	

LUCERNE

Sl.No	Pest	Scientific name	Symptoms	Nature of damage
1	Spotted alfalfa aphid	*Therioaphis maculata*		Nymph and adult suck the sap
2	Lucerne weevil	Hypera variabblis	Leaves with irregular holes. Affected plants remain stunted	Grub and adult cause the damage

SUBABUL

Sl.No	Pest	Scientific name	Symptoms	Nature of damage
1	Jumping lice	*Heteropsylla cubana*	Wilting of growing shoots resulting in die back; plants unable to put forth new growth. Honeydew deposition and sooty mould development	Nymph and adult cause the damage

4.2. PESTS OF STORED PRODUCTS

PRIMARY PESTS

Sl.No	Pest	Scientific name	Symptoms	Nature of damage
1	Rice weevil	*Sitophilus oryzae, S. zeamais, S. granarius*	Presence of irregular holes of 1.5 m diameter on grains of rice, sorghum, wheat, barley, maize before harvest and in store	Both grub and adults cause the damage. Grains are hollowed out; kernels are reduced to mere powder. S. oryzae and S. zeamais starts its attack in field itself. Adults cut circular holes. Heating takes place during heavy infestation, which is known as 'dry heating'
2	Sweet potato weevil	*Cylas formicarius*	Grubs and adults bore into the tubers and make them unfit for consumption	It is an important pest both in the field and in storage. Grubs bore into the tubers and make them unfit for consumption. They also attack the vines. Adult weevils feed on leaves, vines and tubers.
3	Lesser grain borer/ Hooded grain borer/ paddy borer beetle	*Rhyzopertha dominica*	Presence of round tunnel (1 mm) in grains and root crops. In bagged storage irregular messy waste flour spots indicate infestation of the pest. Presence of frass, shelled grains and foul smell.	Heating is very common. Infestation is confined to a small area. Grubs and adults cause damage and are voracious feeders. Adults reduce the grain kernels to mere frass. Grubs eat their way into the grain or feed on the grain dust or starchy material and are capable of attacking grain externally.
4	Cigarette beetle	*Lasioderma sericorne*	Circular, pinhead sized bore holes on processed tobacco	
5	Drug store beetle	*Stegobium paniceum*	Circular pin-head sized bore hole	
6	Pulse beetle	*Callosobruchus chinensis, C. maculatus*	Grains with white cigar like eggs on seed coat and circular exit hole	

Sl.No	Pest	Scientific name	Symptoms	Nature of damage
7	Tamarind beetle	*Pachymeres gonagra*	Circular holes on fruits of tamarind both in tree and storage	
8	Khapra beetle	*Trogoderma granarium*	Stored grains reduced to frass, seed coat chewed up in an irregular manner on all cereals.	Adults are harmless. Grub damages the grain starting with germ portion, surface scratching and devouring the grain. It reduces the grain into frass. Excessive moulting results in loss of market value due to insanitation caused by the cast skin, frass and hair. Crowding of larvae leads to unhygienic conditions in warehouses. Damage is confined to peripheral layers of bags in bulk storage
9	Angoumois grain moth or Grain moth	*Sitotroga cerealella*	Grains with circular emergence hole with characteristic flap/trap door	Only larvae damage grains, adults being harmless. Grains are hollowed out. It attacks both in fields and stores. In stored bulk grain, infestation remains confined to upper 30 cms depth only. Caterpillar enters the grain through crack or abrasion on grain. It feeds inside and remains in a single grain only.
10	Potato tuber moth	*Pthorimaea operculella*	Mine leaves and drooping of twigs of the plants in the field. Black excreta peeping out near the eye buds as a result of tunneling by caterpillars. Rotting and foul-smelling tubers due to bacterial infection caused by caterpillar injuries are the common in the godowns	In the field, the larvae mine into leaves or bore into tender shoots and developing tubers. It is carried over to the storage rooms.

SECONDARY PESTS

Sl.No	Pest	Scientific name	Symptoms	Nature of damage
1	Rust red flour beetle	*Tribolium castaneum*	Heavy infestation causes stinking odour in flour, adversely affecting the dough quality. It is an important pest for mill machiner Presence of grub, adults, moulted skin in the flour, acid odour	Grubs feed on milled products. Flour beetles are secondary pests of all grains and primary pests of flour and other milled products. In grains, embryo or germ portion is preferred. They construct tunnels as they move through flour and other granular food products. In addition they release gaseous quinines to the medium, which may produce a readily identifiable acid odour in heavy infestations.
	Confused flour beetle	*Tribolium confusum*		
2	Flat grain beetle	*Cryptolestus minutas, Laemophloeus pusillus*		Both adults and grubs feed on stored products and are important pests of mills. Grubs feed on germ portion and even on dead insects. Adults are only scavengers, cause heating in grain and flour in case of heavy infestation
3	Saw toothed grain beetle	*Oryzaephilus surinamensis*	Presence of off odour in grain	Adults and grub cause roughening of grain surface and off odour in grain. Grains with higher percentage of broken, dockage and foreign matter sustain heavy infestation, which leads to heating of grain.

Sl.No	Pest	Scientific name	Symptoms	Nature of damage
4	Ricemoth	*Corcyra cephalonica*	Presence of dense webbing of grains and foul smell, which makes the grains unfit for human consumption.	Larva is only responsible for damage. It contaminates foodgrains with frass, moults and dense webbing. In whole grains, kernels are bound into lumps upto 2 kg
5	Fig or Almond or Warehouse moth	*Ephestia cautella*	Before pupation, large number of wandering larvae trails behind silken threads. Such fine threads form the carpets of white sheen over the bags. Matured larva spins a silken cocoon at the junction of two overlapping edges of stacked bags	
6	Indian meal moth	*Plodia interpunctella*	Borehole on the grain	Larva causes serious damage to ear and grain of maize; contaminates the grain with excreta, cast skins, webbings, dead individuals and cocoons; prefers to eat the germ portion and hence grains lose viability.

4.3. INSECT PEST MANAGEMENT

Definition of IPM by FAO (1967)

Integrated Pest Management (IPM) is a system that, in the context of associated environment and population dynamics of the pest species, utilizes all suitable techniques and methods in as compatible a manner as possible and maintains pest populations at levels below those causing economic injury

Cultural methods: Manipulation of cultural practices to the disadvantage of pests

S. No.	Cropping Techniques	Pest Checked
1.	Ploughing	Red hairy caterpillar
2.	Puddling	Rice mealy bug
3.	Trimming and plastering	Rice grass hopper
4.	Pest seed material	Potato tuber moth
5.	High seed rate	Sorghum shootfly
6.	Rogue space planting	Rice brown planthopper
7.	Plant density	Rice brown planthopper
8.	Earthing up	Sugarcane whitefly
9.	Detrashing	Sugarcane whitefly
10.	Destruction of weed hosts	Citrus fruit sucking moth
11.	Destruction of alternate host	Cotton whitefly
12.	Flooding	Rice armyworm
13.	Trash mulching	Sugarcane early shoot borer
14.	Pruning / topping	Rice stem borer
15.	Inter cropping	Sorghum stem borer
16.	Trap cropping	Diamond back moth
17.	Water management	Brown planthopper
18.	Judicious application of fertilizer	Rice leaf folder
19.	Timely harvesting	Sweet potato weevil

Physical methods: Modification of physical factors in the environment to minimise (or) prevent pest problems. Use of physical forces like temperature, moisture and *etc.,* to manage the insect pests.

A. Manipulation of temperature

1. Sun drying the seeds to kill the eggs of stored product pests.
2. Hot water treatment (50 - 55oC for 15 min) against rice white tip nematode.
3. Flame throwers against locusts.
4. Burning torch against hairy caterpillars.
5. Cold storage of fruits and vegetables to kill fruit flies (1 - 2°C for 12 - 20 days).

B. Manipulation of moisture

1. Alternate drying and wetting rice fields against BPH.
2. Drying seeds (below 10% moisture level) affects insect development.
3. Flooding the field for the control of cutworms.

C. Manipulation of light

1. Treating the grains for storage using IR light to kill all stages of insects (eg.) Infra-red seed treatment unit
2. Providing light in storage go downs as the lighting reduces the fertility of Indian meal moth, 3. Light trapping.

D. Manipulation of air: Increasing the CO_2 concentration in controlled atmosphere of stored grains to cause asphyxiation in stored product pests.

E. Use of irradiation: Gamma irradiation from Co^{60} is used to sterilize the insects in laboratory which compete with the fertile males for mating when released in natural condition. (eg.) Cattle screw worm fly, *Cochliomyia hominivorax* control in Curacao Island by E.F.Knipling.

F. Use of greasing material: Treating the stored grains particularly pulses with vegetable oils to prevent the oviposition and the egg hatching. eg. Bruchid adults.

G. Use of visible radiation: Yellow colour preferred by aphids, cotton whitefly: yellow sticky traps.

H. Use of Abrasive dusts

1. Red earth treatment to red gram: Injury to the insect wax layer.
2. Activated clay: Injury to the wax layer resulting in loss of moisture leading to death. It is used against stored product pests.
3. Drie-Die: This is a porous finely divided silica gel used against storage insects.

Mechanical control: Use of mechanical devices or manual forces for destruction or exclusion of pests.

A. Mechanical destruction: Life stages are killed by manual (or) mechanical force.

Manual Force

1. Hand picking the caterpillars
2. Beating: Swatting housefly and mosquito
3. Sieving and winnowing: Red flour beetle (sieving) rice weevil (winnowing)
4. Shaking the plants: Passing rope across rice field to dislodge caseworm and shaking neem tree to dislodge June beetles
5. Hooking: Iron hook is used against adult rhinoceros beetle
6. Crushing: Bed bugs and lice
7. Combing: Delousing method for Head louse
8. Brushing: Woolen fabrics for clothes moth, carpenter beetle.

Mechanical force

1. **Entoleter:** Centrifugal force - breaks infested kernels - kill insect stages - whole grains unaffected - storage pests.

2. Hopper dozer: Kill nymphs of locusts by hording into trenches and filled with soil.
3. Tillage implements: Soil borne insects, red hairy caterpillar.
4. Mechnical traps: Rat traps of various shapes like box trap, back break trap, wonder trap, Tanjore bow trap.

B. Mechanical exclusion

Mechanical barriers prevent access of pests to hosts.

1. Wrapping the fruits: Covering with polythene bag against pomegranate fruit borer.
2. Banding: Banding with grease or polythene sheets - Mango mealybug.
3. Netting: Mosquitoes, vector control in green house.
4. Trenching: Trapping marching larvae of red hairy caterpillar.
5. Sand barrier: Protecting stored grains with a layer of sand on the top.
6. Water barrier: Ant pans for ant control.
7. Tin barrier: Coconut trees protected with tin band to prevent rat damage.
8. Electric fencing: Low voltage electric fences against rats.

Chemical methods

Sucking insects

- Acetamiprid 20 SP – Pride (2 g/lt)
- Imidacloprid 18.5 SL – Tatamida/Tatagold (0.4 ml/lt)
- Thiamethoxam – Actara – 10 g/ha
- Buproferzin 25 SC- Blunt/Applaud- BPH -800 ml/ha
- Acephate 75 SP – Asataf - 1000 g/ha
- Dimethoate 30 EC – Rogor - 1 lt/ha

Chewing insects

- Chlorpyriphos 20 EC – 1 lt/ha
- Quinalphos 25 EC – 1 lt/ha
- Profenophos 50 EC - Profex - 1 lt/ha
- Triazophos – Hostathion - 1 lt/ha
- Diflubenzuron 25 WP – Dimilin – 400 g/ha
- Fipronil 0.5SC – Regent – 1000 ml/ha
- Indoxacarb 14.5 SC – Avaunt – 500 ml/ha
- Thiodicarb 75 WP – Larvin – 1000 g/ha
- Novaluron 10 EC – Rimon – 625 ml/ha
- Spinosad 45 SC – Tracer – 200 ml/ha
- Emamectin Benzoate 5 Sg – Proclaim – 220 g/ha
- Flubendiamide 480 SC – Fame – 100 ml/ha
- Chlorantraniliprole 18.5 SC- Coragen – 500 ml/ha

ETL FOR DIFFERENT CROPS

RICE

Sl. No	Pest	ETL
1	Stem borer	5% dead hearts 1 egg mass / m^2
2	Gall midge	I gall / m^2, 5 affected tiller
3	Leaf folder	1 damaged leaf / hill
4	BPH	5-10 insect/ hill

PULSES

Sl. No	Pest	ETL
1	Gram pod borer	2-3 eggs / plant and 1 matured larvae / 10 plants

COTTON

Sl. No	Pest	ETL
1	Jassids	1-2 Nymphs / leaf
2	Aphids	15-20% infestation
3	Spotted bollworm	10% affected shoots
4	American bollworm	1 egg/plant, 1 larvae / plant and 10% of nose affected bolls
5	Pink bollworm	10% of affected bolls

SUGARCANE

Sl. No	Pest	ETL
1	Shoot borer	15-20% dead hearts
2	Leaf hopper	3-5 nymphs / leaf

CATEGORIES OF PESTS

☞ **Based on occurrence following are pest categories**

Regular pest: Frequently occurs on crop - Close association e.g. Rice stem borer, Brinjal fruit borer

Occasional pest: Infrequently occurs, no close association e.g. Caseworm on rice, Mango stem borer

Seasonal pest: Occurs during a particular season every year e.g. Red hairy caterpillar on groundnut, Mango hoppers

Persistent pests: Occurs on the crop throughout the year and is difficult to control e.g. Chilli thrips, mealy bug on guava

Sporadic pests: Pest occurs in isolated localities during some period. e.g. Coconut slug caterpillar

☞ **Based on level of infestation**

Pest epidemic: Sudden outbreak of a pest in a severe form in a region at a particular time e.g. BPH in Tanjore, RHC in Madurai, Pollachi

Papaya mealy bug in Coimbatore and Erode

Endemic pest: Occurrence of the pest in a low level in few pockets, regularly and confined to particular area e.g. Rice gall midge in Madurai, Mango hoppers in Periyakulam

Parameters of insect population levels

General equilibrium position (GEP)

The average density of a population over a long period of time, around which the pest population tends to fluctuate due to biotic and abiotic factors and in the absence of permanent environmental changes.

Economic threshold level (ETL): Population density at which control measure should be implemented to prevent an increasing pest population from reaching the ETL.

Economic injury level (EIL)

The lowest population density that will cause economic damage

Damage boundary (DB): The lowest level of damage which can be measured. ETL is always less than EIL. Provides sufficient time for control measures.

Major pests of important crops and IPM

Rice

Pest	Symptom	Management
1. Yellow stem borer (*Scirpophaga incertulas*)	Dead heart White ear	*Trichogramma japonicum* Pheromone trapsLight traps
2. Paddy gall midge (*Orseolia oryzae*)	Silver shoots/onion leaf	*Platygaster oryzae* parasitoid Removal and destruction of affected tillers.
3. Brown plant hopper (*Nilaparvata lugens*)	Hopper burn Vector (Ragged stunt virus)	Alternate wetting and drying of the fieldYellow sticky traps
4. Green leaf hopper (*Nephotettix virescens*)	Vector (Rice yellow dwarf, tungro virus)	Yellow sticky traps
5.Leaf fold (*Cnaphalocrosis medinalis*)	White and dried up leaf blades	*Trichogramma chilonis* Light traps

WHEAT

Pest	Symptom	Management
1. Termites *(Odontotermes obesi) Macrotermns obesi)*	-	Drenching with chlorpyriphos
2. Pink stem borer (*Sesamia inferences)*	Dead hearts	-
3. Gujhia weevil (*Tanymecus inducs)*	-	-

Pulses (Red gram, cowpea, black gram and green gram)

- Plume moth - *Exelastis atomosa* (red gram)
- Gram pod fly - *Melanagromysa obtusa*
- Spiny Pod borer - *Etiella zinckinella*
- Pod borer - *Helicoverpa armigera* (NPV 250 larval equivalent/ha, Pheremone trap 10/ha (eg. Helilure), Bird perches 25/ha
- Gram pod borer - *Helicoverpa armigera* Head alone inside the pods, body parts remain outside (Chickpea).

STORED GRAIN PESTS

Pest		Management
Lesser grain borer	*Rhizopertha dominica*	Moisture content of the gram <9%. Malathion (50EC) @ 5ml/20 litres of water / Fenitrothion (50EC) at 5ml/l water (for cleaning storage structures) Malathion 5% dust 2 g/kg seed. Malathion 50 EC 25 ml/quintol
Khapra beetle	*Trogoderma granarium*	
Red flour beetle	*Tribolium castaneum*	
Rice moth	*Corcyra cephalonica*	
Angumois grain moth	*Sitrotroga cerealella*	
Pulse beetle	*Callosobruchus maculates and C. chinensis*	

COTTON

Pest		Management
Cotton jassid (Leaf hopper)	*Amrasca biguttula biguttula*	Release of *Chrysoperla* grubs @ 10000/ha.
Cotton whitefly (Leaf curl virus)	*Bemisia tabaci*	Seed treatment with Imidacloprid 70 WS @ 5 g/kg seed (non hybrid), 10 g/kg seed (hybris)/Thiamethozam (Actara) @ 5 g/kg seed/Acetamiprod 20 SP 20 g/kg seed. Avoid synthetic pyrethroid spraying
Cotton aphid	*Aphis gossypii*	
Spotted bollworms	*Earias vitella* *E. insulana*	
Pink bollworm	*Pectinophora gossypiella*	*Trichogramma chilonis* Pheromone: Gossyplure
American bollworm	*Helicoverpa armigera*	Staking the cotton stalks near the field should be avoided. Destroy opened bolls on the plant.
Red cotton bug	*Dysdercus cingulatus*	

SUGAR CANE

Pest	**Scientific Name**	**Mangement**
Termites	*Odontotermes obesus* *Microtermes anandi*	chloripyriphos @ 0.2% dust in soil
Early shoot borer	*Chilo infuscatellus*	Sturmiopsis inferens Granulosis Virus (GV)
Internode borer	*Chilo sacchariphagus indicus*	Trichigramma chilonis

Top borer	*Scirphophaga excerptalis*	Isotima javensis
Woolly aphid	*Ceratovacuna lanigera*	Dipha aphidivora Micromus sp
or		(either 1000 larvae of *Dipha aphidivora* 2500 larvae of *Micromus* sp. per hectare as soon as the occurrence of woolly aphids is seen)
Leaf hopper	*Pyrilla perpusilla*	*Epiricania melanoleuca*

Coconut Pest		**Parasrtoid**
Rhinocerous beetle	*Oryctes rhinoceros*	*Baculovirus oryctes*
Red palm weevil	*Rhynchophorus ferrugineus*	*Metarrhizium anisopiliae* *Beauveria bassiana and Ferrolure* R
Black headed caterpillar	*Opisina arenosella*	*Bracon brevicornis*
☞ Root feeding: Monocrotophos 10 ml in 10 ml of water		

Legal control methods: Preventing the entry and establishment of foreign plant and animal pest in a country or area and eradication or suppression of the pests established in a limited area through compulsory legislation or enactment

Pests Accidentally Introduced Into India

1. Pink bollworm - *Pectinophora gossypiella*
2. Cotton cushion scale - *Icerya purchasi*
3. Wooly aphid of apple - *Aphelinus mali*
4. SanJose scale - *Quadraspidiotus perniciosus*
5. Potato tuber moth - *Gnorimoschima operculella*
6. Cyst (Golden) nematode of potato - *Globodera sp.*
7. Giant African snail - *Acatina fullica*
8. Subabul psyllid - *Heteropsylla cubana*
9. Bunchytop disease of banana
10. Spinalling whitefly - *Aleyrodicus dispersus*
11. Papaya mealy bug *Paracoccus marginatus*

Foreign pests from which India is free

☞ Mediterranean fruitfly - *Ceratitis capitata*

☞ Grapeavine phylloxera

☞ Cotton boll weevil - *Anthonomos grandis*

☞ Codling moth of apple - *Lasperysia pomonella*

Quarantine

Isolation to prevent spreading of infection

Plant Quarantine

Legal restriction of movement of plant materials between countries and between states within the country to prevent or limit introduction and spread of pests and diseases in areas where they do not exist.

Pest Legislations

- 1905 - 'Federal Insect Pest Act' - first Quarantine act against SanJose scale
- 1912 - 'US Plant Quarantine Act'
- 1914 - 'Destructive Insects and Pests Act' of India (DIPA)
- 1919 - 'Madras Agricultural Pests and Diseases Act'
- 1968 - 'The Insecticides Act'

Different classes of quarantine

Foreign quarantine (Legislation to prevent the introduction of new pests, diseases and weeds from foreign countries)

Plant quarantine inspection and treatments at sea ports of Mumbai, Kolkata, Cochin, Chennai and Visakapattinam and airports of Amritsar, Mumbai, Kolkata, Chennai and New Delhi. Import by post parcel prohibited except by scientists.Import of plant materials prohibited or restricted. Import permits required for importation of plant material. Phytosanitary certificate from the country of origin is required.

Phytosanitary certificate is issued by State Entomologist and Pathologists to the effect that the plant or seed material is free from any pest or disease

a. Fumigation of imported plant material based on need

b. Taking care of pests of quarantine concern in India

Host Plant Resistance (HPR)

Definition: Those characters that enable a plant to avoid, tolerate or recover from attacks of insects under conditions that would cause greater injury to other plants of the same species (Painter, R.H., 1951).

Mechanisms of Resistance

The three important mechanisms of resistance are

- Antixenosis (Non preference)
- Antibiosis
- Tolerance

a. Antixenosis: Host plant characters responsible for non-preference of the insects for shelter, oviposition, feeding, etc. It denotes presence of morphological or chemcial factor which alter insect behaviour resulting in poor establishment of the insect. e.g.

- Trichomes in cotton - resistant to whitefly
- Wax bloom on crucifer leaves - deter feeding by DBM
- Plant shape and colour also play a role in non preference
- Open panicle of sorghum - Supports less *Helicoverpa armigera*

b. Antibiosis: Adverse effect of the host plant on the biology (survival, development and reproduction) of the insects and their progeny due to the biochemical and biophysical factors present in it. Manifested by larval death, abnormal larval growth, etc.

Antibiosis may be due to

- Presence of toxic substances
- Absence of sufficient amount of essential nutrients
- Nutrient imbalance/improper utilization of nutrients

Chemical factors in Antibiosis - Examples

Chemicals present in plants Imparts resistance against

1. DIMBOA (Dihydroxy methyl benzoxazin) Against European corn borer, *Ostrinia nubilalis*
2. Gossypol (Polyphenol) *Helicoverpa armigera* (American bollworm)
3. Sinigrin Aphids, *Myzus persicae*
4. Cucurbitacin - Cucurbit fruit flies
5. Salicylic acid - Rice stem borer

Physical factors in antibiosis

Thick cuticle, glandular hairs, silica deposits, tight leaf sheath, etc.

Examples of resistant varieties in major crops

Pest Resistant varieties

S. No.	Pest	Resistant variety
1	Rice Yellow stem borer	TKM 6, Paiyur 1
2	Brown planthopper	CO 42, IR 36, IR 64
3	Green leaf hopper	IR 50, Ptb 2, CO 46
4	Sugarcane Early shoot borer	CO 312, CO 421, CO 661
5	Internode borer	CO 975, CO 7304
6	Top shoot borer	CO 745, CO 6515
7	Cotton American bollworm	Abhadita
8	Spotted bollworm	Deltapine
9	Stem weevil	MCU 3, Supriya
10	Leaf hopper	MCU 5, K 7, K 8
11	Sorghum Earhead bug	K tall
12	Jasmine Eriophyid mite	Pari Mullai

SPECIFIC MANAGEMENT PRACTICES

- Clipping of top leaf - Rice stem borer
- Rope rouging - Rice case worm
- Alternate wetting and drying - rice BPH
- Trimming and plastering of field bund - Rice grass hopper
- Fishmeal trap - sorghum shoot fly
- Heli lure - H. armigera
- Pherodin SL lure, Blue cloth - S. litura
- Yellow sticky trap - Whitefly & Aphids
- Delta sticky trap - Diamond back moth
- Digging trench around field - Groundnut red hairy caterpillar
- Gossyplure - Pink bollworm
- Earthing up - Sugarcane early shoot borer
- Epiricania melanoleuca (Adult Parasitiod) - Sugarcane pyrilla
- Hooking - Black headed caterpillar, Rhinoceros beetle and longicorn beetles
- Swabbing and plastering - Mango mealy bug
- Fire torch - for hairy caterpillars
- Methyl eugenol lure trap - Fruit fly
- Pairing and prolinage method of insecticide application - Banana Rhizome weevil
- Covering individual fruits with polythene bag - Fruit sucking moth
- Rhino lure - Rhinoceros beetle
- Log trap - Red palm weevil

BIOLGICAL CONTROL OF CROP PESTS

Bio control agents:

- Predators - Australian Lady bird beetles *(Cryptolaemus montroziueri)*, lacewings *(Chrysoperla carnea)*, Ants, spiders, dragonflies, damselflies, ground beetle, Birds, reptiles.
- Wolf spider - *Pardosa pseudoannulata/ Lycosa pseudoannulata*
- Parasitoids - Parasitic wasp & flies
- Entomopathogenic microorganism
  - Bacteria *(Bacillus thuringiensis)*
  - Viruses (Nuclear polyhedrosis virus, Granulosis virus)
  - Protozoa and nematodes

Predations:

- *Rodalia cordinalis* (Lady Bird beetle)
- Vedalia beetle was introduced 1929-30 to control cottony cushion scale on casuarina in India *(Icerya purchasi)*
- Vedalia beetle first introduced from Australia 1888 for control of cottony cushion scale of California
- Australian ladybird beetle *(Cryptolaemus montrouzieri)* introduced in India in 1898 control of Green scale *(Coccus viridis)* but successful in mealy bug, scale (sacforming)
- *Chrysoperla carnea, Mallada boninenesis* = common Chrysopids used in cotton & sunflower - protection from aphids, thnps, red mite, leaf miner, leaf hopper & other soft body insects.
- Predatory mites, *Phyotoseiulus spp and Amblyseius spp* = control phytophagus mites.

Parasitoids:

Trichogramma egg parasitoids

Sugarcane: *Chilo spp; Scirpophaga excerptalis & Acigona steniellus* = @50,000 /ha (INB & ESB) Tissue borer - 5cc/ha

Maize: *Chilo partellus* (stem borer) = @ 75,000 / ha

Rice: *Scirpophaga incertulas* (stem borer) = @ 50,000 - 100,000 /ha

Cotton: Boll worm

- American boll worm *(Helicoverpa armigera)*
- Pink boll worm *(Pectinophora gossypiella)*
- Spotted boll worm *(Earias vitella)*

***Epiricania melanoleuca** - control of sugarcane leaf hopper *(Pyrilla perpusilla)*

Entomopathogens

Entomopathogenic bacteria:

In India, *Bacillus thuringiensis sub sp kurstaki*, isolated from *Helicoverpa armigera* from mysore 1955

1. *Bacillus thuringiensis sub sp kenyae*
2. " *tolworthi Cry 1-Lepidopteran larvae*
3. " *thuringiensis*
4. " *kurstaki*
5. " *aizawai - Cry - 2 lepidoptera & Diptera*
6. " *tenebrionis - Coleopteran - Cry - 3*
7. " *israeliensis - Dipteran- Cry-4*

- Spore less mutant Btk - PIPU 7 = Bangalore = *H. armigera* and safety to silkworm
- *Bacillus thuringiensis* discovered in 1911 from Mediterranean flour moth

Commercial product:

- Agree, biolep, lepidocide, biobit, dipel 8L, Delfin, halt and spicturin
- *Bt. sub sp. israelensis* and *Bt sub spp sphaericus* = Mosquitoes

- ☞ Bt, dosage @ 1-1.25 kg/ha = cotton bollworm complex
- ☞ *Paenibacillus popilliae* @ 0.5 -1kg/ha = white grub. Commercial product: Doom, milk spoe
- ☞ *B. thuringiensis* contain delta endotoxin
- ❖ **Bt endotoxin gene:** The gram positive bacteria *Bacillus thuringiens*is produces a crystal toxin called (delta) endotoxin.

The endotoxin is a stomach poison and kills the lepidopteran insects if consumed.

The gene (DNA fragment) responsible for producing endotoxin is isolated from Bt and cloned into plants like cotton, potato, maize, etc. to produce Transgenic cotton, etc.

Transgenic Bt plants	Targeted insect pests
Cotton	Bollworms, *S. litura*
Maize	European corn borer
Rice	Leaf folder, stem borer
Tobacco, Tomato	Cut worms
Potato, Egg plant	Colarado potato beetle

Baculoviruses:

- o Ha NPV @ 1.5 x10 12 POB/ha (250LE/ha) ---- Pigeonpea
  1% jaggery @ and 0.01 % Teepol
- o Ha NPV @ 3x10 12 POB/ha (500LE/ha) ---- Cotton
  1% jaggery @ and 0.01 % Teepol
- o Sl NPV @ 15x10 12 to 3x10 12 POB/ha ---- Tobacco and ground nut
  1% crude sugar & 0.01 % teepol

Granulosis virus

- o Sugarcane shoot borer *(Chilo infuscatellus)*
- o Diamond back moth

Entomopathogenic fungi:

- o *Beauveria bassiana*= Rice hispa, sugarcane, *tissue* borer
- o *B. brongniarata* and *Metarrhizium anisopliae*= White grub (*Holotrichia consanguinea*)
- o *Verticillinm lecanii* = Aphids, Scales
- o *Hirsutella thompsonii* = Coconut Eriophyiid mite
- o Nomuraea rileyi = *Spodoptera litura*

Entomopathogenic nematodes:

- o *Steinernema thermophilum, S. carpocapsae* ---- White grub
- o *Heterorhabdis bacteriophora* ---- White grub

Protozoa:

- o Vairimorpha = *H. armigera*
- o *Farinocystis tribolii* = Red flour beetle

Plant disease antagonists:

- o *Trichoderma harzianum* ------------ *Pythium spp, Fusarium oxysporum*
- o *T. viride* ------------------- *Scleretium rolfsii, Rhizoctonia solani*
- o *Pseudomonas putida, Phytophthora spp and P. fluorescens*

Antagonists to phytonematode:

a. *Paecilomyces lilacinus* (Fungus) --- *Meloidogyne incognita* (chickpea) and sunflower

b. *Verticillium chlamydosporium* - pigeon pea, cowpea and chick pea nematode

c. *Pasteuria penetrans* (Bacteria)

Antagonists to weeds:

- ☞ Water fern ---- *Crystobagous salviniae*
- ☞ Water hyacinth ----- *Neochetina eichhorniae, N. bruchi*
- ☞ Parthenium --- *Zygogramma bicolorata*
- ☞ Opuntia valgaris ----- *Dactylopius ceylonicus D. opuntiae*
- ☞ Parthenium --- *Puccinia abrupta var partheniicola, Trichoderma virens*

  Water hyacinth ---- *Alternaria eichhorniae*

- o World First classical biological control success goes to India1762 common myna (*Acridotheres tristas*) exported to Mauritius to control the Red locust - *Nomadacris septemfaciata* in sugarcane
- o Trichoderma produces phytotoxin named Viridol

Rice

Pests	Biocontrol agents	Dose
Yellow stem borer *(Scirphophaga incertulas)*	*Trichogramma japonicum* *T. chilonis*	50,000- 1, 00000
Brown plant hopper *(Nilaparvata lugens)*	*Cyrtorhinus lividipennis*	100 adults 50-75 nymphs /m2

SEX PHEROMONES

Chemical nature of sex pheromones

In general pheromones have a large number of carbon atoms (10-20) and high molecular weight (180 - 300 daltons). Narrow specificity and high potency are two attributes which depend on long chain carbon atoms and high molecular weight. But since pheromones are volatile their molecular weights cannot be very high as they cannot be carried by wind.

Butenandt and his coworkers in 1959 isolated 12mg of pheromone from the abdomen of half a million virgin females of silkworm. They named the pheromene as Bombykol. The chemical name is 10, 12 - hexadeca dienol. It is a primary alcohol.

The following are some of the female sex pheromones identified in insects

Sl. No.	Name of the Insect	Pheromone
1.	Silkworm, *Bombyx mori*	Bombykol
2.	Gypsy moth, *Porthesia dispar*	Gyplure, disparlure
3.	Pink bollworm, *Pectinophora gossypiella*	Gossyplure
4.	Cabbage looper, *Trichoplusia ni*	Looplure
5.	Tobacco cutworm, *Spodoptera litura*	Spodolure, litlure
6.	Gram pod borer, *Helicoverpa armigera*	Helilure
7.	Honey bee queen, *Apis sp.*	Queen's substance

Examples of male sex pheromones

Cotton boll weevil - *Anthonomas grandis*, Coleoptera

Cabbage looper, *Trichoplusia ni*, Lepidoptera

Mediterranean fruitfly, *Ceratitis capitata*, Diptera.

Sterile Insect Release Method (SIRM): It is a genetic control method. This is also called Autocidal control since insects are used against members of their own species. E.F. Knipling in 1937 in South East USA used the SIRM technique to control the screw wormfly (Cochliomyia nominivorax) a serious livestock pest.

Methods of sterilizatoin

Chemosterilants: Any chemical which interfere with the reproductive capacity of an insect.

a. Alkylating agents

They inhibit nucleic acid synthesis

- inhibit gonad development
- produce mutagenic effect

(e.g.) TEPA, Chloro ethylamine

b. Antimetabolites

Chemicals having structural similarity to biologically active substances. They interfere with nucleic acid synthesis. e.g. 5-Fluororacil, Amithopterin

4.4 PESTICIDES AND THEIR RESIDUES

Pesticide: Chemicals which are used to kill pests

Trend in pesticides consumption:

- >2.6 million tonnes of active ingredients of pesticide used world wide
- 85% is in agriculture

- India's consumption of pesticides 2% only
- India's third single largest consumer in the world and highest among the south Asian countries
- India - largest producer of pesticide in South Asia

Various generations of insecticides

Generation	Year	Compounds
1. First generation insecticide	1939-1942	BHC and DDT
2. Second generation insecticide	1944-1947	Organophosphates and Carbamate
3. Third generation insecticide	1967	Hormonal insecticides, JH mimic insect growth regulators
4. Fourth generation insecticide	1970s	Synthetic pyrethroids

Pesticide usage comparison

World	%	India	%
Fruits & vegetables	26	Cotton	45
Cereals	15	Rice	22
Maize	12	Vegetables	9
Rice	10	Pulses	4
Cotton	86		

- At present 209 pesticides registered for use in agricultural and public health
- At present 27 pesticides are banned and 10 under restricted use, 7 withdrawn
- DDT and Aluminum phosphide, Lindane, sodium cyanide, Methyl bromide, Methyl parathion, Methoxy ethyl mercury chloride (MEMC)

  * Still only 25-30% total cultivated area under pesticide cover

  *Average consumption of pesticide in India 0.29 kg/ha

  World 0.5 kg/ha

  Japan 12 kg/ha

  Korea 6.6 kg/ha

- Total pesticides used: world
- Herbicides - 36%
- Insecticides- 25%
- Fungicides - 10%

In developing countries

Major = Insecticide

12% = Herbicides

Insecticides

Inorganic : Aluminium phosphide and calcium arsenate

Botanical : Nicotine - Tobacco

Pyrethrin -Chrysanthimum, marigold

Azardirachtin- Neem

Organic :

i. Organochloride - DDT, BHC, Aldrin, Endosulfan

ii. Organo phosphorus

Contact : Dichlorvas, Parathion, Fenitrothine, Malathian

Systemtic : Dimethoate, Monocrotophos, Phosphomidan

iii. Carbamate :

Contact : Carburyl

Systematic : Aldlicarp, carbofuron

Synthetic pyrethroids : Cypermathrin, Fenvalerate, Delta methrin

- Bacteria : *Bacillus thuringiensis*
- Virus : Polyhedral virus, NPV
- Chemosterilant : : Diurea, tepa, m-tepa
- Pheromone : Gyplure, gossyplure
- Repellent : Deet, dimethyl phthalate
- Juvenile tolerance : Farnesol, methoprene
- Acaricide : Dicofol, Dinocap
- Fumigants : Aluminium Phosphide, Methyl Bromide, Phosphine.

Newer molecules

- Imidacloprid
- Thiamethoxam
- Acetamprid
- Indoxacarb

Growth regulators - Buprofezin, Dimilin (Diflubenzuron), Diafenthiuron

Pesticide in environnent:

- International Union of Pure and Applied Chemist (IUPAC)
- Chlorinated hydrocarbon insecticide persists 16 month - 5 years
- Persistent organic pollutants (POPs) ---- alter hormonal system endocrine disrupters

Deposits: Pesticide present on leaves and other surface after application

Residue:

- Pesticide deposits remain on the surface after 1 day after application.
- Pesticide persistence is usually specified in residual half-life (RL50)
- RL50 is a measurement of the time in days required to reduce pesticide residues to half of its initial deposits.

Determination of pesticide residue:

- Gas liquid chromatography (GLC) widely used technique to estimate pesticide residue
- High performance liquid chromatography (HPLC)
- Enzyme linked immuno sorbent assay (ELISA)

Pesticide residue in food:

- * Non-observed adverse effect level (NOAEL)
- *Acceptable or tolerable daily intake (ADI / TDI)
- *Maximum residue limit (MRL)

Toxicity Parameters

Toxicity of a given chemical to an organism can be measured using various parameters as listed below.

1) LD50 or Median lethal dose

LD50 is defined as the amount of insecticide per unit weight which will kill 505 of the particular organism or insect. LD50 usually expressed as mg/kg body weight or g/larva or adult insect.

2) LC50 or Median lethal concentration

Defined as the concentration of insecticide required to kill 50% of the given organism or insect. This is used when the exact dose per insect is not known, but the concentration is known. LC50 is expressed in PPM (1/1,000,000) or Percentage (1/100)

3) LT50 (Median lethal time)

LT50 is defined as the time required killing 50% of the population at a certain dose or concentration.

LT50 expressed in hours or minutes. LT50 is used in field studies and also for testing insect viruses (NPV).

4. KD50: Median knockdown dose	Dose of insecticide or time required to knockdown 50% of the insects. KD50 and KT50 are used for evaluating synthetic pyrethroids against insects. KT50: Median knockdown time
4. ED50: Median effective dose	These terms are used to express the effectiveness of insect growth regulators (IGR)
5. EC50: Median effective concentration	ED50 and EC50 are defined as the dose or concentration of the chemical (IGR) required to affect 50% of population and produce desired symptoms in them.

Pest Resurgence

Tremendous increase in pest population brought about by insecticides despite good initial reduction in pest population at the time of treatment.

Insecticides lead to pest resurgence in two ways.

i. After initial decline, resistant population increase in large numbers

ii. Killing of natural enemies of pest, cause pest increase

e.g. Quinalphos, phorate	-	Cause resurgence of BPH in rice
Carbofuran	-	Leaf folder in rice

Secondary pest outbreak

Application of a pesticide against a major pest, kills the natural enemies of minor or secondary pest. This causes the outbreak of a secondary pest.

e.g. Use of synthetic pyrethroids against bollworms in cotton killed natural enemies of whitefly causing an outbreak of whitefly which was a minor pest till then.

4.5. NEMATODES AND THEIR MANAGEMENT

- Nematodes obligate parasites require to live plant for feeding
- Damage threshold level for nematode pests infective nematode/cm3 soil
- Average losses in developed countries 8.8% developing countries 14.6% (India)
- Molya disease - *Heterodera avenae*
- Ear cockel nematode - *Anguina tritici*
- Root lesion nematode *Pratylechnus coffeae*

Migratory nematode parasites of above ground plant party:

- White tip nematode - *Aphelenchoides besseyi* [Rice- (white tip disease)]
- Bud and leaf nematode - *Aphelenchoides ritzemabosi, A. fragariae* - Chrysanthemum leaves
- Ufra disease - *Ditylenchys angustus* -feeds rice stem, leaves, inflorescences and malformation of tissues
- Red ring disease of palm and coconut - *Rhadinaphelechoides cocophilus*

Sedentary nematode parasites of above ground plant parts:

- Seed gall nematodes = *Anguina tritici* - seed galls on wheat Ear-cockle disease in Wheat
- Tundu disease in Wheat - *A. tritici* + *Corynebacterium tritici*

Ecto parasites of root:

- Xihinema, Longidorus & *Tylenchorhynchns*
- Stubby root & docking disorder = Trichodordus nematode

Semi - endoparasite of underground plant parts:

- Slow decline of citrus = *Tylenchulus semipenetrans*
- Reniform nematode = *Rotylenhus reniformis*

Migratory endoparasites of underground plant parts:

- Root lesion nematode - *Pratylenchns thornel* - wheat, maize sugarcane - *P. penetrans*- Apple, peach, plum
- Burrowing nematode - *Radopholus similis* - Banana, coconut, Tea, pepper
- Black head or toppling disease - Banana
- Yellow disease- black pepper
- spreading decline - *Radoholus citrophilus* - citrus

- ✍ Rice root nematode - *Hirschmaniella oryzae* - Rice
- ✍ Mitimiti disease = *Hirschmanniella miticausa* - *Colocasia esculenta*

Aphlenchoides arachidis - groundnut

Sedentary endoparasites of underground plant parts:

- ✓ Root -knot nematodes = *Meloidogyne spp and M. inocognita*
- ✓ Incentive second - Juvenile stage *M. javanica*
- ✓ False root- knot nematodes- *Nocobbus sp.*
- ✓ Cyst nematodes - *Heterodera avenae* (Wheat molya disease)

  Globodera rostochiensis (Potato cyst)

Disease complexes of nematodes:

- ✍ Black shank in tobacco - *Meloidogyne incognita* + *Phytophtora parasitica*
- ✍ yellow slime rot tundu disease in wheat - *Anguina tritici* + *Clavibacter michiganensis p. var tritici*
- ✍ Cauliflower disease in strawberry - *Aphelenchoides fragariae* + *clavibacter fasciens*
- ✍ Crown - gall disease of raspberry & grapevines - Root -knot reniform nematode

Nemotade as virus vectors:

- o Fan leaf virus of grapevine - *Xiphinema index, X.diversicandatum*
- o Raspbery ringspot virus tobacco black ring virus - *Longidoruss elongatus*
- o Rose hairy roots - *Ratylenchus vulvanus* + *Agrobacteri rhizogene*

Methods of Nematode control

Physical:

a. Cleaning wheat - seed free of ear cockle *(Anguina tritici)* (or)

b. Hot water treatment (40-55°C) with (50 sec- 1 min)

c. Soil solarization

Cultural:

- ✓ Summer ploughing, crop rotation, manipulation of organic matter
- ✓ Dicot plant alternative with wheat, barley, maize & sorghum.

Biological control:

- ✓ Parasitic bacteria - *Pastenria tenetrans*
- ✓ Toxin producing bacteria - *Pseudomonas fluroscense*
- ✓ Nematode parasitic fungi - *Paecilomyces lilacines, Verticillium, Chalmydosporium*
- ✓ Toxin producing fungi - *Trichoderma hazianum, T. viride, Aspergillus niger*
- ✓ Vesicular Arbuscular Mycorrhizae (VAM) - *Glomus fasciculatum, G. mossae*
- ✓ Improving phosphorous availability and also reducing nematode

Chemical control:

- ✓ DD(1,3 dicholoropropene 1,2 dicholoropropane), EDB (Ethelen di-bromide), MBr(Methyl bromide), phorate, Trisulfothion, aldicarb, carbofuran (carbamate) and hydocarbon (O.P)
- ✓ Systemic granular nematicide such as phorate and carbofuran @ 1-2 kg a.i/ha
- ✓ Carbosulfan effective seed treatment 1-3% a.i w/w for Root knot nematode

Entomopathogenic nematode:

- ✓ Steinernema - Xenorahabdus - *Steinernema carpocapse*
- ✓ Heterorhabditis - Photorhabdus - *Heterrhabdus indica*

IMPORTANT NEMATODE PESTS:

Rice:

- ✓ Rice root nematode - *Hirshmanniella oryzae*
- ✓ White tip nematode - *Aphelenchoides bsseyi*
- ✓ Rice - stem nematode - *Ditylenchns angustus* (Ufra - disease)
- ✓ Rice root - knot nematode - *Melidogyne graminicola*
- ✓ Rice cyst nematode - *Heterodera oryzicola*

Wheat:

- ✓ Seed gall nematode - *Anguina tritici* = Earcockle disease - Dip wheat seeds in 5% salt solution
- ✓ Seed gall nematode = *Anguina tritici* - Tundu - Bacteria + *Clavibacter michiganensis* - yellow shimy ear rot
- ✓ Cereal cyst nematode = *Heteodera avenae* = Molya disease

Pulses:

- ✓ Reniform nematode = *Retylenchulus reniformis* (pusa balconi)
- ✓ Root - knot nematode = *Meloidogyne incognita M. javanica*
- ✓ Pigeonpea cyst nematode = *Heterodera cajani*

Vegetables:

Potato

- ✓ Potato cyst nematode = *Globodera rostrochiensis* = Golden nematode

  G. pallida -found in Nilgiris & Kodai hills
- ✓ Kufri swarna - resistant variety
- ✓ Root knot nematode = *Meloidogyne incognita* - Plains
- ✓ Potato, Tomao, Brinjal, chilles *M.hapla* = higher elevations
- ✓ Reniform nematode = *Rotylenchlus reniformis*

Banana: Burrowing nematode - *Radopholns similis*

Citrus: Citrus nematode - *Tylenchulns semipenetrans* - slow decline

Cotton

1. Root knot nematode - *Meloidogyne incognita*
2. Reniform nematode - *Rotylenchulns reniformis*

Rapeseed and mustard: Stunt nematode - *Tylenchorhynchus brassicae*

Sugarcane

- Lesion nematode - *Pratylenchus spp*
- Lance nematode - *Hoplolaimus indicus*

Coffee: Root - lesion nematode - *Pratylenchns coffeae*

Mushroom: *Aphelenchoides composticala* and *A.sacchari*

4.6. PLANT PATHOLOGY

IMPORTANT HISTORIES

- "Mycology' etimologically means "the study of fungi".
- The term 'Pathology' is derived from two greek words 'pathos' and 'logos'.
- 'Pathos' means suffering and 'logos' means the study / to speak / discourse. Therefore if etymologically means "study of suffering".
- The early man celebrated "Robigalia", a special holiday during which they offered scarifies of red dogs and sheep in an attempt to please the rust god "Robigo" and keep him from sending the rust disease to their crop.
- 1866 - ANTON DE BARY (Germany) was the first to indicate the nature of obligate and facultative forms. He laid the foundation of modern experimental plant pathology. He studied about the blight of potato, He is truely regarded as the "FOUNDER OF MODERN MYCOLOGY".
- 1880 - H.M. WARD emphasized the role of environment in the epidemiology of coffee rust. He may be reckoned as he "FATHER OF TROPICAL PLANT PATHOLOGY".
- 1885 - PIERRE MARIE ALEXIS MILLARDET accidentally discovered the Bordeaux mixture for the control of downy mildew of grapes.
- 1920 - E.J.BUTLER initiated an exhaustive study on Indian fungi and the diseases caused by them. He can also be called as the "FATHER OF MODERN PLANT PATHOLOGY IN INDIA". He is the first director of Imperial Mycological Institute (now International Mycological Institute) in England.
- 1876 - ROBERT KOCH of Germany described the theory called "KOCH'S POSTULATES." He established the principles of pure culture technique.
- 1905-1920 E.F.SMITH of U.S.A gave the final proof of the fact that bacteria could bencitants of plant diseases. He also worked on the bacterial wilt of cucurbits and crown gall disease. He is also called as "Father of Phytobacteriology".
- 1886 - ADOLF MAYER - He born in Germany and worked in Holland on Tobacco Mosaic Virus of tobacco. He demonstrated the sap transmission of the disease.
- 1892 - DIMITRII IVANOWSKI (Russia) demonstrated that the agent of tobacco mosaic virus could pass through even those filters that retained bacterial cells.

- 1898 - BEIJERINCK - a Dutch microbiologist and founder of virology proved that the virus inciting tobacco mosaic is not a microorganism. He believed it to be *contagium vivum fluidum* (infectious living fluid)
- 1935 - W.H.STANLEY proved that viruses can be made as crystals. He got Nobel Prize.
- Late blight of potato (*Phytophtora infestans*) caused severe potato losses in much of Northern Europe in the 1840's but it absolutely destroyed the potato crop in Ireland in 1845 and 1846 and caused the great "**Irish famine**". The destruction of potatoes caused widespread famine that resulted in the death of hundreds of thousands of people and the emigration of more than one and half million people from Ireland to the United States.
- In the fag end of the second world war (1943), Bengal had to face a serious famine (**Bengal famine**). One of the reasons to which this famine has been attributed is the loss in the yield of rice due to attack of *Heliminthosporium oryzae* leaf spot disease which had been affecting the crop for last several years.

DISEASE OF FIELD CROPS

RICE *(Oryza sativa)*

Blast [*Pyricularia oryzae* (*Syn : P. grisea*) (*Sexual stage : Magnaporthe grisea*)]

Symptoms - Appear on leaves, leaf sheath, rachis and nodes. The lesions start as small water soaked flecks, soon enlarge and form characteristic spindle shaped spots with grey centre and dark brown margin **(Leaf blast).** The affected nodes may break up and all the plant parts above the infected nodes may die **(Node blast).**

Mode of Spread and Survival - The disease spreads primarily through airborne conidia. The fungus also survives on collateral hosts *viz., Panicum repens, Digitaria magrginata, Brachiaria mutica, Leersia hexandra, Dinebra retroflea, Echinochloa crusgalli* and *Stenotaphrum secondatum.*

Fore casting – Min. night temperature (20-26^{0}C) in association with a high relative humidity 90 % and above lasting for a period of a week or more. In Japan, the first leaf blast model was developed and named as BLAST. Later various models were developed namely, PYRICULARIA, PYRIVIEW, BLASTAM, EPIBLA and P BLAST.

Management- Resistant varieties are Ponmani, ADT40, Co25 in Samba and Co37, Co43, IR20, Co44, ADT36 and ADT39 in Thaladi seasons. Seed treatment with *Trichoderma viride* @ 4g/kg or *Pseudomonas fluorescens* @ 10g/kg of seed. Avoid close spacing of seedlings in the main field. Spray the main field with Edifenphos 250 ml or Iprobenphos 500 ml.

Brown Spot or Sesame leaf spot [Helminthosporium oryzae (Syn : Drechslera oryzae)

(Sexual stage : *Cochliobolus miyabeanus*)]

Symptoms - The fungus attacks from seedling in nursery to milk stage in main field. Lesions (spots) on the coleoptile, leaf blade, leaf sheath, and glume, being most prominent on the leaf blade and glumes. Excess of nitrogen aggravates the incidence.

Mode of Spread and Survival - Airborne conidia infect the plants. The fungus also survives on collateral hosts like *Leersia hexandra, Arundo donux,* and *Echinochlova colonum.*

Management - Slow release nitrogenous fertilizers. Tolerant varieties- Co44, Cauvery. Disease free seeds. Seed treatment - Thiram/Captan at 4 g/kg. Spray Edifenphos 500 ml/ha.

Sheath blight [*Rhizoctonia solani* (Sexual stage : *Thanetophorus cucumeris*)]

Symptoms - The fungus affects from tillering to heading stage. On the leaf sheath oval or elliptical/irregular greenish grey spots are formed. As the spots enlarge, the centre becomes greyish white with an irregular blackish brown or purple brown border.

Mode of Spread and Survival - The pathogen can survive as sclerotia or mycelium in dry soil for about 20 months but for 5-8 months in moist soil. Sclerotia spread through irrigation.

Management - Avoid excess doses of fertilizers. Adopt optimum spacing. Eliminate weed hosts. Spray Carbendazim 250 g / Chlorothalonil 1 kg / Edifenphos 1 lit/ha. Seed treatment with *Pseudomonas fluorescens* @ 10g/kg of seed followed by seedling dip @ of 2.5 kg /ha.

Grain discolouration

Drechslera oryzae, D. rostratum, D.tetramera, Curvularia lunata, Trichoconis

padwickii, Sarocladium oryzae, Alternaria tenuis, Fusarium moniliforme, Cladosporium

herbarum, Epicoccum purpurascens, Cephalosporium sp., Phoma sp., Nigrospora sp.

Symptoms - Dark brown or black spots appear on the grains.

Management - Pre and post-harvest measures for prevention. Spray at bootleaf stage with Mancozeb 1kg/ Iprobenphos 500ml/Carbendazim250g/ha. Store the grain 13.5-14% moisture.

False smut [*Ustilaginoidea virens* (Syn : *Claviceps oryzae - sativa*)]

Symptoms - Fungus transforms individual grains into greenish spore balls of velvety appearance. Due to fructification of pathogen, ovaries are transformed to large green masses.

Favourable conditions - Rainfall & cloudy weather during flowering and maturity periods.

Udbatta disease [*Ephelis oryzae* (Sexual stage : *Balansia oryzae-sativa*)]

Symptoms - Entire ear head is converted to a straight compact cylindrical dark spike like structure. The spikelets are cemented to the central rachis and the size is reduced. No grain is formed. Entire spike covered by greyish stroma and convex pycnidia are immersed in stroma.

Management - Hot water seed treatment at 45°C for 10 min. Removal of collateral hosts like *Isachne elegans, Eragrostis tenuifolia* and *Cynadon dactylon.*

Bunt or Kernel Smut or black smut [*Tilletia barclayana*]

Minute black pustules or streaks are produced in the grains which burst open at the time of ripening. The grains may be partially or entirely replaced by the fungal spores.

Foot rot or Bakanae disease [*Fusarium moniliforme* (Sexual stage : *Gibberella fujikuroi*)]

Symptoms - Infected seedings are lean and lanky, taller than healthy seedlings and die after some time. In the main field, the affected plants have tall lanky tillers. Externally seed-borne.

Management - Treat the seeds with Thiram or Captan or Carbendazim at 2 g/kg.

Bacterial leaf blight [*Xanthomonas oryzae p.v. oryzae*]

Symptoms - Seedlings show circular, yellow spots in the margin, later enlarge, coalesce and cause drying of foliage (**Kresek** symptom). 1-2 weeks after transplanting. Bacterium enters through the cut wounds in the leaf tips, becomes systemic and cause death of entire seedling.

Favourable Conditions - Clipping of tip of the seedling at the time of transplanting.

Mode of Spread and Survival - Pathogen survives in soil and in the infected stubbles and on collateral hosts like *Leersia spp. Plantago najor, Paspalum dictum,* and *Cyanodon dactylon.*

Management - Burn the stubbles. Optimum dose of fertilizers. Avoid clipping of tip of seedling. Avoid flooded conditions. Remove weed hosts. Resistant cultivars like IR 20 and TKM 6. Spray Streptomycin sulphate and tetracycline 300 g + Copper oxychloride 1.25 g/ha.

Tungro disease [Rice Tungro Virus(RTV)]

Symptoms - Yellow discoloration in "Japonica" varieties, orange discoloration in "Indica" varieties. Chemically identified by lodine Test. 10 cm long leaf tip is cut in the early morning before 6 A.M. and dipped in a solution containing 2 g lodine and 6 g Potassium iodide in 100 ml of water for 30 minutes, infected leaves show dark blue streaks.

Pathogen - It is a composite disease caused by two morphologically unrelated viruses: rice tungro bacilliform virus (RTBV) and rice tungro spherical virus (RTSV).

Mode of Spread and Survival - Spherical particles help the transmission of bacilliform virus by the green leaf-hoppers. If the bacilliform virus particles are alone present in the rice plant they will not be transmitted by the leafhopper vector (*Nephotettix virescens, N. nigropictus, N. parvus, N.malayanus* and *Recilia dorsalis*) non-persistent manner. No latent period in the vector and infectivity is retained for a maximum period of 6 days after acquisition of virus.

Management - Summer deep ploughing and burning of stubbles. Destroy weed hosts of virus and vectors. Tolerant cultivars -IR50 and Co45. Spray Phosphomidan 500 ml / Fenthion 500ml/ Monocrotophos 1lit/ha /Neem oil 3% in main field 15 & 30 DAT control leaf hoppers.

Grassy stunt (Grassy stunt virus)

Symptoms - The infected plants stunted,excessive tillering and have an erect growth habit.

Mode of Spread - Transmitted persistent manner by the BPH (*Nilaparvata lugens*).

Rice dwarf (Rice dwarf virus**)**

Symptoms - Stunted growth with chlorotic or whitish specks on leaves. The size of specks varies and form interrupted streaks along the veins and distal part of infected leaves show diffuse yellowing. Plants which infected at early stage produce no ear heads.

Pathogen- Virus particle contains double stranded RNA or of double helical in nature.

Mode of Spread - Virus survive in the gramineous weeds. Transmitted through egg masses of leafhoppers from one generation to another (Trans ovarial transmission).

Management - Destory the weed host which harbour the virus and the vectors. Spray Phosphamidon or Fenthinon 500 ml or Monocrotophos 1 lit/ha.

Ragged stunt (Ragged stunt virus)

Symptoms - stunting of plants, vein swelling, delay in flowering, production of nodal branches and incomplete emergence of panicles are the main diagnostic symptoms.

Mode of Spread and Survival - The virus is transmitted in a persistent manner by the brown planthopper, *Nilaparvata lugens* but not through seed.

Rice yellow dwarf [*Mycoplasma Like Organism* (*MLO*)]

Symptoms - Stunted and have yellowish green to whitish green leaves. There is excessive tillering and leaves became soft and droop slightly. Plants are usually sterile.

Mode of Spread and Survival - Transmitted by *Nephotettix virescens* and

N. nigropictus with a latent period of 25-30 days. It survives on several grass weeds.

Management - Summer deep plough and burning stubbles. Resistant varieties IR62 & IR64.

SORGHUM *(Sorghum bicolor)*

Downy Mildew *Peronosclerospora sorghi*

Symptoms - Systemic invades the growing points of young plants, either through oospore or conidial infection and as the leaves unfold. The first few leaves partially infected with green/ yellow colouration of the infected portion. Abundant downy white growth on the lower surface of the leaves, consists of conidiospores and conidia. Normally 3 or 4 leaves develop the chlortic downygrowth. As the infected bleached leaves mature they become necrotic and the interveinal tissues disintegrate, releasing the resting spores (oospores) and leaving the vascular bundles loosely connected to give shredded leaf symptom.

Favourable Conditions – Max. sporulation at 100% relative humidity. Optimum temp. for sporulation is 21-23°C. Light drizzling & cool weather is highly favourable.

Mode of Spread - The primary - soil borne oospores persist. Secondary - air-borne conidia.

Management - Crop rotation. Moderately resistant variety -Co25 and Co26. Seed treatment with Metalaxyl (Apron 35 SD) 6 g/kg of seed. Spray Metalaxyl 500 g or Mancozeb 1 kg/ha.

Leaf blight [*Exerohilum turcicum* (Syn : *Helminthosporium turcicum*)]

Symptoms – It causes seed rot and seedling blight of sorghum.

The straw coloured centre becomes darker during sporulation.

Mode of Spread and Survival - The fungus is found to persist in the infected plant debris. Seed borne conidia are responsible for seedling infection.

Management - Use disease free seeds. Treat the seeds with Captan or Thiram at 4 g/kg.

Rectangular Leaf spot [*Cercospora sorghi*]

Symptoms - Small leaf spots which enlarge to become rectangular lesions. The colours of the spots vary from red, purple, brown or dark depending upon the variety.

Mode of Spread and Survival - Spreads through air-borne and seed-borne conidia.

Management - Disease free seeds. Treat the seed with Captan or Thiram at 4 g/kg.

Rust [*Puccinia purpurea*]

Symptoms - Purplish pustules (uredosori) appear on both surfaces of leaf which rupture to release reddish powdery masses of uredospores. Teliospores develop later.

Mode of Spread and Survival - Uredospores survive short time in soil and infected debris. **Management** - Remove the alternate host *Oxalis comiculata.*

Grain smut/Kernel smut / Covered smut / Short smut (*Sphacelotheca sorghi*)

Symptoms - Individual grains replaced by smut sori at a particular part of the head or entire inflorescence. The sori are oval or cyclindrical and are covered with a tough white cream to light brown skin (peridium). Ratoon crops exhibit higher incidence of disease.

Loose smut/Loose kernel smut (*Sphacelotheca cruenta*)

Symptoms - Affected plants detected before the ears come out. They are shorter than the healthy plants with thinner stalks and marked tillering. The ears come out much earlier than the healthy. The sorus is covered by a thin membrane.

Long smut (*Tolyposporium ehrenbergii*)

Symptoms - This disease restricted to small proportion of the florets. The sori are long, more or less cylindrical, elongated, slightly curved with a thick creamy-brown covering membrane (peridium). The peridium splits at the apex to release black mass of spores.

Head smut (*Sphacelotheca reiliana*)

Symptoms - The head is completely replaced by a large gall (sorus). The galls are at first covered by a whitish grey membrane of fungal tissue, ruptures.

Management for all smuts – Seed treatment -Captan /Thiram at 4 g/kg. Disease free seeds. Follow crop rotation. Collect the smutted ear heads in cloth bags and dip in boiling water.

Ergot or Sugary disease (*Sphacelia sorghi*)

Symptoms - Secretion of honey dew from infected florets, the drops of honey dew which had fallen on the soil. Long, straight or curved, cream to light brown, hard sclerotia develop. **Favourable Conditions -** High rainfall and high humidity during flowering season.

Mode of Spread and Survival - Primary source is germination of sclerotia. Secondary spread through air and insect- borne conidia. Rain splashes also help in spreading the disease.

Management - Adjust the date of sowing, crop does not flower during Sep.-Oct.. Spray Ziram (or) Zineb (or) Mancozeb (or) Carbendazim at 1 kg/ha at emergence of earhead (5-10 per cent flowering stage) followed by a spray at 50 % flowering.

Head mould/Grain mould/Head blight (More than thirty two genera of fungi)

Symptoms - Rains occur during flowering and grain filling stages, severe grain moulding can occur. Genera are *Fusarium, Curvularia, Alternaria, Aspergillus* and *Phoma. Fusarium semitectum* and *F.moniliforme* develop a fluffy white or pinkish colouration. *C. lunata* colours the grain black. Symptom varies depending upon the organism involved and the degree of infection.

Management - Adjust the sowing time. Spray Mancozeb 1 kg/ha. or Captan 1 kg + Aureofungin-sol 100 g/ha a week later and during milky stage.

CUMBU *(Pennisetum americanum)*

Downy mildew (*Sclerospora graminicola*)

Symptoms - Systemic infection and symptoms appear on the leaves and the inflorescences. First symptoms appear in 3 to 4 leaf stage. Light green to light yellow colour on the upper surface of leaves and the corresponding lower surface bears white downy growth.

Mode of Spread and Survival - Oospores viable in soil for 5 years to cause the primary infection on seedling. Secondary spread is through sporangia (rainy season).

Management - Deep ploughing. Rogue out infected plants. Adopt crop rotation. Excess seed rate. Resistant varieties like X5, WCC-75 and Co7. Treat seeds with Metalaxyl (Apron 35SD) at 6g/kg. Spray Mancozeb 1kg or Metalaxyl+Mancozeb (Ridomil MZ) at 2kg/ha on 20DAS.

Smut (*Tolyposporium penicillariae*)

Symptoms - Pathogen infects few florets and transforms them into plump sacs (sori) containing black powder (smut spores). Early stage, the sori are larger and greener than mature they become dark brown and are easily broken and release black smut spore balls.

Favourable Conditions - High humidity and successive cropping with cumbu.

Rust (*Puccinia penniseti*)

Symptoms - Leaf soon becomes covered by uredosori which appear on upper surface. The pustules may be formed on leaf sheath. stem and even on peduncles. Later, telial formation takes place on leaf blade, leaf sheath and stem. While brownish uredinia get exposed at maturity, the black telia remain covered by the epidermis for a longer duration.

Favourable Conditions - Closer spacing. Presence of abundant brinjal plants and other species of *Solanum* viz., *S.torvum, S. xanthocarpum* and *S. pubescents.*

Management - Spray with Wettable Sulphur 2.5 kg / Captafol 1 kg or Mancozeb 1.25 kg/ha.

Ergot or Sugary disease (*Claviceps fusiformis*)

(as in Sorghum)

MAIZE *(Zea mays)*

Downy mildew/Crazy top (*Peronosclerospora sorghi*)

Symptoms - White downy growth on the lower surface of leaf also on the chlorotic streaks. Downy growth also occurs on bracts of green unopened tassel. Proliferation of auxillary buds on the talk of tassel as well as the cobs is very common (Crazy top).

Management - Deep ploughing. Crop rotation with pulses. Adjust the time of sowing. Rogue out infected plants. Treat the seeds with Metalaxyl compound (Apron 35SD) at 6g/kg. Spray Metalaxyl + Mancozeb 2kg or Mancozeb 1.25 kg/ha on 20th day after sowing. Resistant varieties and hybrids are Co1, CoH1 and CoH2.

Leaf blight [*Helminthosporium maydis (Syn : H. turcicum*)]

Symptoms - Fungus affects the young stage maize plant. Small yellowish round to oval spots are seen on the leaves. The spots gradually increase in area into bigger elliptical spots and are straw to grayish brown colour in the centre with dark brown margins.

Mode of Spread and Survival - It is seed-borne fungus. It also infects sorghum, wheat, barely, oats, sugarcane and spores of the fungus are also found to associate with seeds of green gram, black gram, cowpea, varagu, Sudan grass, Johnson grass and Teosinte.

Management - Treat the seeds with Captan or Thiram at 4 g/kg. Spray Mancozeb 1.25 kg/ha.

Rust (*Puccinia sorghi*)

Symptoms - As in sorghum

Mode of Spread - Presence of alternate hosts (*Oxalis corniculata* and *Euchlaena Mexicana*).

Management - Remove the alternate hosts. Spray Mancozeb at 1.25 kg/ha.

Head smut (*Sphacelotheca reiliana*)

Symptoms - Large smut sori replace the tassel (partially or wholly) and the ear.

Favourable Conditions - Low temperature. it also infects the sorghum crop and *vice versa.*

Mode of Spread and Survival - Spores retains its viability for two years. The fungus is externally seed-borne and soil-borne. Major source of infection is soil-borne chlamydospores.

Charcoal rot [*Macrophomina phaseolina (Rhizoctonia bataticola)*]

Symptoms - First exhibit the wilting symptoms. Stalk - grayish streaks and become weak. Pith become shredded and grayish black minute sclerotia develop abundantly on the vascular bundles. Shreading of interior of the stalk and root bark and disintegration of root system. **Favourable Conditions -** High temperature and low soil moisture (drought)

Management - Long crop rotation with non- host of the fungus. Irrigate the crops at the time of earhead emergence to maturity. Treat the seeds with Carbendazim or Captan at 2 g/kg. Grow disease tolerant varieties viz., SN-65, SWS-8029, Diva and Zenit.

RAGI *(Eleusine coracana)*

Blast (*Pyricularia grisea*)

(As in Rice) Presence of collateral hosts like Tenai, bajra, wheat, barley and oats.

Seedling blight or Leaf blight [*Helminthosporium nodulosum* (Sexual stage : *Cochliobolus nodulosum*)]

Wilt [*Sclerotium rolfsii* (Sexual stage : *Corticium rolfsii*)]

Symptoms - Fungus attacks basal stem portion and later the leaf sheath and culm. The infected portion becomes soft and dark brown in colour.

Mode of Spread - Soil sclerotia and spreads through irrigation water and implements.

Management - Spot drench with Copper oxychloride at 0.25 per cent. Crop rotation.

RED GRAM (*Cajanus cajan*)

Wilt (*Fusarium oxysporum f. sp. Udum*)

Symptoms - Gradual withering and drying of plants. Yellowing of leaves and blackening of portion of stem appear, starting from collar to branches which gradually result in drooping and premature drying of leaves, stems, branches and finally death of plant. Vascular tissues exhibit brown discolouration and resulting in partial wilting.

Favourable Conditions - Soil temperature of 17-25°C. Continuous cultivation.

Mode of Spread and Survival - Primary spread is by soil-borne chlamydospores and seed contaminant. The secondary spread in the field is through irrigation water and implements.

Management - Treat the seeds with *Trichoderma viride* at 4 g/kg. Avoid successive cultivation of redgram in the same field. Crop rotation with tobacco. Adopt mixed cropping of sorghum. Grow resistant lines like ICP 8862, ICPL 88046, ICPL 227, BWR 254, DPPA 85-14, DPPA 85-15 and GPS 52.

Dry root rot [*Macrophomina phaseolina* (Sclerotial stage : *Rhizoctonia bataticola*)]

Symptoms - Occurs both young seedlings and grown up plants. Reddish brown discoloration at collar region and sudden death of the plants occur in patches. The lower leaves show yellowing, drooping and premature defoliation. The bark near the collar region shows shredding. The plant can be easily pulled off minute dark sclerotia seen in the shredded bark. **Favourable Conditions -** Prolonged drought followed by irrigation. High temp. (28-35°C).

Management - Treat the seeds with Carbendazim or Thiram at 4 g/kg or pellet the seeds with

Trichoderma viride at 4 g/kg. Apply heavy doses of FYM or green leaf manure like *Gliricidia maculata* at 10 t/ha or Apply Neemcake at 250 kg/ha.

Powdery mildew (*Leveillula taurica*)

Symptoms - White powdery growth on the lower surface of leaves. The corresponding areas in upper surface show pale yellow discoloration.

Mode of Spread and Survival - Fungus survives in the soil as cleistothecia and ascospores.

Management - Spray Carbendazim 500g or Wettable sulphur 1.5 kg/ha at the initiation of the disease and repeat after 15 days.

BENGAL GRAM *(Cicer arietinum)*

Root rot [*Rhizoctonia solani* (Sexual stage : *Thanatephorus cucumeris*)]

Symptoms - Brown lesion develop on the stem near ground level. The infected seedling slowly turn yellow and petioles and leaflets show drooping, leads to complete drying of the plant. The black dead tissues contain pinkish mycelial growth and sclerotia.

Favourable Conditions - Heavy soils with poor drainage and warm humid weather.

Mode of Spread and Survival - Sclerotia survives for several years in soil. Secondary spread through farm implements, irrigation water and rain splash.

Management - Treat the seeds with Carbendazim or Thiram at 2 g/kg or seed pelleting with

Trichoderma viride at 4 g/kg or *Pseudomonas fluorescens* @ 10g/kg of seed. Apply FYM at 10 t/ha. Resistant genotypes -ICCC 32, ICCC 42, ICCC 12223, ICC 11322 & H82-2.

Wilt (*Fusarium oxysporum f.sp. ciceri*)

Symptoms – Vascular browning, black streaks on the stem and root portion below the bark.

Favourable Conditions - High soil temperature (Above 25^{0}C), high soil moisture and presence of weed hosts like *Cyperus rotundus, Tribulus terrestris* and *Convolvulus arvensis.*

Mode of Spread and Survival - Primary infection through chlamydospores. The weed hosts also serve as a source of inoculum. The secondary spread is through irrigation water.

Management - Treat the seeds with Carbendazim or Thiram at 2 g/kg or Carbendazim 1 g+Thiram 1g/kg or treat the seeds with *Trichoderma viride* at 4 g/kg *Pseudonomas fluorescens* @ 10g/kg of seed. Apply heavy doses of organic manure or green manure. Grow resistant cultures like ICCC 42, H82-2, ICC 12223, ICC 11322, ICC 12408, P 621 and DA1.

Ascochyta blight (*Ascochyta rabiei*)

Symptoms - On leaflets, the lesions are round or elongated, bearing irregularly depressed brown spots, and are surrounded by a brownish red margin.

Mode of Spread - Infected plant debris as pycnidia. Externally and internally seed-borne.

Rust (*Uromyces ciceris-arietini*)

Mode of Spread and Survival - Fungus survives as uredospores in the legume weed *Trigonella polycerata.* The spread is through wind-borne uredospores.

Management - Destory weed host. Dust Sulphur at 20 kg/ha. Spray Wettable sulphur 1 kg or

Carbendazim 500 g/ha.

BLACK GRAM *(Vigna mungo)* and GREEN GRAM *(Vigna radiata)*

Powdery mildew (*Erysiphe polygoni*)

Symptoms - Small, irregular powdery spots appear on the upper surface of the leaves. Leaves turn yellow and finally shed. Often pods are malformed and small with few ill-filled seeds.

Favourable Conditions - Warm humid weather (late kharif and rabi seasons).

Mode of Spread and Survival - Primary infection ascospores. Secondary -air-borne conidia.

Management - Spray Carbendazim 500g or Wettable sulphur 1.5 kg or Tridemorph 500 ml/ha. Grow resistant varieties like LBG17, PDU10, ICI12/2 and PLU 322.

Anthracnose [*Colletotrichum lindemuthianum* (Sexual stage : *Glomerella lindemuthianum*)]

Symptoms - Dark brown to black sunken lesions on the hypocotyl area and cause death of the seedlings. Small angular brown lesions appear on leaves. Minute water soaked lesion appears on the pods & becomes brown and enlarges, necrotic areas with black dots(Acervuli).

Mode of Spread and Survival - It is seed-borne and cause primary infection. Secondary spread by air borne conidia.

Management - Remove and destroy infected plant debris in soil. Treat the seeds with Carbendazim at 2 g/kg. Spray Carbendazim 500g or Mancozeb 1 kg/ha.

Leaf spot (*Cercospora canescens*)

Symptoms - Small, circular spots develop on the leaves with grey centre and brown margin.

Mode of Spread and Survival - The fungus survives on diseased plant debris and on seeds. The secondary spread by air-borne conidia.

Management - Remove and burn infected plant debris. Spray Mancozeb at 1 kg/ha.

Rust [*Uromyces phaseoli typical* (Syn: *U.appendiculatus*)]

Symptoms - Small, round, reddish brown uredosori mostly on lower surface. They may appear in groups and several sori coalesce to cover a large area of the lamina.

Pathogen - Orange coloured *cupulate aecia* develop later on the lower surface of leaves. The aeciospores are unicellular and elliptical.

Dry root rot [*Rhizoctonia bataticola* (Pycnidial stage : *Macrophomina phaseolina*)]

(As in Red gram)

Yellow mosaic (*Yellow mosaic Virus*)

Symptoms - Initially small yellow patches or spots appear on green lamina. The young leaves are the first to show the symptoms.

Favourable Conditions - Summer sown are highly susceptible. The weed hosts *viz., Croton sparsiflorus, Acalypha indica, Eclipta alba* and *Cosmos pinnatus* and legume hosts.

Mode of Spread and Survival - Spreads through wind-borne white fly (*Bemisia tabaci*).

Management - Rogue out the diseased plants. Remove the weed hosts peiodically. Increase the seed rate (25 kg/ha). Grow resistant black gram variety like VBN-1, PDU 10, IC12/2 and PLU 322. Follow mixed cropping by growing two rows of maize (60 x 30 cm) or sorghum (45 x 15 cm) or cumbu (45 x 15 cm) for every 15 rows of black gram or green gram.

Leaf crinkle (*Leaf crinkle Virus*)

Symptoms -Later crinkling and curling of the tips of leaflets. Petioles and internodes are shortened. Infloresence, malformed, small size flower buds & fail to open.

Favourable Conditions - Weed hosts - *Aristolochia bracteata* and *Digera arvensis.*

Mode of Spread and Survival - The virus is seed-borne also transmissible. White fly (Bemisia tabaci) is the vector, helps in the secondary spread.

Management - Use increased seed rate (25 kg/ha). Rogue out the diseased plants. Cultivate seed crop during rabi season. Remove weed hosts. Spray Monocrotophos or Methyl demeton on 30 and 40 days after sowing at 500 ml/ ha.

Leaf curl (*Tomato Spotted Wilt Virus*)

Symptoms - Chlorosis of lateral veins near the leaf margins. Leaves are brittle and vein necrosis on the under surface of the leaves, produce a few small and malformed pods.

Favourable Conditions - The weed hosts *viz., Acanthospermum hispidum, Ageratum conyzoides, Amaranthus viridis, Calotropis gigantea, Lagasca mollis, Trianthema portulacastrum, Cassia tora, Cleome gynandra, Solanum nigrum* and *Datura metal* and other legume hosts.

Mode of Spread and Survival - Virus is transmitted by thrips *viz., Frankliniella schultzii, Thrips tabaci* and *Scirtothrips dorsalis.* It survives in weed hosts, tomato, petunia and Chilli.

Management - Rogue out infected plants. Remove the weed hosts. Spray Monocrotophos or Methyl demeton at 500 ml/ha on 30 and 45 days after sowing.

COWPEA *(Vigna unguiculata)*

Wilt (*Fusarium oxysporum f. sp. tracheiphilum*)

Symptoms - Shedding /withering of leaves. Brownish, purple discoloration of the cortical area, often extends throughout the plant.

Favourable Conditions - Temperature of 20-25^{0}C and moist humid weather.

Management - Treat the seeds with Carbendazim or Thiram at 2 g/kg or treat the seeds with

Trichoderma viride at 4 g/kg. Spot drench with Carbendazim at 0.5 g/litre.

Bacterial leaf spot (*Xanthomonas vignicola*)

Symptoms - Red lesion at cotyledon leads to withering. In adult plants yellow circular spots appear on leaves, enlarge and become angular and encircled by a yellow halo. The affected twigs show twig blight and infected pods contain shrivelled seeds.

Favourable Conditions - Heavy rainfall, high relative humidity

Mode of Spread and Survival - The bacterium is seed-borne. The wind, rain splash and insects help in the secondary spread.

Management - Disease free seeds.

GROUNDNUT *(Arachis hypogaea)*

Tikka leaf spots

Early leaf spot : *Cercopora arachidicola* (Sexual Stage) : *Mycosphaerella arachidis*)

Late leaf spot : *Phaeoisariopsis personata* Syn : *Cercospora personata* (Sexual stage) : *Mycosphaerella berkeleyii*)

Symptoms - all above ground parts of the plant. Lesions on petiole, stem and pegs. Spotted leaves shed prematurely.

Mode of Spread - Primary infection ascospores. Secondary spread by wind blown conidia.

Management - Remove and destory the infected plant debris. Eradicate the volunteer plants. Keep weeds under control. Treat the seeds with Carbendazim at 2g/kg. Spray Carbendazim 250g /Mancozeb 1 kg or Chlorothalonil 1 kg/ha. Moderately resistant varieties like ALR.1.

Rust (*Puccinia arachidis*)

Symptoms - Small brown to chestnut dusty pustules (uredosori) on lower surface of leaves. Corresponding to the sori, small, necrotic, brown spots appear on the upper surface of leaves.

Mode of Spread - Fungus survives on the collateral hosts like *Arachis marginata, A. nambyquarae* and *A. prostrate.*

Management - Avoid monoculturing. Remove volunteer plants and reservoir hosts. Spray Mancozeb 1 kg or Wettable sulphur 2.5 kg or Tridemorph 500 ml or Chlorothalonil 1 kg/ha. Grow moderately resistant varieties like ALR.1

Collar rot or seedling blight or crown rot (*Aspergillus niger and A. pulverulentum*)

Symptoms - Fungus is seed-borne and soil-borne. The disease usually appears in 3 phases.

i. Pre-emergence rot: Seeds are attacked prevents to germinate by soil-borne conidia.

ii. Post-emergence rot: Attacks the seedling and cause circular brown spots on the cotyledons. Brown discoloured spots appear on collar region.

iii. Crown rot: Adult plants show crown rot symptoms. Large lesions develop on the stem below the soil and spread upwards along the branches causing drooping of leaves and wilting.

Favourable Conditions - Deep sowing, high soil temp. (30-35°C) and low soil moisture.

Mode of Spread and Survival - Pathogen survive in plant debris. Primary source is contaminated seeds. The fungi are carried on the seed surface or under the testa.

Management - Select good quality seeds. Treat the seeds with Carbendazim 2 g/kg.Avoid deep sowing. Destroy the crop debris by burning.

Dry Root rot (*Macrophomina phaseolina*)

(As in Red gram)

Favourable Conditions - Prolonged rainy season at seedling stage and low lying areas.

Peanut spotted wilt / bud necrosis / groundnut ring mosaic [*Tomato spotted wilt virus* (*TSWV*)]

Symptoms - Emerging leaves are small, rounded or pinched inwards and mottling. Necrotic spots and irregular lesions on leaves and petioles. In advanced conditions, the necrosis of buds occurs. Drastic reduction of flowering. Seeds are abnormally small and wrinkled.

Mode of Spread and Survival - Perpetuates in the weed hosts *viz., Bidens pilosa, Erigon bonariensis, Tagetes minuta* and *Trifolium subterraneum.* It transmitted by thrips *viz., Thrips tabaci* and *Frankliniella sp..*

Management - Extraction of 10 Per cent Antiviral principle

Dried sorghum or coconut leaves are cut and powdered. To one kg of leaf powder, two litres of water is added and heated to 60°C for one hour. It is then filtered through muslin cloth and diluted to 10 litres and sprayed. Five hundred litres of extract is required to cover one ha.

GINGELLY *(Sesamum indicum)*

Root rot or stem rot or charcoal rot (*Macrophomina phaseolina* Syn :Cercospora personata (Sclerotial stage : *Rhizoctonia bataticola*)

(as in Red gram)

Leaf blight (*Alternaria sesame*)

Symptoms - Small, circular, reddish brown spots on leaves with concentric rings.

Pathogen - Conidia are transverse and longitudinal septa.

Wilt (*Fusarium oxysporum f.sp. sesame*)

Powdery mildew [*Erysipha cichoracearum* (Syn : *Oidium acanthosperml*)]

(as in other crops)

Phyllody (*Mycoplasma Like Organism*)

Symptoms - The floral parts are transformed into green leafy structures, grow profusely. Bear cluster of leaves and malformed flowers at the tip. The flower is rendered sterile.

Mode of Spread and Survival - Survives on *Brassica campestris* var. *toria, B. rapa, Cicer arietinum, Crotalaria* sp., *Trifolium* sp., *Arachis hypogaea* and some weed hosts. The disease is transmitted by jassid, *Orosius albicinctus.* Nymphs are incapable of transmitting the MLO. Vector population is more during summer and less during winter months.

Management - Remove reservoir and weed hosts. Avoid growing sesame near cotton, groundnut and legumes. Spray Monocrotophos /Dimethoate at 500ml/ha to control jassids.

SUN FLOWER (*Helianthus annuus*)

Root rot or charcoal rot [*Rhizoctonia bataticola* (Pycnidial : *Macrophmina phaseolina*)]

Symptoms - Seed-borne and primarily causes seedling blight and collar rot in the initial stages. Drooping of leaves and death occurs in patches. The bark of the lower stem and roots shreds and associated with sclerotia. Dark coloured, minute pycnidia on lower stem portion.

Mode of Spread and Survival - The pathogen survives in soil and seed-borne Wind-borne conidia cause secondary spread.

Management - Treat the seeds with Carbendazim at 2 g or Thiram at 4g/kg or with *Trichoderma viride* at 4g/kg. Spot drench with Carbendazim at 1g/litre.

Leaf blight (*Alternaria helianthi*)

Symptoms - The fungus produces brown spots on the leaves, but the spots can also be seen on the stem, sepals and petals. The lesions on the leaves are dark brown with pale margin surrounded by a yellow halo.

Head rot (*Rhizopus sp.*)

Symptoms - Water soaked lesions on the lower surface, which later turns brown and head become soft and pulpy. The larvae and insects which attack the head pave way for the entry of the fungus which attacks the inner part of the head. The seeds are converted into a black powdery mass. The head finally withers and droops down with heavy fungal mycelial nets.

Favourable Conditions - Prolonged rainy at flowering and damages caused by insects.

Mode of Spread – It survives as a saprophyte in debris and. Spread by wind blown spores.

Management - Treat the seeds with Thiram or Carbendazim at 2g/kg. Control the caterpillars feeding on the heads.

Rust (*Puccinia helianthi*)

(as in other crops)

CASTOR (*Ricinus communis*)

Seedling blight (*Phytophthora parasitica*)

Symptoms - Circular, dull green patch on both the surface of the cotyledonary leaves and causes rotting. The infection moves to stem and causes withering and death of seedling.

Favourable Conditions – Continuous rainy weather, low temp. (20-25°C), ill drained soils.

Mode of Spread and Survival - Chlamydospores and oospores act as primary source of infection. The fungus also survives on other hosts like potato, tomato, brinjal, sesamum etc. The secondary spread takes place through wind-borne sporangia.

Management - Remove and destroy infected plant residues. Avoid low-lying and ill drained fields for sowing. Treat the seeds with Thiram or Captan at 4g/kg.

Rust (*Melampsora ricini*)

Symptoms - Minute, orange-yellow coloured, raised pustules appear with powdery masses on the lower surface of the leaves and the corresponding yellow.

Mode of Spread - It attacks *Euphorbia obtusifolia, E.geniculata,* and *E.marginata.*

Management - Rogue out the self-sown. Spray Mancozeb at 1kg/ha or dust "S" at 25kg/ha.

Leaf blight (*Alternaria ricini*)

(as in other crops)

Brown leaf spot (*Cercospora ricinella*)

Symptoms - The disease appears as minute brown specks surrounded by a pale green halo. The spots enlarge to greyish white centre portion with deep brown margin.

Mode of Spread and Survival - The fungus remains as dormant mycelium in the plant debris. The fungus mainly spreads through wind borne conidia.

Management - Remove the infected plant debris. Spray Mancozeb at 1kg/ha.

COCONUT *(Cocos nucifera)*

Bud rot (*Phytophthora palmivora*)

Symptoms - Indication seen on central shoot of the tree (spindle). Heart leaf shows discolouration. Drooping and breading off the heart leaf. Entire base of the crown rotten emitting a foul smell. The central shoot comes off easily. Subsequent wilt and death of tree.

Favourable Conditions - High rainfall and humidity (above 90%), low temperature (18-20^{0}C) and wounds caused by Rhinoceros beetles.

Mode of Spread and Survival - Dormant mycelium in infected tissues and also survives as chlamydospores and oosepores in crop residues. Spread is through air-borne sporangia and zoospores. Rainfall also helps in spreading the diseases.

Mangement - Remove and burn affected trees (beyind recovery). If in early stage, remove the infected tissue thoroughly and protect with Bordeauex paste. Prophylactic spray with 1% Bordeaux mixture to all the healthy plams in before onset of monsoon rains.

Basal Stem Rot / Thanjavur wilt / Bole rot (*Ganoderma lucidum*)

Symptoms - Yellowing, withering and drooping of the outer fronds which remain hanging around the trunk. The wilting plants also show bleeding patches near the base of the trunk. A brown gummy liquid oozes out from the cracks in the tree which result in the death of outer tissues. The fungus produces fruiting body (Bracket) along the side of the basal trunk.

Favourable Conditions - Sandy loam and sandy soils, water logging during severe rains, low soil moisture content during summer months and damages caused by weevils and beetles.

Management - Remove and burn severely infected trees. Isolate the diseased trees by digging a trench all around to check further spread. Irrigate the palms at least once in a fortnight during summer months. Apply heavy doses of farm yard manure or compost for green manure at 50 Kg/tree/year along with 5 kg of neem cake. Drench the soil near the tree with 40 litres of 1 per cent Bordeaux mixture at quarterly interval for thrice a year and repeat after 2-3 years. Apply Aureofunginsol 2g+Copper sulphate 1g in 100 ml of water or Tridemorph 2ml/100 ml of water through stem injection or root feeding.

Stem bleeding (*Ceratocystis paradoxa*)

Symptoms - Exudation of reddish brown fluid from the cracks in stem. Area underneath the bark gets decayed and the bleeding patch extends further up. Tree is not killed.

Favourable Conditions - Copious irrigation or rainfall followed by drought, shallow loamy soils or laterite soil with clay or rock layer beneath the soil, damages by *Diocalandra* and *Xyleborus* beetles.

Mode of Spread - Fungus survies in the infected debris and soil as perithecia and conidia.

Management - Scoop out the diseased tissue and apply molten coal tar followed by swabbing Bordeaux paste. When stem bleeding is observed in association with *Ganoderma,* follow root feeding or stem injection technique. Irrigate during summer.

Kerala wilt / Root wilt disease (Mycoplasma Like Organism)

Symptoms - "Flaccidity" of leaves (curve abnormally inwards). Leaves are Yellowing, marginal necrosis and fan-like appearance. Wilting of leaves and shedding of buttons and immature nuts. The mature nuts are small with thin kernel. The root wilt palms highly susceptible to leaf rot disease (*Bipolaris halodes*).

Pathogen- Transmitted by vegetative vector (Dodder) to Periwinkle (*Catharanthus roseus*).

Favourable Conditions - Sandy and sandy loam soils, severe floods and lace wing bug (*Stephanitis typia*).

Management - Remove all severely infected palms. Spray the leaves with 0.01 per cent Monocroptophos or Thiodan alternatively. Grow green manure crops like *Pueraria phaseloides* in basin and incorporate at the time of fertilizer application. Control the leaf rot by spraying 1%Bordeaux mixture or 0.3% Mancozeb. Avoid water logging in rainy seasons. Raise the intercrops like Tapioca, elephant yam and yam.

Grey leaf blight (*Pestalotia palmarum*)

Symptoms - Centre of spots turns to greyish white with dark brown margins with a yellow halo. Complete drying and shrivelling of the leaf blade, giving blighted or burnt appearance.

Management - Remove and burn the infected, fallen leaves. Apply heavy doses of potash. Improve the drainage conditions. Spray the crown with 0.25 per cent copper oxychloride or 1 per cent Bordeaux mixture before the onset of rains.

COTTON (*Gossypium spp.*)

Wilt (*Fusarium oxysporum* f.sp. *vasinfectum*)

Symptoms - Cotyledons turn yellow and then brown. Base of petiole shows brown ring, wilting and drying of the seedlings leaves loose their turgidity, gradually turn brown, droop and finally drop off.

Favourable Conditions - Soil temperature of 20-30°C, hot and dry periods followed by rains, heavy black soils with alkaline reaction, high dose of nitrogen and phosphatic fertilizers, wounds caused by nematode (*Meloidogyne incognita*) and grubs of Ashweevil *(Myllocerus pustulatus).*

Mode of Spread and Survival - Externally and internally seed-borne.

Management - Acid-delinting with Carboxin or Chlorothalonil at 4 g/kg. Apply heavy doses of FYM at 100/ha. Mixed cropping with non-host plants. Resistant varieties of *G.hirsutum* and *G. barbadense* like Varalakshmi, Vijay Pratap, Jayadhar and Verum.

Verticillium wilt (*Verticillium dahlia*)

Symptoms - Drying of leaf margins & between veins("Tiger stripe /Tiger claw" appearance).

Favourable Conditions - Low temperature 15-20°C, low lying and ill-drained soils, heavy soils with alkaline reaction and heavy doses of nitrogenous fertilizers.

Mode of Spread and Survival – Infects other hosts like brinjal, chilli, tobacco and bhendi. The fungus can survive in the infected debris and in soils as micro sclerotia upto 14 years.

Management - Treat the delinted seeds with Carboxin or Carbendazim at 4 g/kg. Crop rotation by paddy or lucerne or chrysanthemum for 2-3 years. Resistant varieties are Sujatha, Suvin and CBS 156 and tolerant variety like MCU 5 WT.

Root rot [*Rhizoctonia bataticola* (Pycnidial stage : *Macrophomina phaseolina*)]

(as in other crops)

Anthracnose (*Colletotrichum capsici*)

Symptoms - Small reddish circular spots on the cotyledons and primary leaves. The lesions develop on the collar region, stem may be girdled, causing seedling to wilt and die. Small water soaked, circular, reddish brown depressed spots appear on the bolls. The lint is stained to yellow or brown, becomes a solid brittle mass of fibre.

Favourable Conditions - Prolonged rainfall at the time of boll formation and close planting.

Mode of Spread and Survival - Pathogen perpeptuates on the rotten bolls and debris in the soil. Secondary spread by air-borne conidia. Weed hosts viz., *Aristolachia bractiata* and *Hibiscus diversifolius.*

Management - Treat the delinted seeds with Carbendazim or Carboxin of Thiram or Captan at 2g/kg. Rogue out the reservoir weed hosts.

Grey or Areolate mildew [*Ramularia areola* (Sexual stage : *Mycosphaerella areola*)]

Symptoms - Irregular to angular pale translucent lesions.

Favourable Conditions - Wet humid conditions during winter cotton season.

Management - Rogue out the self-sown cotton plants. Grow the resistant varieties like Sujatha and Varalakshmi.

Boll rot - It is a complex disease caused by *Fusarium moniliforme, Colletotrichum capsici, Aspergillus flavus, A. niger, Rhizopus nigricans, Nematospora nagpuri* & *Botryodiplodia* sp.

Bacterial blight (*Xanthomonas campestris* p.v *malvacearum*)

Symptoms - Bacterium attacks all stages. Symptoms are,

i) Seedling blight; ii) Angular leaf spot; iii) Vein blight or vein necrosis or black vein blackening of the veins and veinlets; iv) Black arm; v) Square rot / Boll rot.

Favourable Conditions - Rain followed by bright sunshine.

Mode of Spread and Survival - Bacterium survives on infected, dried plant debris in soil for several years. Other hosts like *Thumbergia thespesioides, Eriodendron anfructuosum* and *Jatropha curcus.*

Management - Delint seeds.

SUGARCANE *(Saccharum officinarum)*

Red rot [*Colletotrichum falcatum* (Sexual stage: *Physalospora tucumanensis*)]

Symptoms - Third or fourth leaf. Internodes of a stalk splitting it longitudinally and reddening of the internal tissues. Emits acidic-sour smell. Advances, the stalk becomes hollow and covered white mycelial growth. Minute red spots on the centre of the mid-rib.

Favourable Conditions - Monoculturing, ratoon cropping and injuries.

Mode of Spread and Survival - Sett-borne. Fungus persists in soil. Collateral hosts are *Sorghum vulgare, S. halepense* and *Saccharum spontaneum.*

Management - Adopt crop rotation. Select disease free setts. Grow resistant varieties like Coc 62198, Coc 90003, Coc 91061, Co 6806 and Co 7704.

Smut (*Ustilago scitaminea*)

Symptoms - Central shoot is converted into a long whip-like, dusty black structure. This structure is covered by a thin, white papery membrane. On maturity it ruptures and millions of tiny black smut spores (teliospores) are liberated and disseminated by the wind.

Favourable Conditions - Monoculturing, ratooning and dry weather.

Management - Disease free Setts. Remove and destory the smutted clump (Collect the whips in a thick cloth bag/ polythene bag and immerse in boiling water for 1hr to kill the spores). Treat the setts in hot water at 55-60OC for 10 minutes or 52°C for 30 minutes. Grow resistant varieties like Co 7704, Co 449, Co 6806, Co 62175, Co 7807, Co 8139, Co 62198, Coc 8001.

Sett rot or Pineapple disease (*Ceratocystis paradoxa*)

Symptoms - Affected tissues reddish colour which turns to brownish black in the later. A characteristic pineapple smell is associated with the rotting tissues.

Mode of Spread and Survival - Insects like cane borer *(Diatraea dyari)* helps in the spread.

Pathogen also survives on coconut, cocoa, mango, papaya, coffee, maize and arecanut.

Management - Soak the setts in 0.05% Carbendazim oor 0.25% Emisan for 15 minutes. Use long setts having 3 or 4 buds. Provide adequate drainage during rainy seasons.

Rust [*Puccinia erianthi (Syn : P. melanocephala* and *P. kuehnii*)]

Symptoms - Minute, elongated, yellow spots (uredia), on both the surfaces of young leaves.

Favourable Conditions - Temperature of 30°C, relative humidity between 70 and 90 %, high wind velocity and continuous cloudiness.

Mode of Spread - Collateral hosts are *Erianthus fulvus* and *Saccharum spontaneum.*

Management - Remove the collateral hosts. Spray Tridemorph 1 kg or Mancozeb 2 kg/ha.

Gummosis (*Xanthomonas campestris* p.v. *vasculorum*)

Symptoms - Longitudinal stripes or streaks on mature leaves. Dull yellow bacterial ooze comes out from the cut ends.

Mode of Spread and Survival - The bacterium remains viable in the soil. Primary transmission through naturally affected diseased setts or through soil-borne contamination. Secondary spread through wind splashed rain, harvesting implements, animals and insects.

Management - Remove and burn the affected clumps. Select setts from disease free areas. Avoid growing collateral hosts like maize, sorghum and cumbu near the sugarcane fields.

Red stripe (*Pseudomonas rubrilineans*)

Symptoms - Stripes appear as water soaked, long, narrow chlorotic streaks and become reddish brown in few days.

Favourable Conditions - Continuous rationing, prolonged rain and low temperature (25°C)

Mode of Spread and Survival - Bacterium survives on sorghum, cumbu, maize, ragi and other species of *Saccharum.*

Management - Select healthy setts. Aviod growing collateral hosts near the fields.

Mosaic (*Sugarcane Mosaic Virus*)

Symptoms - Chlorotic or yellowish stripes alternate with normal green portion of the leaf. Elongated necrotic lesions are produced on the stalks and stem splitting occurs.

Mode of Spread and Survival - Virus remains viable on several other hosts like *Zea mays, Sorghum vulgare, Pennisetum americanum, Eleusine indica, Setaria lutescens, Echinochloa crusgalli, Stenotaphrum secondatum, Digitaria didactyla etc.,*. Spreads from other aphids *viz.*, *Melanaphis sacchari* and *M. indoscchari*.

Management - Treat the setts in hot water as follows : 52°C for 20 minutes on the first day, 57.3^{o}C for 20 minutes, on the second day and 57°C for 20 minutes on the third day or Aerated Steam Therapy (AST) at 56°C for 3 hrs.

Grassy shoot [*Mycoplasma Like Organism* (*MLO*)]

Symptoms - The production of lanky tillers from affected shoots. Plants appear bushy and 'grass like' due to reduction in the length of internodes premature and continuous tillering.

Mode of Spread and Survival - It transmitted by aphids viz., *Rhopalosiphum maidis, Melanaphis sacchari* and *M. indosacchari.* In addition, leaf hopper, *Proutista moesta* also involves in the transmission. Sorghum serves as a natural collateral host.

Management - Hot Water Treatment (HWT) at 50°C for 21/2 hr. Hot Air Treatment (HAT) at 4°C for 8 hr or Aerated Steam Therapy (AST) at 50°C for 1 hr or Moist Hot Air Treatment (MHAT) at 54°C for 2 hr.

Ratoon stunting [*Rickettsia Like Organism* (RLO)]

Symptoms - Stunted growth, reduced tillering, thin stalks with shortened internodes.

Mode of Spread - Maize, sorghum, sudan grass and *Cynodon* are collateral hosts.

Management - Select the setts from disase free fields.

TOBACCO *(Nicotiana tabacum)*

Damping off (*Pythium aphanidermatum*)

Symptoms - Fungus attack the seedling. Water soaked minute lesions appear on the stems near the soil surface, soon girdling the stem leading to toppling over of the seedlings (Post-emergence damping off).

Favourable Conditions - Over crowding of seedling, ill drained nursery beds, heavy shade.

Mode of Spread - The fungus survives in the soil as oospores and chlamydospores.

Management - Prepare raised seed beds. Drench the seed bed with 1 per cent Bordeaux mixture or 0.2 per cent Copper oxychloride, two days before sowing. Avoid overcrowding.

Black shank (*Phytophthora parasitica* var. *nicotianae*)

Symptoms - Seedlings show black discolour on stem near soil level and blackening of roots.

Favourable Conditions - Frequent rainfall, high soil moisture and high population of rootknot nematodes (*Meloidogyne incognita* var. *acrita*).

Mode of Spread– Survive as saprophyte on organic wastes and infected crop residues in soil.

Management - Burn the seed beds with paddy husk. Provide drainage in nursery. Drench the nursery beds with 1 per cent Bordeaux mixture or 0.2 %, two days before sowing.

Frog eye spot (*Cercospora nicotianae*)

Symptoms - Lower leaves as small grey spots with brown border. White centre, surrounded by grey and brown portions, surrounded by a black margin, resembling the eyes of a frog.

Mode of Spread and Survival - Seed-borne. The secondary spread is wind-borne conidia.

Management - Avoid excess nitrogenous fertilization. Adopt optimum spacing. Spray the crop with 0.4 per cent Bordeaux mixture.

Powdery mildew (*Erysiphe cichoracearum* var. n*icotianae*)

(as in other crops)

Wild fire (*Pseudomonas tabaci*)

Symptoms - Black spots with a yellow halo spreads, causing withering and drying of leaves.

Favourable Conditions - Close planting, humid wet weather and strong winds.

Mode of Spread and Survival - Survives in the infected crop residues in the soil.

Management - Remove and burn the infected crop residues in soil. Avoid closer planting.

Mosaic [*Tobacco Mosaic Virus* (TMV)]

Symptoms - Appears a characteristic light and dark green pattern, the dark green areas are usually associated with the veins. The leaves are narrowed, puckered, thin and malformed, Later, dark brown necrotic spots develop under hot weather (Mosaic burn/Mosaic scorching).

Mode of Spread and Survival - Viable in plant debris. It has wide host range. Sap-transmissible and enters the host through wounds. Seed-transmitted in tobacco but tomato seeds transmit the virus. In the field, the virus is transmitted by contact.

Management - Remove and destroy infected plants. Keep the field free of weeds which harbour the virus. Wash hands with soap and running water before or after handling the plants or after weeding. Spray the nursery and main field with leaf extracts of *Bougainvillea or Basella alba* at 1 litre of extract in 150 litres of water, two to three times at weekly intervals. Adopt crop rotation by growing non-host plants for two seasons. Grow resistant varieties like TMV RR2, TMV RR 2a and TMV RR3.

Leaf curl [*Niccotiana Virus 10* (*Ruga tabaci*)]

Symptoms - Three forms of leaf curl expression are observed. Enation on the veins.

Mode of Spread and Survival - Whitefly, *Bemisia tabaci* is the vector.

Management - Avoid growing solanaceous crops like tomato near tobacco fields. Spray Methyldemeton at 0.1 to 0.2 per cent to control the vectors.

Phaneroganic parasite

Broom rape (*Orobanche cernua* var. *Desertorum*)

Symptoms - The affected tobacco plants are stunted and show withering and drooping of leaves to wilting.

Parasite - It is a total root parasite. It is an annual, fleshy flowering plant with a short, stout stem, 10-15 inches long. The stem is pale yellow or brownish red in colour and covered by small, thin, brown scaly leaves and the base of the stem is thickened. White-coloured flowers appear in the leaf axils.

Mode of Spread and Survival - The seeds of the parasite remain dormant in the soil for several years. The dormant seeds are stimulated to germinate by the root exudates of tobacco and attaches itself to the roots by forming haustoria. Later, it grows rapidly to produce shoot and flowers. It attacks brinjal, tomato, cauliflower, turnip and other cruciferous crops.

Management - Rogue out the tender shoots of the parasite before flowering and seed set. Spray the soil with 25 per cent Copper sulphate. Spray 0.1 per cent Allyl alcohol. Apply few drops of kerosene directly on the shoot. Grow decoy or trap crops like chilli, motherbean, sorghum or cowpea to stimulate seed germination and kill the parasite.

Phanerogamic parasite [*Witch Weed* (*Striga asiatica and Striga densiflora*)]

It is a partial root parasite and occurs mainly in the rainfed sorghum. It is a small plant with bright green leaves, grows upto a height of 15-30 cm. It always occurs in clusters of 10- 20/host plant. *S. asiatica* produces red to pink flowers while. *S. densiflora* produces white flowers.

Management - Regular weeding and interculture operation during early stages of parasite growth. Spray Fernoxone (sodium salt of 2, 4-D) at 450g/500 litre of water.

DISEASE OF HORTICULTURE CROPS

MANGO

Anthracnose (*Colletotrlchum gloeosporioides*)

Symptoms : Leaf spots, blossom blight, wither tip, twigs blight and fruit rot. Spots appear on fruits. The affected areas usually crack and sink slightly.

Mode of survival and spread - Contact with diseased fruit during transport and storage.

Favourable conditions : Temperature of 25°C and R.H 95-97%

Management - Spray carbendazim or Topsin (0.1%) or Chlorothalonil (0.2%), Blitox and Dithiocarbomate at 14 days intervals until harvest. Before storage, treat with hot water, (50-55°C) for 15 minutes or dip in Benomyl solution (500ppm a.i.) or Thiobendazole (1000ppm) for 5 minutes or expose them to ammonia, sulphur dioxide and Carbon dioxide gases.

Powdery mildew [*Oidium mangiferae (Acrosporum mangiferae)*]

Symptoms - Appearance of wefts of white mycelium on the affected parts. The affected fruits do not grow in size and may drop before attaining pea size.

Favourable conditions - Warm temperature with heavy morning dew and cloudy weather.

Management - Dusting the plants with fine sulphur (250-300 mesh) at the rate of 0.5 kg/tree. The first application may be soon after flowering, second 15 days later.

Mango malformation : (*Fusarium moliliforme* var. *subglutinans*)

Symptoms - Floral malformation and vegetative malformation.

Management - Spraying 100-200ppm NAA during October. Pruning of diseased parts. This is followed by the spraying of Carbendazim (0.1%) or Captof (0.2%).

Stem rot (*Botrydiplodia theobromae*)

Symptoms - The dark epicarp around the base of the pedicel in the initial stage rapidly turns the whole fruit completely black within two or three days.

Management - Dip mangoes in 6 percent borax solution at 43°C for 3 minutes.

Die-back (*Botryodiplodia theobromae*)

Symptoms - Discolouration and darkening of the bark some distance from the tip. **Management -** Prune and destroy infected twigs and spray Carbendazim or Thiophanate Methyl (0.1%) or Chlorathalonil (0.2%) as fortnightly interval during rainy season.

Red-rust (*Cephaleuros mycoides*)

Symptoms - The algae attacks foliage, bark and twigs. Initially the spots are greenish-grey in colour and variety in texture, but later on the surface bears reddish-brown appearance.

Management - Bordeaux mixture (0.6%), or Copper oxychloride 0.25%

Black Spot (Xanthomonas campestris pv. mangiferae indicae)

Symptoms - Small, water soaked lesions appear in groups towards the tip of the leaf blade.

Infected fruits may show crack in the skin and badly affected ones drop prematurely.

Management - Orchard sanitation. Spray Agrimycin-100, Plantomycin (200 ppm).

GRAPE VINE

Downy mildew (*Plasmopara viticola*)

Symptoms - Irregular, yellowish, translucent sports on the upper surface of the leaves. Correspondingly on the lower surface, white, powdery growth on fungus appear.

Management - Bordeaux mixture 1%, metalaxyl+mancozeb 0.3 % copper oxychloride 0.25%.

Powdery mildew (*Uncinula necator*)

Symptoms - Powdery growth mostly on the upper surface of the leaves.

Management - Spraying wettable sulphur 0.2%/Carbendazim 0.1%/Karathane 0.07%

Bird's Eye Spot/Anthracnose (*Gloeosporium ampelophagum*)

Symptoms - Visible on leaves, stem, tendrills and berries. 'shot hole'. Sunken spots with ashy grey centre and dark margin, resembling that of a bird's eye are produced.

Management - Remove infected twigs. Spraying Bordeaux mixture 1%

CITRUS

Gummosis (*Phytophthora parasitica, P. Palmivora*)

Symptoms - Dark staining of bark. Girdling and finally death of the tree. Lateral profuse exudation of gum from the bark of the trunk.

Favourable conditions - Water logged areas and heavy soils.

Management - Injuries should be avoided. Bark of trunk should be coated with ZnSO4, CuSO4, lime (5:1:4) with the consistency of a wash.

Powdery mildew (*Oidium tingitaninum*)

Symptoms - White powdery on young twigs and leaves.

Management - Apply 350 mesh sulphur dust in the early morning.

Canker (*Xanthomonas campestris* pv *citri*)

Symptoms - Lesions are typically circular with yellow halo; appear on both sides of leaf, also produced on twigs, they are girdled and die. On fruits, canker lesions reduce market value.

Mode of survival and spread - Wind and rain splashes. Injury caused by leaf miner.

Management - Streptomycin sulphate 500-1000 ppm; or Phytomycin 2500 ppm or Copper

oxychloride 0.2% at fortnight intervals.

Tristeza or quick decline (*Tristeza Virus*)

Symptoms - Lime is susceptible both as seedling or buddling on any root stock. Roots decay, twigs die back. Fine pitting of inner face of bark of sour orange stock. Tree stunted and dies.

Fruits are small in size.

Mode of spread - *Toxoptera citricida* (aphid) is the vector. Not seed-borne.

Management - Avoid susceptible root stocks. Pre-immunised with mild strain of tristeza.

Exocortis of scaly butt (*Exocortis Viroid*)

Symptoms - Affects only Rangpur lime, Trifoliate orange and citrange root stocks. Vertical cracking and scaling of bark, root stock portion, yellow blotching twigs, extreme stunting.

Management - Use virus-free certified budwood; use tolerant stocks like rough lemon.

Greening [*Liberobactor asiaticum* (Phloem limited Bactria- like- organism)]

Symptoms - Stunting of leaf, sparse foliation, twig die back, predominantly greened, worthless fruits. Green circular dots on leaves. Fruits small, lopsided with curved columella.

Mode of spread - Infected budwood; psyllid vector-*Diaphorina citri*

Management - Control psyllids with insecticides. 500 ppm tetracycline spray.

BANANA

Panama wilt disease (*Fusarium oxysporum* f. sp. *cubense*)

Symptoms - Yellowing extends upwards and finally heart leaf alone green Longitudinal splitting of pseudostem. Reddish brown radiating mycelial strands in c.s of corms.

Management - Variety Poovan shows resistance, Rasthali and monthan susceptible to the disease. Capsule application of carbendazim @ 40 mg/capsule/tree on the capsule is applied in the corm by making a hole 45° of 10 cm depth. Flooding the fields. Raise paddy.

Sigatoka leaf spot (*Cercospora musae*)

Symptoms - First symptoms are seen on the leaves. Yellow elongated spots on leaves. Spots surrounded by yellow halo. Number of spots cause drying of leaves.

Survival and spread - The fungus survives in the infected banana leaves.

Management - Removal of the affected leaves, Avoid night irrigation. Spraying copper oxychloride 2kg or Carbendazim 500g or Mancozeb 1.25kg

Anthracnose (*Gloeosporium musarum*)

Symptoms : The skin at the distal ends of the fingers turn black. The fungus produces masses of conidia (pinkish coat). The entire fruit, stalk and buds are affected. Fruits get shrivelled.

Management - Spraying copper oxychloride 0.25% or Bordeaux mixture 1%.

Banana bunchy top (*Banana bunchy top Virus*)

Symptoms - Plants are extremely stunted. Leaves are reduced in size and chlorotic. Leaves stand upright and become brittle. Many leaves are crowded at the top.

Spread –Primarily suckers. Secondarily through aphid (*Pentalonia nigronervosa* var *typical*).

Management - Disease free suckers. Spraying dimethoate 500ml or monocrotophos 250ml/ha to control the vectors. Infected plants destroyed in situ using 4ml of 2, 4, D (50g in 400 ml of water).

Infectious chlorosis (*Cucumber mosaic virus*)

Symptoms - Chlorotic or yellow green leaves, upward curling of leaves, mottling of leaves.

Spread - Through infected sucker and *Aphis gossypii*

Control - Destroy infected plants. Use disease free suckers. Arrest vector by systemic insecticide. Dry heat treatment of sucker at 40°C for 1 days.

Moko wilt (*Pseudomonas solanacearum*)

Symptoms - Rapid wilting and collapse of the leaves. Discolouration in vascular region. Blackening of the suckers. Discolouration in fruits.

Spread - Poorly drained soil. Primary infection through injured roots.

Control - Eradicate infected plant. Expose soil to sunlight. Fallowing and crop rotation. Disinfection if implements. Providing good drainage.

Tip over or Head rot (*Erwinia carotovora*)

Symptoms - Small sword leaves. Pseudostem easily come out. Oozing visible on the edge of corm and pseudostem confirm bacterial infection.

Spread - Severe during summer. Through soil and infected sucker.

Control - Plant disease free suckers. Remove infected plants and destroy. Drench methoxy ethyl mercuric chloride (Emisan-6) 0.1%. Avoid drenching (Emisan-6) after shooting. Poovan is resistant Robusta is susceptible to this disease.

GUAVA

Wilt (*Fusarium Oxysporum f. sp. psidi*)

Symptom - Leaves turn pale green. Premature shedding of leaves. Fruits produced in the affected plants are very small. Diseased plants fail to produce new flush and flowers. Plants dries within a year. Rotting of roots at the basal region Bark turn light brown.

Favourable conditions - Trees of age 5 years are normally affected. 10-20 years are highly susceptible. Alkali soils favours the pathogen.

Mode of spread - Survives as saprophytes and chlamydospores dormant for several years.

Management - Removal of infected branches, uproot wilted plants. Add green manure, oil cake. Avoid water logging. Resistant varieties are Banarasi Doalka, Nasik and white guava.

Stem Canker (*Physalospora psidi*)

Symptom - Causes craks in branches wilting of branches.

Mode of spread - Primary spread through conidia and secondary spread through ascospores.

Management - Removal of dead stem paint the cut end with bordeaux paste.

Red rust (*Cephaleuros virescens*) (as in Mango)

DISESASES OF JACK

Rhizopus fruit rot (*Rhizopus atrocarpi*)

Symptom - Fruit rot begins in stalk and covered by mycelium. The fruits are mumified.

Mode of Spread - The pathogen lives in the debris.

Management - Improve sanitation. Spray Bordeaux mixture 0.4% /copper oxychloride 0.2%.

PINE APPLE

Leaf and fruit rot (*Ceratocystis paradoxa*)

Symptoms - Spots produced on leaves and fruits. The spots are grey with dark margins and later turning olive brown or white. On the water soaked lesions appear which turn yellowise, then dark. The tissue becomes soft and rotting soon set in when the ripened fruits are affected a characteristic smell or odors develops.

Management - The diseased plant must be destroyed. Disease free suckers.

Root rot (*Phytopthora parasitica*)

Symptom - Yellowing of leaves which later turns above brown coloured. Fruit development is arrested and entire plant wilts from the tip downward. The fruit exhibits spongy texture

When the plant is pulled out the roots appear dark and in the process of decaying.

Favourable conditions - Excess moisture.

Mode of spread - Through irrigation water (zoospores)

Management - Prophylactic treatment with 1% Bordeaux mixture (or) metalaxze + mancozeb 0.2% to 0.4%.

POMEGRANATE

Bacterical leaf spot (*Xanthomonas campestris* pv. *punicae*)

Symptom - Dark coloured irregular spots 2-5 mm in diameter. The leaves often distorted and malformed. Premature dropping of leaves. The normal growth of the plant in affected. Raised spots are seen on the fruits which are irregular.

Mode of Spread - The bacteria infect through wounds and stomatal openings.

Survival - Survive in the soil.

Management - (Streptomycin + Tefracycline) Agrimycin-100 100 ppm.

VEGETABLES

CRUCIFEROUS VEGETABLES

Cabbage and Cauliflower

Club root (*Plasmodiophora brassicae*)

Symptoms - Infected roots enlarge to form 'clubs' (Hyperplasia and hypertrophy). Wilting of plants during night time. Due to secondary infection the clubs decay.

Favourable conditions - Low lying areas, ill drained soils with a soil temperature of 15-25°C. Acid soils are highly favourable.

Survival - It persists in the soil and crop refuse.

Mode of spread-Farm implements. Irrigation. Manure from cattle fed on diseased root crops.

Management - Crop rotation. Improving drainage facilities. Adding lime. Seed treatment with carbendazim 2g/kg. Nursery drenching with carbendzim or chlorothalanil 0.1%. Dip seedlings 0.8% carbendazim for 20 min. Spot drenching carbendazim/chlorothalonil 0.1%.

Downy mildew (*Peronospora parasitica*)

Symptoms - Production of white fluffy growth on the lower side of the leaves, stem etc. Yellow areas on the upper side and mildew on lower side. Severe infection at the seedling stage results in complete death of seedlings.

Management - Seed treatment with metalaxyl (Apron 35 S.D.) 6g/kg. Spraying Metalaxyl + mancozeb combi 4g/litre or Mancozeb 0.25%.

CUCURBITS

Powdery mildew (*Erysiphe cichoracearum*)

Symptoms - White powdery coating on all parts of the plant except fruits.

Management: Destroy affected plant parts. Spraying Tridemorph at 0.05%.

CARADAMOM

Katte diseases (*Virus*)

Symptoms: Spindle shaped chlorotic flecks (2.5mm) apper on the leaves; these flecks progress into light green, discontinuous stripes which run parallel to the vein. Mosaic symptoms are also frequently mixed.

Mode of spread: Non-persistant virus transmitted by the aphid *pentalonia nigronervosa.*

Management: Healthy seedlings.

COFFEE

Leaf rust (*Hemiliea vastatrix*)

Symptoms: Small yellowish, translucent oily spots appear on 12 week old young leaves. Spots expand, a powdery coating of spores on the under surface. Finally the spots turn bright orange to red with a yellow rusted band.

Favourable conditions: Rainy weather and periods with mist or dew.

Management: The blossom spray 0.5 percent Bordeaux mixture in February March pre monsoon spary with 0.03% a.i plantvax in May-June.

Terminologies :

Virus - A virus is a nucleoprotein that is very small to be seen with a light microscope multiply only in living cells and has the ability to cause disease.

Viroids - Viroids are small low molecular weight ribonucleic acids that can infect plant cells replicate themselves and cause disease.

Phytoplasma - They lack cell wall, are bounded by a unit membrane and have cytoplasm ribosomes and strands of nuclear material.They are pleiomorphic.

Endemic disease - The word "endemic" means "prevalent in, and confined to, a particular country or district" and is applied to disease. A disease is classified as endemic when it is

constantly present in a moderate to severe from and is confined to a particular country or district. E.g. Club root disease of cabbage is endemic in the Nilgiris district.

Epidemic or Epiphytotic disease - A disease usually occurs widely but periodically in a destructive form is referred as epidemic or Epiphytotic disease. E.g. Powdery mildew disease in grapevine.

Sporadic disease - Sporadic disease is one which occur at very irregular intervals and locations and in relatively fewer instances. In reality, sporadic disease belongs to the epidemic group.] E.g. Udubatta disease in rice.

Pandemic disease - These occur all over the world and result in mass mortality. ' E.g. Damping off disease in Tomato.

Plant Hyphae - It is a tubular, transparent filament, usually branched, composed of an outer cell wall and a cavity filled with protoplasm including cytoplasm. Hyphae are divided into compartments or cells by cross walls called septa and are generally characterised as septate (with cross wall) or coenocytic (aseptate - without cross wall).

Dolipore - It is a pore in the cross wall/septum through which cytoplasm is in continuity among adjacent cells. The region of the septum around the dolipore is swollen and the swollen rim on either side is capped by a membrane which is called as "parthenosome" or verschluss band or septal pore cap.

Mycelium - The hyphal mass (or) network of hyphae constituting the body (thallus) of the fungus is called as mycelium.

Eucarpic thallus - The thallus is differentiated into vegetative part which absorbs nutrients and a reproductive part which forms reproductive structure. Such thalli are called as eucarpic. (eg) *Pythium aphanidermatum*

Holocarpic - The thalli does not show any differentiation on vegetative and reproductive structure. After a phase of vegetative growth, it gets converted into one or more reproductive structures. Such thalli are called as *'holocarpic'* (eg) yeast, *Synchytrium* sp.

Monokaryotic mycelium or uninucleate - Mycelium contains single nucleus that usually forms part of haplophase in the life cycle of fungi.

Dikaryotic mycelium (binucleate) - Mycelium contains pair of nuclei (dikaryon) which denotes the diplophase in the life cycle of fungi.

Homokaryotic mycelium - The mycelium contains genetically identical nuclei.

Heterokaryotic mycelium - The mycelium contains nuclei of different genetic constituents.

Modification of fungal hyphae - In most fungi, during certain stages of its life cycle the mycelium become organised in to loosely or compactly wooven tissues which are distinguishable from the normal hyphae composing a thallus.

1) **Plectenchyma : (Gr. plekein =to weave + enchyma = infusion) -** Hyphae may have an altered shape and are closely packed together, easily indistinguishable from true tissues of higher plants. Such pseudo tissue are called as plectenchyma.

The plectenchyma is of 2 types.

   a) **Prosenchyma** (Gr. Pros = towards + enchyma = infusion). If the altered hyphal elements are visible as hyphae and mostly linear cells lie parallel to one another forming a loose tissue and such tissue is termed as prosenchyma

   b) **Pseudoparenchyma (Gr. Pseudo - false + parenchyma - a type of plant tissue)** In some fungi, modified hyphal elements are oval or isodiametric in shape and closely packed.

2) **Rhizomorph** (Gr. rhiza - root + morphe - shape) In many root invading (or) wood destroying fungi, hyphae aggregated longitudinally and form long root like strands of compact masses. These strands are called as rhizomorph have a thick hard cortex and a growing tip. These are the organs of mycelial migration and food transport and it provides a great inoculum potential to the invading hyphae in either colonization of a dead substrate or infection of a host. (eg) *Armillariella mellea*.

3) **Sclerotium (Pl. Sclerotia)** It is a hard, compact resting body, made up of mass of mycelium produced in some fungi after a vigorous active growth. They are generally black or purple outside and greyish white inside. They are irregularly shaped and possess rich amount of food reserve (eg) *Rhizoctonia solani Sclerotium*

rolfsii, Macrophomina phaseolina. Sclerotia may vary from the size of pinhead to 1 cm in diameter. The fungus *Polyporus* produces the largest size of sclerotium as man's head and size as much as 15 kg.

4) **Stromata -** Any compact hyphal aggregation in or on which fructification are formed are called as stromata. *Acervuli, sporodochia* are nothing but fertile stromata bearing sporophores and spore. Some kind of sclerotia also produce stromata during its germination as mentioned in *C. fusiformis*.

5) **Sporogenous -** Sclerotia on germination gives rise to a spore are called as sporogenous (eg) *Botrytis cinerea*.

6) **Chlamydospore -** Fungi like *Fusarium* sp produce swollen or thickened resting structures by the rounding up of cells (resting cells) and are termed as *chlamydospores*. They are produced either singly or in chains and may be terminal or intercalary in the hyphae.

   The rhizomorph, selerotium and chlamydospores are some of the resting spores of fungi that remain dormant over long period and germinate when the favourable condition appears.

7) **Gemnae -** They are similar in structure to chlamydospore but they are not thick walled, occur either singly or chains and are generally borne terminally eg *Mucor* sp.

8) **Appressorium (Pl. Appressoria) (L - apprimere - to press against)** These are localized swellings of the tip of germ tube or older hyphae that develop in response to contact with the host. In simple these are special structures for attachment in the early stage of infection. From these a minute infection peg usually grows and enters the epidermal cell of the host.

9) **Haustoria -** They are Nothing but organ for absorption. It is the lateral outgrowth of intercellular or superficial hyphae which will help to absorb food and nutrients from the host. They are of different shapes and size ranging from knob like structures to simple, lobed, branched, coiled and they are able to penetrate only in the cell wall and not in the plasma membrane.

MYCORRHIZA - It is one kind of mutualism between the fungi and roots of higher plants. The name 'mycorrhiza' literally means "fungus root" and it was coined by Frank in 1885. They are quite common in nature and majority of the plant species are known to have some form of mycorrhizal system in their root. ie fungal hyphae like in association with the roots either on the periphery or in the cells of certain layers of the cortex.

1. **Ectomycorrhiza -** These mycorrhiza forms an external fungal sheath or mantle round the root. The mantle is of considerable thickness and may be differentiated into layers *Pseudoparenchymatous layer*) differing in compactness and commonly two in number. The hyphal branches from the sheath grow intercellularly into the cortical parenchyma of the host tissue in the form of network known as "Hartig net" named after the German botanst "Hartig". Hyphal branches from the sheath also moves to the soil. These kind of mycorrhiza is noticed in most of the forest trees like pine, fir, oak etc. Most of the ectomycorrhizal fungi belong to basidiomycotina and some to ascomycotina (eg) *Amanita* spp, *Boletus* sp, *Pisolithus tinctorius, Marasmius scorodonius*.

2. **Endomycorrhiza -** These mycorrhiza does not form external sheath or mantle but the hyphae penetrate both between (inter cellular) and with in (intracellular) tissues of cortex and epidermis. The endomycorrhiza may have septate (mostly basidiomycotina) or aseptate mycelium (mostly zygomycotina). Most common aseptate endo mycorrhiza are known as vesicular - arbuscular mycorrhiza (VAM). After germination the aseptate hyphae colonize the cortex, both inter and intracellularly and it branches repeatedly in a dichotomous manner to form a complex branched haustorium called as arbuscules. The hyphae within the cells or intercellular spaces also swell to form thick walled vesicles, that contains fat globules due to the formation of vesicles and arbuscules.

3. These mycorrhiza are named as VAM. *Endogone* sp. *Gigaspora* sp, *Glomus* sp. *Acaulospora* sp. *Gigaspora margarita, Glomus mosseae, G. fasciculatus, Ectendomycorrhiza*. They (Have characters of both endo and ectomycorhiza. *Endogone* sp. *Gigaspora* sp, *Glomus* sp. *Acaulospora* sp. *Gigaspora margarita, Glomus mosseae, G. fasciculatus, Ectendomycorrhiza*. They (Have characters of both endo and ectomycorhiza. In this case, they is usually form a mantle or sheath. The external hyphae penetrate between the epidermal and

cortical cell constituting the Hartig net just like ectomyconrhiza and also penetrate into the active living cell similar to endomycorrhiza. (eg) *Coenococcum geophilum.*

In all the three types, the hyphae of the fungal symbionts permeate the soil and obtain scarce and relatively immobile nutrients especially phosphorus but also nitrogen, potassium, Cu, Zn and supplied more effectively to the plants than the root hairs of the concerned plant.

Phanerogamic parasites

Parasite	Type of Parasitism		Host plants
Broomrape (*Orobanche* sp.)	Root parasite	Complete	Tobacco, Egg plant, Safflower, Tomato
Witchweed (*Striga* sp.)		Partial	Mustard, Cumin, Sorghum, Sugarcane, Rice, Kodomillet
Dodder (*Cuscuta* sp.)	Stem parasite	Complete	Citrus, Ornamentals
Loranthus, *Dendrophthoe* sp.		Partial	Forest trees, Mango, Citrus, Jack, Pomegranate, Neem

Plant Diseases introduced to India

Diseases	Year of Introduction	Introduced from
Leaf rust of coffee	1879	Sri Lanka
Late blight of potato	1883	Europe
Flag smut of wheat	1906	Australia
Downy mildew of grape	1910	Europe
Downy mildew of cucurbits	1910	Sri Lanka
Powdery mildew of rubber	1938	Malaya
Fireblight of apple &pear	1940	England
Bunchy top of banana	1940	Sri Lanka

Aphid Vectors	Viral Diseases
Aphis gossypii	Bean mosaic virus, CMV,
Aphis maidis	Sugarcane mosaic virus
Myzus persicae	Potato virus Y, Bean mosaic virus, CMV,
Pentalonia nigronervosa	Banana bunchy top, Katte disease of cardamom
Toxoptera citricidus	Citrus tristeza
Leafhopper Vectors	**Viral Diseases**
Nephotettix virescens	Rice dwarf
Nilaparvatha lugens	Rice tungro

White fly Vectors	Viral Diseases
Bemisia tabaci	Cassava Mosaic Virus, Cotton Leaf Curl, Bhendi leaf curl,
Bemisia gossypiperda, Bemisia goldingi	Cotton leaf curl
Thrips Vectors	**Viral Diseases**
Frankliniella schultzei	Tomato spotted wilt virus
Mite Vectors	**Viral Diseases**
Aceria cajani	Redgram sterility mosaic
Fungal Vectors	**Viral Diseases**
Olpidium brassicae	Lettuce big vein
Spongospora subterranea	Potato mop top
Synchytrium endobioticum	Potato virus X
Nematode Vectors	**Viral Diseases**
Paratrichodorus	Tobacco rattle virus
Xiphinema index	Grapevine fan leaf virus

Plant Disease management

Bordeaux mixture – 1 kg of copper sulphate is dissolved in 50 litre of water. Similarly 1 kg of lime of dissolved in 50 litre of water. Copper sulphate solution is slowly added in to lime with constant stirring. (1:1:100)

Bordeaux paste – 1:1:10

Burgundy mixture - 1 kg of $CuSo_4$ and 1 kg of sodium carbonate in 100 lit of water.

Cheshnut compound – mixing 2 parts of $CuSo_4$ and 11 parts of ammonium carbonate.

Spraying solution (Volume based classification)

Spray fluid	Droplet		
	Litre used per acre	**Size**	**Coverage per day**
High volume (Knapsack, Rocker)	200-400	150µ	2.5ac
Low volume (Power sprayer, Mist blower)	90-100	70-150 µ	5-6ac
Ultra low volume	81	20-70µ	20ac

Commercial formulation required = % of spray fluid X Wt of volume required for appl./ % of active ingredient in commercial formulation.

Edible mushrooms

Button mushrooms – *Agaricus bisporus*

Oyster mushrooms – *Pleurotus sajar caju, P. Citrinopileatus* (Co.1), *P. djamor* (MDU1), *P. ostreatus* (Ooty1), *P. eous* (APK1), *P. florida, P. Platypus.*

Paddy straw mushroom – *Volvariella volvaceae*

Milky mushroom – *Calocybe indica* (APK2)

FUNGICIDES - The word 'fungicide' originated from two latin words, viz., 'fungus' and 'caedo'. The word 'caedo' means 'to kill.' Thus the fungicide is any agency/chemical which has the ability to kill the fungus.

Fungistat - Some chemicals do not kill the fungal pathogens. But they simply arrest the growth of the fungus temporarily.

Antisporulant - Some other chemicals may inhibit the spore production without affecting the growth of vegetative hyphas and are called as 'Antisporulant'.

Protectant - As the name suggests, protectant fungicides are prophylactic in their behaviour. Fungicide which is effective only if applied prior to fungal infection is called a protectant, eg., Zineb, Sulphur.

Therapeutant - Fungicide which is capable of eradicating a fungus after it has caused infection and there by curing the plant is called chemotherapeutant. eg. Carboxin, Oxycarboxin antibiotics.

Eradicant - Eradicant are those which remove pathogenic fungi from an infection court (area of the host around a propagating unit of a fungus in which infection could possibly occur). eg. Organic mercurials, lime sulphur, dodine etc. These chemicals eradicate the dormant or active pathogen from the host.

EPIDEMIOLOGY - Epidemiology is a study of outbreak and spread of diseases in a population.

Compound Interest – It introduced by van der Plank in 1963. The term Compound Interest diseases refer to those diseases in which the incubation period and sporulation period of the pathogen is short. Thus, there are many generations of the pathogen in the life of the crop if pathogens spreading by means of air disseminating propaguies, such as rusts of cereals and late blight of potato.

Simple Interest Diseases - There is only one generation of the pathogen in the life of the crop. The primary inoculum is seed or soil borne and secondary infection rarely occurs during the season e.g., Diseases caused by soil borne fungi such as wilts and root rots and seed and soil borne smuts such as loose smut of wheat, covered smut of barley, sorghum, etc.

METHODS OF SURVEY

(i) **Fixed plot survey -** Select two villages and fix two fields per village. Select five sampling plots per field of one acre and it should be three metres away from the bunds. The size of the sample plot should be 1 Sq, m. Select 20 plants or leaves perplot and observe forthe disease intensity using the standardulated using the following formula, disease score chart having 0-9 grades. Work Out the percent disease index(PDI)usingthestandardformula. Survey the field every week and prepare the report.

(ii) **Roving survey -** Select four villages and in each village, fix two fields. Observe and score 100 plants or leaves in each field by walking across, starting from south west comer to North East corner. Work out the PDI using the standard formula. Survey the fields every week and prepare the report.

DISEASE SURVEILLANCE REPORTS

(i) **White Report -** It contains the disease/pest surveillance detail in a particular block/division/district for a week.

(ii) **Yellow Card -** It should be sent when the disease/pest occurrence attained half the level of Economic Threshold Level (ETL).

(iii) **Red Card -** It should be sent when the disease/pest occurrence exceeded the Economic Threshold Level.

PLANT QUARANTINE - The term 'Quarantine' literally means a 40 day period. 'Plant Quarantine' is the use of exclusion or other regulatory actions as a control strategy against the entry of exotic new pests and pathogens along man-made pathway into a new area.

Embargo - Some of the plant materials have been totally banned for import into India irrespective of any certificate. Potatoes cannot be imported to India from any wart disease *(Synchytrium endobioticum)* and/or golden nematode infested areas. Sugarcane cuttings from Australia, Fiji and Philippines cannot be imported. Import of cocoa from Africa and Sri Lanka is totally banned.

Domestic Quarantine - Within India, domestic quarantine has been established. To prevent the spread of Bunchy top of banana, an Act has been promulgated which prevent the movement of suckers, stems and materials of banana plant used for packing and wrapping from the states of Assam, Kerala, West Bengal and Orissa to other states.

4.7. APICULTURE

Apiculture is the science and culture of honeybees and their management

- ☞ First attempts to keep movable - frame hive -1882 - Bengal
- ☞ 1911 Rev Newton devised hive for Indian bee - south India called Newton bee - hive
- ☞ **Indian bee journal** published by All India beekeepers association located at PUNE
- ☞ ICAR established Beekeeping Research station 1945-Punjab and 1951- Coimbatore

Bee hive products: Honey, beewax, royal jelly, bee venom

Honey bee = Pollinating and Producing honey

Species	Scientific Name	Honey yield (kg/hive/year)
Indian bee	*Apis cerana indica*	12-15
Italian/ western bee	*Apis mellifera*	30-40
Rock bee	*Apis dorsata*	50
Little bee	*Apis florae*	200-250g
Dammer bee / Mosquito bee / Stingless bee	*Mellipona irridipennis*	100g

- ☞ Honey from Stingless bee is having more medicinal properties

Italian bee

- ☞ Widely distributed and commercially reared honey bee species in the world
- ☞ Apis mellifera - the best & introduced honeybee to our country

Colony organization

- ☞ Division of labour = Queen (1), worker (1000 to 30,000) and drone (few hundreds)

Queen

- Queen - Perfectly developed female
- Fertilized egg = Queen & workers
- Unfertilized egg = Drone

Worker

- Life span - 6 weeks
- Imperfectly developed females
- Maintenance and welfare of the colony

Ist half ---- Nurse bee - Indoor duties; royal jelly secretion, feeding brood, secreting bees wax, Building combs, cleaning and fanning,

"Guard bee" protect colony from other colony bees

IInd half= "Searcher and gatherer" - Locating source and collecting pollen, nectar, propolis & water

Drone: Functional male - meant especially for to mate with queen

Communication behavior

Dancing language = Karl von Frisch (1967)

Round dance:- Circle of comb, when food source is within 100 m distance

Wagtail dance:- More than 100 m and also which direction. It says both distance and direction

Sickle dance:- Intermediate between round and wagtail dance

Zig Zag Dance:- If the field contains poison

Bee Space is also called as crawl space

Bee Pasturage

Plants that yield pollen and nectar are collectively called bee pasturage or bee forage.

Plants which are good source of nectar

1. Tamarind
2. Neem
3. Soapnut tree
4. Eucalyptus
5. Pungam
6. Moringa
7. *Prosopis juliflora*
8. *Glyricidia maculata*
9. *Tribulus terrestris*

Plants which are good source of pollen

1. Sorghum
2. Maize
3. Millets like Cumbu, Tenai, Varagu, Ragi
4. Roses
5. Pome granate
6. Sweet potato
7. Tobacco
8. Coconut
9. Castor
10. Date palm

Plants which are good source of Pollen and Nectar

1. Banana
2. Citrus
3. Apple
4. Berries
5. Pear
6. Plum
7. Peach
8. Guava
9. Sunflower
10. Safflower
11. Mango

Honey

- A sweet, viscous fluid - Produced by honey bees
- Collected as nectar from nectaries at base of flower
- Also collected from extra floral nectaries (nectar secreted by parts other than flowers)
- Collected also from fruit juice, cane juice

Collection and ripening of honey

- Bee draws nectar by its tongue (proboscis)
- Regurgitated by field bees
- Collected by hive bees - Deposited in cells in comb
- Nectar contains 20-40% sucrose
- Invertase converts sucrose into dextrose (glucose) and levulose (fructose)
- Invertase is present in nectar itself and in saliva of honey
- Ripening of honey is by action of enzyme and by evaporation of water by fanning of wings

Composition of fully ripened honey percent (Approx.)

Levulose - 41.0

Dextrose - 35.0

Sucrose - 1.9

Dextrins - 1.5

Minerals - 2.0

Water - 17.0

Undetermined (Enzymes, Vitamins, Pigments, etc.) 1.6

Pigments Carotene, Chlorophyll, Xanthophyll

Minerals include Potassium, Calcium, Phosphorus, Sodium, Magnesium, Manganese, Copper, Sulphur, Silica and Iron.

Vitamins Vit B1 (Thiamine), B2 (Riboflavin), Nicotlinic acid, Vit. K, Folic acid, Ascorbic acid and Pantothenic acid.

Physical properties of honey

- Honey is hygroscopic. If exposed to air it absorbs moisture
- Honey is a viscous fluid. Heating of honey reduces viscosity
- Specific gravity of pure honey is 1.35 - 1.44 gms/cc
- Refractive index of honey - Helps to find moisture content measured using refractometer

Purity test for honey

- Measure specific gravity of honey using hydrometer
- If the specific gravity is between 1.25-1.44 it is pure honey

Aroma and flavour of honey

- Acquired from the nectar of the flower
- Lost if heated or exposed to air for long time

Colour of honey

- Depends on the nectar of flower (plant species)
- Darker honey has stronger flower
- Lighter honey has more pleasant smell

Fermentation of honey

- Honey containing high moisture can ferment
- Sugar tolerant yeast present in honey cause fermentation
- Fermentation more at 11-21°C
- Fermentation lends to formation of alcohol and carbondioxide
- Alcohol later converted into acetic acid
- Fermented honey sour in taste due to acidity
- Heating honey to 64°C for 30 min destroys yeast and prevents fermentation

Crystallization or granulation of honey

- This is a natural property of honey (particularly at low temperature)
- Dextrose present in honey granulate and suffer down
- Levulose and water remain top - More prone to fermentation
- High ratio of Levulose/Dextose (L/D) - Less granulation
- High ratio of Dextrose/Water (D/W) - More granulation

Royal jelly is a milky white secretion produced from the hypopharyngeal glands of young worker bees.

It is rich in vitamins B and C, and is lacking in vitamin E.

It's contains 10-hydroxy decanoic acid which has an antibiotic and anti-tumor activity.

Propolis is a resinous substance collected by bees from the leaf buds and bark of trees.

Bee diseases:

Brood diseases are more serious than adult diseases. Brood diseases are

- Thai Sac brood
- American foulbrood (AFB) is a bacterial disease of larvae and pupae.
- European foulbrood (EFB) is a bacterial disease of larvae.
- Chalkbrood is a fungal disease of larvae.
- Nosema is a protozoan disease of adult bees.

Viral disease - Thaisac brood disease (TSBV)

- Strain of this virus first reported in *Apis cerana indica* - Thailand 1976
- In India 1978 from Meghalaya

Bacterial disease: Temperature & subtropical = *Apis cerana indica* ---- Singh

- American font brood -- *Bacillus larvae*
- European font brood --- *Melissococcus pluton*

Protozoan disease Tropical India = Rahman and Rahman (1996)

- *Apis mellifera* = most serious disease caused by Nosema apis

Bee mites:

- Acarapis woodi mite of adult bee --- Apis cerana

The microscopic mites enter the tracheae (breathing tubes) of young bees. Inside the tracheae, mites block air exchange and pierce the walls of the tubes to suck blood.

- Varroa jacobsoni --- Present in Europe next in India --- *Apis cerana*
- Miticides (Apistan®, Checkmite+®, Amitraz®)

Bee enemies:

Greater wax, moth, *Galleria mellonella*, wasps, ants, brids, lizards and cockroaches

4.8 SERICULTURE

India has unique distinction of producing five known varieties of silk host plant

Name	Species	Host
Mulberry silkworm	*Bombyz mori*	Mulberry
Muga	*Antheraen assamensis*	Som and Soaln-Endemic to India
Tropical tasar	*Antheraea mylitta*	*Terminaha arjuna, S.robusta, T. tomertasa*
Oak tasar silk	*Antheraea frithii*	Oak, *A.perhys, A. roylei, A. Proyle,*
Eri silk worm	*Philosamia ricini*	Castor

- India occupies second position among silk producing countries next to china and mulberry silk dominates 90% of silk production
- In India more than 98% mulberry silk produced form Karnataka, A.P, W.B, T.N and J&K
- Tukra disease and mealy bug (Maconelliceccns hirsutus) are major problems in mulberry
- Central Silk Board (CSB), Bangalore was started in 1949
- Central Sericultural Research & Training Institute (CSR & TI), Mysore
- Central Silk Technology Research Institute (CSTRI), Bangalore was started in 1983
- Centarl Tasar Research and Training Institute - Ranchi

Silkworms:

Univoltine - One generation per year; good quality cocoon

Bivoltine - Two generation per year; Inferior quality

Multivoltine - More than three generations per year; short filament cocoon

- CSR & TI- Central Silk Board and Training Institute - Mysore
- Silkworm egg expressed by = DFL (Disease free egg)
- Silkworm rearing house should be maintained with the 27-280 C temperature & 85-90% relative humidity
- Young stage of silkworm is called chawki worms
- Matured worms spin cocoons on montages/ chandrike
- **Female:** sexual markings appear as a pair of milky white spots on each of the eighth and ninth segments. Pair of spots in eighth segment is known as ISHIWATA'S fore glands while in ninth segment is known as ISHIWATA'S HIND GLANDS.
- **Male:** Small milky white body HEROLDS GLAND appears at the centre of the ventral side between the eighth and ninth segments.

Silkworm Diseases

- Pebrine - *Nosema bombycis* - Protozoa / microsporidian - Transovarial transmission
- Grasserie - Nuclear polyhedrosis virus - viral disease
- Flacherie - Cytoplasmic polyhedrosis virus- viral disease
- Muscardine - *Beauveria bassiana* - Fungal
- Aspergillasis - *Aspergillus flavus; A. orytae*

Silkworm Pests:

Uzifly - *Exorista bombycis* is an endolarval parasitoid of silkworm

- Ecto-pupal parasitaid Nesolynx thymns used as a bio-control agents and uzitrap release by CSR &TI in Mysore and uzicide are also used to control Uzifly

Process of stifling: The silk moth emerges from the cocoon with in 8 to 10 days from the date of commencement of spinning the cocoon. Such pierced cocoons are not useful for reeling. There fore, in order to preserve the cocoons before reeling, the pupae inside the cocoons should be killed without delay.

Points to remember

- ✓ In India, first proposed the concept of integrated pest control Dr. S. Pradhan
- ✓ Egg parasotoid - *Trichogramma chilonis* - Bollworms, leaf folder
- ✓ T. japanicum - Rice stem borer, leaf folder
- ✓ Larval parasitoid -*Bracon brevicornis*- shoot borer (sugarcane, cotton and Rice)
- ✓ Pupal parasitoid -*Trichospilus pupivora* - coconut black headed caterpillar
- ✓ Papaya mealybug parasitoid- *Acerophagus papaya*
- ✓ Predators - *Chrysoperla carnea or Cryptolaemus montrouzieri* on sucking pests
- ✓ Green leaf hopper - Vector= Rice yellow dwarf, Rice tungro virus
- ✓ Cotton whitefly *Bemisia tabaci* - leaf curl virus
- ✓ Sesamum phyllody - *Orosius albicinctus*
- ✓ Pea nut bud necrosis - Ground nut thrips
- ✓ Pigeon pea sterility mosaic virus - *Aceria cajani*
- ✓ *Helicoverpa armigera* - NPV - 250 LE / ha
- ✓ LE - Larval equivalent
- ✓ POB - Poly occlusion bodies
- ✓ Hadda beetle - *Henosepilachna vigintioctopunctata* only cocinellid pest
- ✓ Mustard sawfly - only Hymenopteran pest
- ✓ *Epiricania molanoleuca* - only Lepidopteran parasitoid
- ✓ thuringiensis contain delta endotoxin
- ✓ Total pesticides registered - 209 (2007)
- ✓ Banned - 27 (2007)
- ✓ Ear cackle nematode - *Anguina tritici*
- ✓ Tundn disease - *A. tritici. f corynebacterium tritici* - wheat
- ✓ Wheat molya disease - *Heterodera avenae*
- ✓ Kufri swarna - resistant for Potato cyst nematode
- ✓ Late blight of potato *(Phytophthora infestans)* in Europe - Irish famine - 1845
- ✓ Brown leaf spot of rice *(Helminthosporium oryzee)* - India (Bengal famines) - 1943
- ✓ Banana bunchy top - vector - *Pentalonia nigronervosa*
- ✓ Loranthus (partial stem parasite) - fruits, avenue & forest plants
- ✓ Orabanche (total root parasite) - roots of Brinjal, tomato, & other solanaceous cricifers plants

- ✓ Striga (partial root parasite) - roots of sugarcane, maize & millets
- ✓ Cuscuta (total stem parasite) variety of plants except cereals
- ✓ Pre planting - fluchloralin
- ✓ Pre emergence - atrazine, pendimethalin
- ✓ Post emergence - 2,4-D glyphosphate, cladinatop
- ✓ Selective - Atrazine in maize, isoproturon in wheat
- ✓ Non selective - paraquat, glyphosphate
- ✓ Destruction of lantana - by *Telenomia scrubulosa*
- ✓ Water hyacinth - by *Neochetina bruchi, N. eichhorniae*
- ✓ *Parthenium hysterophorus* - by mexican beetle *(Zygograma bicolorata)*
- ✓ Salvinia moesta - by *Cyrtobagous salviniae*
- ✓ 1845, late blight of potato - *Phytophthora infestans*
- ✓ 1943, Bengal famine - *Helminthosporium oryzae*
- ✓ Father of Modern Plant Pathology - Anton de Bary
- ✓ Gene for gene concept, 1946 - Flor
- ✓ Founder of Virology - Beijerinck
- ✓ First Indian scientist, collected & identified fungi in India - KR Kirtikar
- ✓ "Fungi and Disease in Plants" was written by - EJ Butler
- ✓ "Fungi and Plant Diseases" was written by - BB Mundakur
- ✓ Bacteria variability - Conjugation, Transformation and Transduction
- ✓ JC Luthra, solar heat treatment of wheat seeds for- Loose smut
- ✓ Done most of his work on rust disease in India - KC Mehta
- ✓ Common scab of potato controlled by reducing pH below 5.2 - *Streptomyces scabies*
- ✓ *Plasmodiophora brassicae* club root disease of crucifers - raising pH of soil
- ✓ Citrus greening disease is mainly confine to - North India
- ✓ Mycoplasma like organisms (MLO) causes - Brinjal little leaf, Rice yellow dwarf, Sugarcane grassy shoot, Sesamum phyllody, Sandal spike disease
- ✓ Virus - can not synthesis protein by own enzymes.
- ✓ Father of Plant Pathology in India - EJ Butler
- ✓ Father of Indian Mycology - EJ Butler
- ✓ The bacterial colony- Ooze out test

- ✓ Green ear / downy mildew of pearlmillet first reported in India by - EJ Butler
- ✓ The downy mildew of pearlmillet is primarily a - soil borne
- ✓ Loose smut of wheat - Internally seed borne
- ✓ Covered smut of barley - Externally seed borne
- ✓ Loose smut of barley - Internally seed borne
- ✓ Early blight of potato - Soil borne disease
- ✓ Wilt disease of sugarcane was first reported in India - Bihar
- ✓ The incidence of black scurf of potato is more in - sandy soil
- ✓ Sheath blight of rice is more during - Active tillering stage
- ✓ Cell was lack microorganisms - MLO
- ✓ Father of Microbiology - Leuvanhoeck
- ✓ MLO first discovered by - Doi
- ✓ Plant pathology written by - Agrios
- ✓ Plant disease written by - RS Singh
- ✓ MLO diseases are transmitted by - Leaf hopper
- ✓ In north India and central India, the black rust inoculum cause from - South India (Uredospores produced in Nilgiri and Palaney hills during off season of the crop)
- ✓ Renowned scientist Dr. NE Borlaug belongs to - Plant Pathology
- ✓ Floral abnormalities - downy mildew of mustard, green ear of bajra, white rust of crucifers
- ✓ Entomopathogenic fungi - *Beauveria bassina and Metarrhizium anisopliae*
- ✓ Application of potash increases - Disease resistance in plants
- ✓ Viruses contain - either DNA or RNA, never both
- ✓ Hypotrophy - decrease in size, hypertrophy - increase in size, hyperplasia - increase in number
- ✓ Sucidal germination take place in - Striga
- ✓ Bordeaux mixture, Millardet at Bordeaux in France- 1885 (downy mildew of grapes)
- ✓ Major genes contribute disease resistance - Vertical resistance
- ✓ Partial resistance against all the races of pathogen - Horizontal resistance
- ✓ Etiology/Aetiology -deals with cause of plant diseases & nature of causal agent
- ✓ Who discovered the downy mildew for the first time in India - EJ Butler
- ✓ Yeast - single cell fungi
- ✓ *Aspergillus* - weed fungi for lab

- ✓ AP Michelli - Father of mycology
- ✓ Plantvax - rust fungi
- ✓ Vitavax - smut fungi
- ✓ Calaxin - powdery mildew
- ✓ ELISA technique can not be detect - viroids
- ✓ Fungi (eukaryotic, spore bearing, achlorophyllous) reproduce - sexually & asexually.
- ✓ Bacteria - primitive organism, prokaryotes (nuclear material is not separated from the cytoplasm by a nuclear membrane).
- ✓ Virus - a nucleoprotein, electron microscope, multiply only in living cells.
- ✓ Viroids - small low molecular weight ribonucleic acids.
- ✓ PLO lack cell wall, are bounded by a unit membrane, pleiomorphic.
- ✓ The names of the bacteria are governed by - International Code of Nomenclature of Bacteria (ICNB)
- ✓ Fastidious vascular bacteria called as Rikettsia Like Organism or RLO
- ✓ Complete root parasite - broomrape (*Orabanche sp.*) - tobacco, egg plant, safflower and tomato
- ✓ Partial root parasite - witch weed (*Striga sp.*) - mustard, cumin, sorghum, sugarcane, rice and kodo millet
- ✓ Complete stem parasite - dodder (*Cuscuta sp.*) - onion, alfalfa, flax, clover and ornamentals
- ✓ Complete stem parasite - *Cassytha* sp. - citrus and ornamentals
- ✓ Partial stem parasite - giant mistletoe (Loranthus) - forest trees, mango, citrus, jack, pomegranate and neem.
- ✓ Button mushroom - *Agaricus bisporus*
- ✓ Oyster mushroom - *Pleurotus sp.*
- ✓ Paddy straw mushroom - *Volvariella volvaceae*
- ✓ Milky mushroom - *Calocybe indica*
- ✓ Spiroplasma - lack the cell wall, spiral / helical in shape, resistant to penicillin but sensitive to tetracycline.

SEED BIOLOGY AND TECHNOLOGY

5

V. Alex Albert

5.1. SEED AND SEED PRODUCTION

Origin of Plentiful Yield is the Seed - *Parashara 400BC*

Good Seed in Good Soil Yield Abundantly - *Manu Smriti 200BC*

Botanically seed is a matured ovule containing an embryo in arrested state of development with a food reserve (endosperm) and protective coat (seed coat). In albuminous seeds the endosperm is present eg. rice, wheat, castor (cereals). Whereas, exalbuminous seeds the endosperm is absent eg. onion and pea (pulses).

Types of fruit/ seed

i. *Simple fruit* is derived from a single fruit eg. paddy and sorghum

ii. *Aggregate fruit* is collection of simple fruits eg. strawberry and polyalthia

iii. *Multiple/Composite fruit* is composed of the ovaries of several flowers eg. jack, pineapple, fig and mulberry

iv. *Caryopsis* is a single seeded fruit with fused fruit and seed wall, eg., rice, wheat and sorghum

v. *Achene* is a single seeded fruit where pericarp can be removed from the matured seed eg. sunflower and safflower

vi. *Schizocarp* is a dry fruit which is separated into two or more units at maturity eg. coriander and carrot

vii. *Nut* is a fruit with hard shell like fruit wall eg. cashew nut and walnut

viii. *Berry* is a fleshy pericarp present in tomato and grapes

ix. *Drupe* is made up of stony endocarp, thick, leathery, fleshy mesocarp and a thin exocarp eg. mango and plum

x. *Pome* is made up of thick outer fleshy layer, papery inner pericarp forming a multi-seeded core, eg., apple and pear

xi. *Hesperiduim* is a berry like fruits with papery internal separations and a leathery separable rind eg. lemon and orange

xii. *Silique* is a fruit with two locules, split away from the base at maturity eg. mustard

xiii. *Samara* is achene fruit with thin, flat wing like appendage eg. apple and ash tree

Pod-legumes	Sorosis - jack, pineapple
Capsule - bhendi, gingelly	Single seeded berry - date, arecanut
Regma - castor, jatropha	Single seeded nut/drupe - litchi
Balusta - pomegranate	Ettaeria of berries - annona

Amphisarca - bael Pepo - cucumber Utricle - ragi Syconium - fig	Ettaeria of drupes - raspberry Ettaeria of achene - strawberry Ettaeria of follicles - periwincle

Appendages of seed

- Awn is thorn like projection at tip of the paddy seeds
- Aril is coloured flesh mass as additional integument of some nutmeg seeds
- Hairs are thread like appendages present on the surface of the cotton seeds
- Wings are papery structure attached to the side of the seed moringa coat
- Caurancle is spongy outgrowth of the outer integument at micropyle region-castor, rubber and tapioca

Xenia : effect of foreign pollen on the seed coat colour in cross pollinated crops like maize and datepalm. The gene present in sperm cell contributes in the expression of endosperm colour

Polar nuclei	Sperm nucleus	Endosperm	Xenia effect
YY	YY	YYY	Yellow
YY	y	YYy	Medium yellow
Yy	y	Yyy	Light yellow
yy	y	yyy	white

SEED PRODUCTION : It is the generation system of seed multiplication *viz.,* nucleus seed followed by breeder seed, foundation seed and certified seed. The ISI colour standards for certification tag are

Certification stage	Tag colour	ISI no.	Size (cm)
Breeder seed	Golden yellow	356 (IS:5-1978)	12 x 6
Foundation seed	White	-	15 x 7.5
Certified seed	Azure blue	104	15 x 7.5
Truthfully labeled seed	Opal green	ISC no.275 of IS 5	15 x 10

Minimum isolation distance recommended for certified seed production

Distance (m)	Crops
3	Rice, wheat, oat, finger millet, groundnut, soybean
5	Blackgram, cowpea, greengram, potato, sweet potato, tapioca
10	Napier grass, guinea grass, buffel grass, dinanath grass
20	Pointed gourd, coccinia, rice bean
25	Mustard (SI), lettuce, tomato

30	Cotton, jute
50	Mustard (SC), sesame, true potato seed
100	Hybrids-rice, wheat, sorghum, redgram, brinjal, tomato
150	Seed crop infected by loose smut-wheat, barley, oat
200	Sunflower, safflower, brinjal, bhendi, hybrids-maize, sorghum, pearl millet
300	Castor, asparagus, celery
400	Sorghum-presence of Johnson grass, forage sorghum
500	Onion, bitter gourd, bottle gourd, cucumber, watermelon
600	Maize inbred-different kernel colour, texture
800	Carrot
1000	Cabbage, cauliflower, sugarbeet, hybrids-watermelon, bottlegourd
1600	Hybrids-cabbage, cauliflower, radish, turnip

Seed Multiplication Ratio (SMR)=No. of seeds produced(Seed Yield)/No. of seeds sown(Seed Rate)

Seed Replacement Rate (SRR)=Actual quantity of quality seed used for sowing/Total quantity of seed required to cover the area

Field inspection :

Field counts : The number of counts taken and the method of taking counts vary from crop-to-crop. Normally in all crops five counts are taken for an area upto 5 acres, and an additional count taken for each additional five acres as given below:

Area (ac)	No. of counts
<5	5
5-10	6
10-15	7
15-20	8
20-25	9

The tillering habit and planting space the number of plants/heads comprising a single count differs from crop to crop. The prescribed number of plants to be covered per count is given below

- ❖ Closely spaced tillering crops like rice, wheat, sorghum bajra, finger millet is 1000 plants
- ❖ Closely spaced non-tillering crops like soybean, jute, lucerne, mesta is 1000 plants
- ❖ Medium spaced crops such as beans, bengalgram, blackgram, greeengram, mustard, sesame, safflower is 500
- ❖ Widely spaced crops like maize, cotton, castor, potato, groundnut, sunflower, brinjal, tomato, bhendi is 100 plants

Objectionable weeds of seed crops

Crop	Common name of the weed	Botanical name of the weed
Rice	Wild rice	*Oryza sativa var fatua*
wheat	Wild morning glory	*Convolvulus arvensis*
Mustard	Mexican poppy	*Argemon mexicana*
Sunflower	Orabanche	*Orabanche cumana*
Napier grass	Johnson grass	*Sorghum halopense*
Bhendi	Wild okra	*Abelmoschus moschatus*
Fenugreek	Senji	*Meliliotus sp.*

Physiological maturity : A stage at which the seed has maximum dry matter accumulation, germination and vigour

Harvestable maturity : A stage at which commercial harvest has to be followed, usually at minimum moisture level. Relative term representing a stage of development when a plant possesses the prerequisites for utilization by consumer for a particular purpose

Crop	Maturity Indices
Rice	Seed colour turn from straw yellow to golden yellow
Wheat	Deposition of red colour in the seed
Sorghum	Duncan layer formation
Maize	Duncan layer formation
Cumbu	Leaves turn from green to yellow
Redgram	Dried brown colour pods
Blackgram	Pod colour turn green to black
Greengram	Pod colour turn green to black
Soybean	Shredding of leaves
Bengalgram	Pod become creamy colour
Sunflower	Receptacle become lemon yellow, seed become black
Groundnut	Inner layer of pod turn to black
Sesame	Seed become black in the main stem 10th capsule

SYNTHETIC SEEDS : Synthetic or artificial seeds are the living seed like structure derived from somatic embryoids *in vitro* culture and encapsulated with hydrogel as artificial endosperm. Synthetic endosperm containing carbon sources, nutrients, growth regulators and antimicrobial agents. Encapsulation is done by various types of hydrogel which are water soluble.

Gel concentration (w/v) (%)	Complexing agent (μM)
Sodium alginate (0.5-5)	Calcium chloride (30-100)
Carragenan (0.2-0.8)	Potassium chloride (500)
Locust beam gum (0.4-1)	Ammonium chloride (500)

Steps in synthetic seed production

- ➢ Establishment of callus culture
- ➢ Induction of somatic embryogenesis
- ➢ Maturation of somatic embryos
- ➢ Encapsulation of somatic embryos
- ➢ Evaluation of embryoid and plant conversion
- ➢ Planting in green house/field

Advantages of synthetic seeds

- ✓ As clonal propagation to replace traditional seed production
- ✓ As replacement to hand pollinated hybrid seed
- ✓ As a carrier for beneficial microorganism, pesticides and growth regulators
- ✓ Planting efficiency of vegetatively propagated crops would be increased
- ✓ Useful in tree crops where breeding cycle is long
- ✓ Useful in germplasm conservation

HYBRID SEED PRODUCTION : It is the process of producing seeds by crossing plants belonging to different lines of same species

Genetic male sterility : It is defined as the failure of plants to produce functional anthers, pollen, or male gametes and it is used in cotton, red gram and tomato

Cytoplasmic male sterility : It is entirely controlled by cytoplasmic factors or by interaction between cytoplasmic and nuclear factors.

Line breeding

- ✓ **One line breeding :** Stubble breeding in rice also called as single line breeding
- ✓ **Two line breeding :** TGMS and PGMS mostly followed in rice at China
  - TGMS- Temperature Sensitive Male Sterility
  - PGMS-Photosensitive Male Sterility
- ✓ **Three line breeding :** by using male sterility systems
  - **A-Line,** It is male sterile line, used as female line in hybrid seed production
  - **B- Line**, It is isogenic line of a line except for the male sterility gene, this is used as male line (due to male fertile in nature)
  - **R- Line**, it is restorer line it restore the fertility

Chemically induced male sterility : Also called as CHA'S (Chemical Hybridizing Agents). eg. malic hydrazide, ethrel and arsenates. Gibberilic acid is used to increase panicle exertion of CMS lines.

Male sterile cytoplasm source used in different crops

Rice - WA (Wild Abortive) based CMS lines have imperfect panicle exertion.

Wheat - *T. tymphovi*

Maize - Texas cytoplasm; cytoplasm; S- cytoplasm

Sorghum - khafir 60 (Milo cytoplasm)

Pearl millet - Tift 23 A, 732-A, L111

5.2. SEED PROCESSING AND STORAGE

Seed Processing

Improvement in physical purity of seed lot by removal of undesirable, undersized and damaged seeds by means of mechanical device with minimum damage and loss to the seed.

Basic steps in seed processing are

- Receiving, conditioning and pre-cleaning
- Cleaning, separating and upgrading
- Treating, drying
- Packing, storage/shipping

Process :

- Preconditioning is to prepare the seed lot for basic cleaning
- Scalper is a rough cleaner to remove large size particle and trash
- Huller scarifier to remove hull/husk, buffel grass and alfalfa
- Debearder is the process of removing the awn in oat and barley
- Pebble mill is used to remove the hairs and fuzz in cotton
- Hammer mill is used to extracting the seeds from hard fruits
- Sheller as used for shelling maize
- Blending is mixing or intermingling of two or more dissimilar lots to produce overall uniformity in the final seed lot. Pearson square law is an easy way to determine the proportion of different lots that can be blended together to give a final lot of desired quality
- Basic seed cleaning is to remove contaminant and made in to finished product
- Sorting is a process of separation of cleaned product in to various quality factors based on size, identity, shape texture and colours
- Grading is the process of classification of cleaned product based on commercial value and usage

Parts in air screen cleaner

- Screen for cleaning and grading
- Crushing roll to separate lumps and onion bulblets
- Tapper/Knocker is to jar loose materials wedged
- Shoe is a shaking section helps to flow the seed
- Fan is to remove chaff and dust

Functions of typical 4 screen cleaner

Screen no.	Function
1	Scalping
2	Grading
3	Close scalping
4	Fine grading

Screen section

Seed shape	Top screen	Bottom screen
Round	Round	Oblong
Oblong	Oblong	Oblong
Lens	Oblong/rectangle	Round

Seed processing machinery

Physical character used for separation	Machine
Width/thickness	Air screen cleaner/width separator
Length	Disc separator/intended cylinder separator
Shape	Spiral separator/inclined draper
Surface texture	Velvet roll mill, inclined draper, magnetic separator
Weight	Gravity separator, aspirator, pneumatic separator, blower
Electrical conductivity	Electrostatic separator
Colour	Electronic colour sorter
Others	Picker belt, polisher, vibrator

Conveying equipments : to move the seed into, through or away from the processing plant in horizontal, vertical or inclined directions

1. Bucket elevator-vertical or near vertical direction
2. Belt conveyor-horizontal direction
3. Vibrating conveyor-horizontal and near horizontal short distance
4. Pneumatic conveyor-vertical and horizontal short distance
5. Screw/auger conveyor-vertical and horizontal direction
6. Chain conveyor-horizontal 45° angle inclined plane
7. Lift truck-serve as conveyor and storage unit-useful in large no. of lots handled

SEED STORAGE

The purpose of seed storage is to maintain seed vigour and viability in terms of germination from harvest until planting.

Types of seed storage (based on storage period)

1. Short term (microbiotic)-harvest to next planting-6 to 9 months
2. Medium term (mesobiotic)- harvest to second planting-upto 18 months
3. Long term (macrobiotic)- harvest to next planting-upto 10years

Types of seed storage

1. Conditioned storage with the relative humidity and temperature of 10 to 11% at 20°C or 12 to 13% at 25° C

2. Cryogenic storage is stored in liquid nitrogen at -196° C
3. Hermetic storage is storing seeds in sealed containers
4. Containerized storage is storing seeds in bins or specially constructed rooms

Drying : The process of removal of moisture from the seed to a safe moisture limit at which the rate of deterioration is slow down (in order to maintain seed viability and vigour)

Higher the seed moisture, lesser should be the drying temperature and *vice versa*

Moisture content (%)	Drying temperature
<10	110° F (43.3° C)
10-18	100° F (42.2° C)
18-30	90° F (32.2° C)

Orthodox seeds : seeds can be dried to safe moisture content and stored without loss of viability for specific period-cereals, pulses, oilseeds. or desiccation tolerant seeds.

Recalcitrant seeds : Seeds looses its viability when dried below a critical moisture level (20-50% of wet weight basis)-coffee, coconut, rubber, mango and jack or Thermo sensitive or desiccation intolerant seeds.

Equilibrium moisture content : A point at which moisture content of the seed attains equilibrium with the surrounding atmosphere at given temperature, RH and atmospheric pressure.

Tempering : It is the time interval provided in-between two successive drying processes in order to stabilize the moisture content of seed by diffusion. So that drying will be uniform in all parts of the seed.

Harrington's rule of thumb

1. For every 1% decrease in moisture content, the life span of the seed is doubled-applicable between 5-14% of seed moisture content.
2. For every 5° C (10° F) drop in temperature, the storage life of seed is doubled-applicable between 0-15° C (50°-100° F)
3. Sum total of RH and temperature (F) should not exceed a numerical sum of 100 with not more than 50% contributed by RH.

Classification of packaging materials / containers

1. Moisture and vapour previous containers: allows entry of water in the form of vapour and liquid-cloth bag, gunny bag and paper bag
2. Moisture resistant containers: allows the entry of water in the form of vapour but not in liquid form-urea bag and 300 gauge thickness polythene bag
3. Moisture and vapour proof containers: Never allows the entry of moisture in the form of vapour or liquid-rigid plastics-aluminum foil pouch and 700 gauge thickness polythene bag

Moisture content of seed (%)	Expected storage life (year)
11-13	0.5
10-12	1
9-11	2
8-10	4

Seed Deterioration : Seed deterioration is an irreversible and degenerative change occurs in the seed after it reaches maximum quality level. Three general concepts of seed deterioration are: 1. Inexorable, 2. Irreversible, 3. Variable among seed population

Causes of seed deterioration

1. Lipid peroxidation
2. Degeneration of functional tissue
3. Enzyme degradation and inactivation
4. Formation and activation of hydrolytic enzymes
5. Genetic deterioration
6. Accumulation of toxic compounds

5.3. SEED TESTING

Frerich Nobbe is known as Father of seed testing

Primary sample : randomly selected sample and drawn from the top, middle and bottom of the container

Composite sample : combined and mixed sample of various primary samples

Submitted sample : repeated halving or by abstracting portion from the composite sample

Working sample : sample taken from submitted sample by the seed analyst for conducting various tests. It contains at least 2500 seeds but subject to maximum of 1000 grams

Minimum number of primary samples for seeds stored in containers

Container	Sampling intensity
Upto 5	One for each, minimum 5
6-30	Sample 5 containers, one for every 3 containers
31-400	Sample 10 containers, one for every 5 containers
>400	Sample 80 containers, one for every 7 containers

Minimum number of primary samples for seeds stored in bulk

Quantity (kg)	Sampling intensity
Upto 500	Minimum 5
501-3000	One for every 300kg, not <5 samples
3001-20000	One for every 500kg, not <10 samples
>20000	One for every 700kg, not <40 samples

Sampling equipments

1. Stick or sleeve type trier
2. Bin sampler
3. Nobbe trier
4. Sampling by hand

Dividing methods used for getting working sample

1. Boerner (conical) divider
2. Soil (riffle) divider
3. Centrifugal (gamet) divider
4. Random cup method
5. Spoon method
6. Hand halving method
7. Modified halving method

Physical purity is determination of different components of physical purity namely pure seed, other crop seed, weed seed and inert matter.

Pure seed : seeds of kind/species stated by the sender or found to predominant in the test

Crop	Inseparable other crops
Wheat	Barley, gram, oats, triticale
Barley	gram, oats, triticale, wheat
Oats	Barley, gram, wheat, triticale
Triticale	Barley, gram, oats, rye

Minimum standards for physical purity

Physical purity (%)	Crops
99	Bhendi
98	Rice, wheat, oat, barley, maize, sorghum, bajra, bengalgram, blackgram, greengram, cowpea, redgram, castor, sunflower, safflower, cotton, brinjal, chilli, tomato, gourds, beans, cabbage, cauliflower, onion, potato
97	Finger millet, mustard, sesame, jute, celery
96	Groundnut, asparagus, spinach, beet root
95	Carrot, amaranthus
90	Stylo, marvel grass
80	Buffalo grass, guinea grass

Genetic purity : Trueness to type or variety. Seeds conforming to characters of the variety as described by the breeder.

Genetic purity standards

Seed class	Genetic purity (%)
Breeder seed	>FS std. (100)
Foundation seed (FS)	99.0
Certified seed	98.0
Hybrids (general)	95.0
Hybrids-brinjal, cotton, tomato and watermelon	90.0
Hybrid castor	85.0

Tests for the determination of genuiness of varieties

- Phenol colour reaction in wheat
- Seed coat peroxidase test in sorghum
- Ferrus sulphate colour reaction in paddy
- NaOH or potassium dichromate in pea
- Alkaloid test in lupinus

Germination test : It determines the maximum germination potential or viability of the seed. The test is conducted in the germination medium such as germination paper, sand and soil.

Seedling evaluation : Seedlings which have reached a stage when all essential structures can be accurately assessed shall be removed from the test at the first or any other intermediate counts.

Categories of seedlings

1. **Normal Seedlings :** Normal seedling is one which shows the capacity for continued development into mature plant when grown in good quality soil and under favourable conditions of water supply, temperature and light. According to the ISTA (1985) seedlings to be classified as normal seedling, must conform to one of the following categories:

   a)Intact seedlings : Seedlings with all their essential structures, well developed complete in all proportion and healthy.

   b) Seedlings with slight defects : Seedlings showing certain slight defects of their essential structures provided they show an otherwise satisfactory and balanced development comparable to that of intact seedlings of the same test.

   c) Seedlings with secondary infections : Seedlings which are seriously infected by fungi or bacteria are classified as normal, if it is evident that the parent seed is not the source of infection, and if it can be determined that all the essential structures were present.

2. **Abnormal seedlings :** An abnormal seedling is one which does not have the capacity to develop into a normal plant when grown in the soil under favourable conditions because one of more of the essential structures is irreparably defective. Three major classes of abnormal seedlings are damaged Seedlings, deformed or unbalanced seedlings and decayed seedlings.

Multigerm seed units : Seeds which are capable of producing more than one seedling. Several types of seed units can produce more than one seedling eg. clusters of *Beta vulgaris*, fruits of *Tectona grandis*, polyembryonic seeds. In such cases only one normal seedling is counted for determining the germination percentage.

Ungerminated seed : Seeds which have not germinated by the end of test period when tested under the ideal conditions. They are hard seeds, fresh seeds and dead seeds.

Minimum standards for germination

Germination (%)	Crops
90	Maize
85	Wheat, oat, barley, bengalgram, mustard
80	Rice, horsegram, sesame, safflower, jute, lucerne
75	Sorghum, bajra, finger millet, cowpea, blackgram, greengram, redgram, french bean, lablab
70	Sunflower, castor, soybean, groundnut, tomato, brinjal, cabbage, onion, radish
65	Cotton, bhendi, cauliflower
60	Chilli, carrot, beetroot, watermelone, gourds
50	Dinanath grass, setaria grass
40	Stylo, marvel grass
30	Buffel grass
20	Guinea grass
15	Daraf grass

Seed viability : It is the ability of the embryo to germinate and is affected by a number of different conditions.

Viability test : Tetrazolium test, commonly known as the TZ test for seed viability (2, 3, 5 triphenyl tetrazolium chloride 0.2-1%)

Vigour : Sum total of those properties of the seed which determines the potential level of activity and performance of seed or seed lot during germination and seedling emergence. According to ISTA seed vigour is "the sum of those properties that determine the activity and performance of seed lots of acceptable germination in a wide range of environments.

Seed vigour testing

i. **Direct vigour tests :** Assessed by Hiltner (brick gravel) test, paper piercing test, accelerated ageing test and cold test.

ii. **Indirect vigour tests :** Tested by dry weight of seedlings, speed of germination, seedling length measurement, exhaustion test, electrical conductivity of seed leachate, tetrazolium test, glutamic acid decarboxylase activity (GADA) test, cool test and respiration.

Seed health test : Seeds may harbour pathogens both inside and on their surfaces. Seeds may be infected with viruses, bacteria, fungi and nematodes. The seed health test is to determine the insects, nematode and plant pathogens present in the seed sample or the planting material.

Fungi : rice bunt-NaOH test, wheat loose smut-lactophenol test

Bacteria : ELISA, SSEM, radio immuno assay, double diffusion, radial diffusion methods

Nematode : direct seed examination, lactophenol

Designated disease of seed crops

Crop	Designated disease	Casual organism
Wheat	Loose smut	*Ustilago tritici*
Bajra	Green ear	*Sclerospora graminicola*
Sorghum	Grain smut	*Sphacelotheca sorghi*
Cowpea	Anthracnose	*Colletotrichum lindemuthianum*
Greengram	Halo blight	*Pseudomonas phasiolicola*
Sunflower	Downy mildew	*Plasmophora halstedii*
Sesame	Leaf spot	*Cercospora sesame*
Brinjal	Phomopsis wilt	*Phomopsis vexans*
Chilli	Die back	*Colletotrichum capsici*
Cluster bean	Aschocyta blight	*Ascochita sp.*
French bean	Bean mosaic	*Macrosiphum pisi*

Varietal purity test is done by grow out test (GOT).

Moisture content determination

1. Reference methods: Done by using phosphrous pentoxide and Karl Fischer titration.
2. Routine methods : Done by using hot air oven and moisture meter

i. Low constant temperature of 103° C for 17 hours for the crops like finger millet, soybean, groundnut, castor, mustard, sesame, cotton, jute, brinjal, chilli and onion

ii. High constant temperature for 130° C for 1 hour for the crops likes rice, wheat, maize (4 hours), sorghum (2 hours), bajra, cowpea, bengalgram, blackgram, greengram, redgram, peas, tomato, bhendi, carrot, beet, gourds, coriander, lablab and cucumber

Submitted sample size is required for moisture estimation is 50g for species that need not be ground and 100g for species that need to be ground.

5.4. SEED GERMINATION

The emergence and development from the embryo of those essential structures which are indicative of the ability to produce a normal plant under favourable conditions. Availability of sufficient amount of light, moisture, temperature and oxygen is the favourable condition for seed germination.

Phases of seed germination

- Imbibitions and enzyme activation (active metabolism)
- Initiation of embryo growth
- Protrusion of radicle
- Seedling establishment

i. Hypogeal Germination: Cotyledons remain below the soil level and the plumule is carried upward by elongation of the epicotyl, eg. rice wheat, maize, pea, bengalgram and redgram.

ii. Epigeal germination: Cotyledons are raised above the ground level due to the elongation of hypocotyls, eg. dicots and onion.

Quiescence stage is non dormant seed, unable to germinate due to unfavourable environment conditions.

Dormancy : Inability of viable and mature seed to resume growth immediately after harvest in a favourable environment condition.

i. Primary dormancy: The most common form of dormancy and is a condition in which essential germination components not available to the seed or due to the inherent property (inhibitors) of seed.

ii. Secondary dormancy: Due to the exposure to one or more unfavourable conditions, i.e. adverse temperature or light and excessive moisture.

Types of secondary dormancy

- Thermo dormancy is seed sensitivity to heat or cold
- Skotodormancy is brought about when fully hydrated, mature seeds experience gravely sub-optimal germination conditions such as anoxia, darkness, excessively strong light, maximal and minimum temperatures for the completion of germination and water stress
- Photodormancy is light sensitivity affects germination of some seeds

Dormancy breaking : Dormancy can be broken to induce the germination by following treatments

Scarification : It is a process that allows water and gases to penetrate into the seed; it includes methods to physically break the hard seed coats or soften them by chemicals, such as soaking in hot water or poking holes in the seed with a pin or rubbing them on sandpaper or cracking with a press or hammer.

Stratification : It is also called moist-chilling, breaks down physiological dormancy and involves the addition of moisture to the seeds so they imbibe water and they are then subjected to a period of moist chilling to after-ripen the embryo (Cold 0-5° C and Warm 40-50° C).

Leaching : It is the process of soaking in water removes chemical inhibitors in some seeds that prevent germination

Growth regulators : Endogenous dormancy may be due to presence of germination inhibitors. Application of low level of growth regulators (i.e. Gibberellins, Cytokinins and Ethylene etc) may break the seed dormancy.

5.5. SEED ENHANCEMENT TECHNIQUES

Seed hardening : It is the process of hydration and dehydration of seeds to induce drought tolerance. The chemicals used are KCl, $CaCl_2$, NaCl and KH_2PO_4

Priming : It is the controlled hydration of seeds to proceed first phases of seed germination (pregerminative metabolic activity) but prevents radicle protrusion

Osmo-priming : Soaking the seeds in osmotic solution using PEG 6000 & 8000, D-mannitol

Halo-priming : Soaking the seeds in salt solution of KNO_3 and K_2PO_4

Bio-priming : Coating the seeds with biological agents like bacteria

Solid matrix priming : Solid material as carrier for osmotic solution-vermiculite

Hydro-priming : Soaking the seeds in excess water

Drum-priming : Subjecting the seeds to mist in a rotating drum

Pre sowing seed treatments : Seed treatments given prior to sowing such as hardening, priming, scarification, fungicide treatment, bacterial inoculation etc.

Seed invigouration : Improvement of seed performance by any physio-chemical treatments resulting in improvement of germination, better field performance and storability

Seed fortification : Impregnation of bioactive chemicals (for invigourating the seeds) through water or diluted solutions of micronutrients, growth regulators, vitamins and plant extracts

Infusion : Impregnation of bioactive chemicals through organic solvents for invigourating the seeds-acetone, ethanol, petroleum ether etc.

Seed stimulation : Dry seed treatment with magnetic field, electric field and radio frequency waves

Designer seeds

Seed coating is a general terminology for the application of finely ground materials containing suspended/ dissolved solids to form continuous layer covering the natural seed coat with or without changing the original size and shape of the seed.

1. **Seed pelleting :** It is the process of enclosing the seed with small quantity of ingredients along with filter materials to produce a spherical or elliptical unit of uniform size to facilitate precision planting. Filler materials used are gypsum, limestone, dolomite, rock phosphate, vermiculite, peat, poultry manure, dried blood, wood charcoal, milk powder, yeast extract, biofertilizer and talc. Leaf powders such as arappu, pungam, notchi, prosopis, neem, moringa and tamarind. Adhesives used are gum acacia, gelatin, methyl cellulose, polyvinyl alcohol, polyvinyl acetate, polyelectrolyte or dextrin. Low cost binding materials are rice gruel, maida paste, starch and sago gruel.
2. **Film coating :** It is a sophisticated process of application of precise amount of active ingredients along with a liquid material directly on to the surface without obscuring the shape-biodegradable polymer
3. **Seed colouring :** Application of precise amount of dyes or pigments directly on to the seed to improve its brand identity and marketability
    - Chemical dyes: Congo red, pink CM, bromocresol green, jade green, sky blue, turkuoise blue, direct chenbaga
    - Botanical dyes: turmeric-lemon yellow, margold-golden brown, opuntia-red, beet root-teak, annatto-dawn glow, basella-leaf brown, henna, hibiscus & jamun with seed-sand stone, jamun without seed-light brown

Seed tablet : A composite of seed and solid materials formed by compression in a tablet press such as issued in the pharmaceutical industry

Seed tapes : Narrow bands of degradable materials that contain seeds spaced as per crop requirements. It is sown by laying the tape in the field and covering with soil.

5.6. SEED LEGISLATION IN INDIA

- ✓ Seed act was enacted by the parliament on 29 Dec 1966
- ✓ Seed Rules notified in 1968
- ✓ Seed Act come whole county - 2nd Oct, 1969
- ✓ Essential commodities act -1955
- ✓ Functions of central variety release committee - 1969
- ✓ Notifications of kinds (or) varieties as provided for in the speed act was started in 1969

- ✓ Central seed laboratory located in New Delhi
- ✓ Seed control order issued - 1983
- ✓ Packing commodities order - 1975
- ✓ Standards of weights and measurements act - 1976
- ✓ Consumer Protection act - 1986
- ✓ National Seed policy - 2002
- ✓ Protection of Plant Variety Rules - 2003
- ✓ Seed bill - 2004 (To replace the seed act 1966)

International Seed Testing Association (ISTA) founded in 1924. ISTA certificate used are

- *Orange certificate* : Official sampling and testing is done in same country.
- *Green certificate* : Official sampling in one country and analyzed in another country.
- *Blue certificate* : This is not officially sampled.

Indian Seed Industry

- National Seed Corporation established in 1963
- State Farm Corporation of India Limited established in 1969
- National Seed Research and Training Centre (NSRTC) established at Varanasi in 2005
- Protection of Plant Varieties and Farmers Right Act (PPV & FR Act) enacted in India (2001)

Comparison of Seed Act 1966 and 2004

S. No	Particular	Seed Act 1966	Seed Bill 2004
1.	Scope	It includes agriculture and horticulture	It includes agriculture, horticulture, forestry, plantation crops, medicinal and aromatic plants.
2.	Category of seed permitted for sale	Certified seed	Certified seed
3.	License for selling seed by seed dealer	Compulsory	Compulsory
4.	Requirements for seed multiplication	Released and Notified Varieties	Released and Notified Varieties
5.	Registration with PVP authority	Not required	Compulsory
6.	Criterion for registration	Not applicable	Novelty, distinctness, uniformity and stability
7.	Certification	By state seeds certification agency	Self certification is permitted
8.	Period of protection as per UPOV Act 1978	Not applicable	For field crops 15 years and for trees and vines 18 years as per UPOV act 1978

9.	Compensation of loss by use of certified seed	Through Consumer Court	Through Consumer Court
10.	Farmers exemptions	Already there	Allowed
11.	Purity and germination standards	Adopted	Adopted
12.	Penalty for violation of act	Rs.100-1000 and six month imprisonment	Rs. 5000 - 50,000

Seed Industry

- Tamil Nadu State Seed Development Agency was established on 10.04.2015
- State Seed Farms-41, State Seed Certification Agencies-22, State Seed Corporations-16, NSC-1
- Notified State Seed Testing Laborataries-(107-124-136)
- Central Seed Testing Laboratory established at Varanasi(for all seeds) and referral Lab at Nagpur(only for Bt cotton seeds)
- ISTA accredited Lasbs-5, Member labs-18
- Private sector and public sector share in the seed market is 74:26 by its value
- A total of 8187 varieties in 164 crop species are notified as per Seeds Act 1966 viz., Cereals(03-2165), Pulses and Millets(16-3194), Oilseed(16-1605), Fibres (06—624), Vegetables(10-414, Sugar crop(1-99), cash crop(1-2), Flowers & Fruits(2-5), Forage crops(21-79)
- 6 Rice varieties of neighboring countries Binadhan 8,10,11,12 of Bangladesh and Sukhanham 5,6 of Nepal were also notified as per Seed Act 1966 for cultivation in West Bengal, Tripura, Assam, Uttar Pradesh and Bihar
- A total of 109 varieties of 20 crop species were enlisted in OECD list of varieties for varietal certification
- PPV&FRA registration certificate was issued for 2444 varieties of 140 crop species upto Nov.2016

Web resources:

- http://seednet.gov.in
- http://www.worldseed.org
- https://www.seedtest.org
- http://plantauthority.gov.in
- http://nsai.co.in
- http://nsrtc.nic.in
- http://www.agricoop.nic.in
- http://122.15.179.102/tanseda

CROP PHYSIOLOGY

6

N. Sridharan

6.1. PLANT WATER AND MINERAL NUTRITION

Plant water

Water form the major constituent of living (cells) things and the cells originated in a highly aqueous medium and all the vital processes of the life are carried out in it. Besides, water predominately acts as a source of hydrogen to plants and is released by the photolys is of water during photosynthesis.

In living tissue, water is the medium for many biochemical reactions and extraction process. Inorganic nutrients, photosynthesis, bases and hormones are all transported in aqueous solution. Soil nutrients are available to plant roots only when it get dissolved in water. It is essential for life and plays a unique role in virtually all biological process.

Water absorption

Plants absorbed water through root hairs which are in contact with soil water and form a root hair zone a little behind the root tips.Root hairs are tubular hair like prolongations of the cells of the epidermal layer of the roots. When roots elongate, the older root hairs die and new root hairs are developed so that they are in contact with fresh supplies of water in the soil.

Mechanism of water absorption

1. Active absorption of water : In this process the root cells play active role in the absorption of water and metabolic energy released through respiration is consumed active absorption may be of two kinds.

- **i. Osmotic absorption:** Water is absorbed from the soil into the xylem of the roots according to osmotic gradient.
- **ii. Non-osmotic absorption:** Water is absorbed against the osmotic gradient.

2. Passive absorption of water : It is mainly due to transpiration, the root cells do not play active role and remain passive.

Water potentials : Every component of a system possesses free energy capable of doing work under constant temperature conditions. For non-electrolytes, free energy/mole is known as chemical potential. The chemical potential of water is called as water potential (ψ). For a pure water, the water potential is 0. The presence of solute particles will reduce the free energy of water or decrease the water potential. Therefore it is expressed in negetative value.

ψ_w = $\psi_w + \psi_s + \psi_p$

ψ_S = is the solute potential or osmotic potential

ψ_p = pressure potential or turgor potential

ψ_w = is the matric potential. Matric potential can be measured for the water molecules adhering on the soil particles and cell wall.

In plant system, the matric potential is disregarded.

Therefore, $\psi_w = \psi_s + \psi_p$

Osmotic pressure : Osmotic pressure in a solution results due to the presence of solutes and the solutes lower the water potential. Osmotic pressure has positive sign and osmotic potential has negative sign (ψ_s)

Turgor pressure : In plant cell, the turgor pressure results due to the presence of water molecules is turgor pressure. The potential created by such pressures is called presume potential (ψ_p)

Diffusion : The movement of particles or molecules from a region of higher concentrations to a region of lower concentration is called as diffusion.

Importance of diffusion

- ❖ It is an essential step in the exchange of gases during respiration and photosynthesis
- ❖ During passive salt uptake, the ions are absorbed by diffusion
- ❖ It is important in stomatal transpiration as the last step in the pollen, where diffusion of water vapour from the interrelation space into the outer atmosphere occurs through open stomata.

Osmosis: The diffusion of solvent molecules into the solution through a semi-permeable membrane is called as osmosis.

Endo-osmosis

If a living plant cell is placed in water or hypotonic solution whose osmotic potential is lower than cell sap, water enters into the cell sap by osmosis and the process is called endo-osmosis. As a result of entry of water with the cell sap, a pressure is developed which press the protoplasma against the cell wall and the become turgid. This pressure is called an **turgor pressure.**

Exosmosis : The plant cell is placed in hypertonic solution (whose osmotic potential is higher than cell sap) the water comes out from the cell sap into the outer solution and the cell becomes flaccid. This process is known as exosmosis.

Imbibition : Certain substances if placed in a particular liquid absorb it and swell up. These substances are called as imbibants and the phenomenon is imbibition, certain force of attraction is existing between imbibants and the involved substance.

Plasmolysis : When a plant cell or tissue is placed in a hypertonic solution water comes out from the cell sap into the outer solution of exosmosis and the protoplasm begins to contract. The protoplasm separate from the cell wall and assures a spherical form and this phenomenon is called plasmolysis. Water enter into the cell sap, the cell becomes turgid and the protoplasm again assumes it normal shape and position. This phenomenon is called deplasmolysis.

Diffusion pressure deficit (DPD) : The difference between the diffusion pressure of the solution and its solvent at a particular temperate and atmosphere conditions is called as diffusion pressure deficit.

Transpiration : It is the process by which moisture is carried through plants from roots to small pores on the underside of leaves, where it changes to vapour and is released to the atmosphere.

Types of transpiration

Stomatal transpiration : Stomata are natural epidermal openings on leaf thought which 80 - 90% of total transpiration takes place in plants. 80-90% (Blue light responsible for stomata opening and glucose 1 phospahte enzyme responsible for stomata open and close)

Cuticular transpiration : A small amount of water(20%) is lost in form of water vapour through the epidermal cuticle of stem and leaf by simple diffusion.

Lenticular transpiration : Very minute permanent openings present on the bark of trees and woody fruits called lenticels. Through these pores nearly 0.1% of total water lost is transported.

Apoplast : Water moves from cell to cell through *cell walls* until it reaches the endodermis

Symplast : Water moves from cell to cell in the cytoplasm via the plasma membranes and plasmodesmata.

Vascular pathway : Water moves through the cell surface membrine, cytoplasm and tonoplast of the vacuoles

Anti-transpirants

Type	Mechanism	Examples
Stomata closing	It reduces water loss through stomatal closing	Phenyl mercuric acetate (PMA), Atrazine in low concentrations and ABA
Film forming	Retard moisture loss due to formation of thin film as physical barrier	Mobileaf, hexadecanol, silicone, oils, waxes
Reflecting	It reflects the radiation and thus reduce leaf temp and vapour pressure gradient from leaf to atmosphere and ultimately transpiration is reduced	Celite and Kaolin
Growth retardants	It reduces shoot growth and increase root growth and thus enable the plants to resist drought. It also induces stomatal closure	Cycocel (CCC)

Guttation : It is the process by which loss of water in liquid form from the uninjured leaf or stem of the plant, principally through water stomata.

Difference between Transpiration and Guttation

Transpiration	Guttation
Loss of water from living tissues of aerial parts of the plants in the form of water vapour	Water loss in liquid form
Transpiration occurs mostly through stomata. It may also takes place through cuticle and lenticels	It occurs only through hydathodes (water stomata)
It takes place throughout the day, its rate being maximum at noon.	It takes place only early in the morning when root pressure and the rate of water absorption are higher
Loss as pure water	Loss as salt and sugars

MINERAL NUTRITION

- The chemical compounds required by an organism are termed as nutrients
- Nutrition may be defined as the supply and absorption of chemical compounds needed for plant growth and metabolism
- For plant growth and metabolism, 17 elements are essential. They are C, H, O, N, P, K, Ca, S, Mg, Fe, Mn, Zn, B, Cu, Mo, Cl and Ni

  These essential elements are classified into two groups

  1. Major elements (Macro nutrients)
  2. Minor elements (Micro nutrients or Trace elements)

Major elements : The essential elements which are required by the plants in comparatively large amounts are called as major elements or macro nutrients.

They are C, H, O, N, P, K, Ca, S, Mg.

Minor elements: The essential elements which are required in very small amounts or traces by the plants are called as minor elements or micronutrients or trace elements.

They are Fe, Zn, Mn, B, Cu and Mo.

Based on the mobility, elements are also classified into three types.

1. Mobile elements : N, P, K, S and Mg
2. Immobile elements : Ca, Fe and B
3. Intermediate in mobility : Zn, Mn, Cu, Mo

6.2. PHOTOSYNTHESIS AND RESPIRATION

Photosynthesis

- Photosynthesis means synthesis with the help of light i.e. plant synthesize organic matter in the presence of light (carbohydrates)
- Photosynthesis is sometimes called as carbon assimilation.

This is represented by the following equation.

$$6CO_2 + 12H_2O \xrightarrow[\text{Green pigments}]{\text{sun light}} 6O_2 + C_6H_{12}O_6\text{-}$$

Sugar (carbohydrates)

- During the process of photosynthesis, the light energy is converted into chemical energy and is stored in the organic matter which is usually the carbohydrates and $O\text{-}_2$ form the end products of photosynthesis.
- One molecule of glucose for instance, contains about 686 K.Cal. of energy.
- CO_2 and water constitute the raw material for this process.

Photosynthetic apparatus : The chloroplast in green plants constitute the photosynthetic apparatus. Typically, the chloroplast of higher plants are discoid in shape, 4-6μ in length and 1-2μ thick. Internally, the chloroplast is filled with a hydrophilic matrix called as stroma in which are embedded grana. Each grana consists of 5-25 disk shaped grana lamellae (thylakoids) placed one above the other like the stack of coins. Each grana lamella of thylakoid encloses a space, the loculus and the thylakoid membrane consists of alternating layer of lipids and proteins.

Chlorophyll and other photosynthetic pigments are confined to grana. The chlorophylls are the site of photochemical reactions.

Photosynthetic pigments : Photosynthetic pigments are of three types namely Chlorophylls, Carotenoids and Phycobillins.

- Chlorophylls and carotenoids are insoluble in water and can be extracted only with organic solvents (acetone, petroleum ether, alcohol).
- Phycobillins are soluble in water
- Carotenoids include carotenes and xanthophylls.

1. Chlorophylls : Chlorophylls are magnesium porphyrin compounds. The porphyrin ring consists of four pyrrol rings joined together by CH bridges. Molecular formula for *chlorophyll a* and *chlorophyll b* are $C_{55}H_{72}O_5NH$ Mg and $C_{55}H_{70}O_6N_4$ Mg respectively.

2. Carotenoids : (Yellow or orange pigments)

Carotenes: Carotenes are hydrocarbons with a molecular formula $C_{40}H_{56}O$.

Xanthophylls (carotenols) : They are similar to carotenes but differ in having two oxygen atoms in the form of hydroxyl or carboxyl group. Mole. formula $C_{40}H_{56}O_2$.

Phycobillins : They are red and blue pigments.They also contain four pyrrol rings and lack of Mg and the phytol chain.

Distribution of photosynthetic pigments in plant kingdom

Pigments	Plant kingdom
Chlorophylls	
Chlorophyll –a	All photosynthesizing plants except bacteria
Chlorophyll –b	Higher plants and green algae
Chlorophyll –c	Diatomes and brown algae
Chlorophyll –d	Red algae
Backeriochlorophylls a, b, c, d & e	Purple and green bacteria
Carotenoids	
Carotenes	Higher plants and algae
Xanthophylls	Higher plants and algae
Lutein	Green leaves and Green and Red algae
Violaxanthin	Green leaves
Fucoxanthin	Brown algae
Phycobilins	Blue algae
Phycocyanins	Blue green algae
Phycoerythins	Red algae
Allophycocyanin	Blue – green and Red algae

Source sink relationship

Source : Leaves and other green tissues that produce photosynthesis (carbohydrates) are called as sources

Sink : Organs which utilize the photosynthates for their growth and then store the photosynthates are called sink.

RESPIRATION

The energy stored in carbohydrate molecules during photosynthesis is released during cellular oxidation of carbohydrates into CO_2 and H_2O is called as respiration. In respiration the oxidation of various organic food substances like carbohydrates, fats, proteins etc, may take place. Among these glucose is the commonest.

$$C_6H_{12}O_6 + 6O_2 \rightarrow 6CO_2 + 6H_2O + \text{Energy (686 k cal)}$$

Mechanism of Respiration

Glycolysis (Embden-Mayer Hof–Paranas Pathway) : It is the metabolic pathway that converts glucose into pyruvate. The free energy released in this process is used to form the high-energy compounds of ATP (adenosine triphosphate), $FADH_2$ and NADH (Nicotinamide adenine dinucleotide) and it occurs in cytoplasm.

Metabolism : The sum total of catabolic and anabolic reaction occuring at any time in a cell is called metabolism

Anabolism : The phase of metabolism in which simple substances are synthesized into the complex materials of living tissue.

Catabolism : The metabolic breakdown of complex molecules into simpler ones, often resulting in a release of energy.

Aerobic respiratin-Krebs' cycle (TCA cycle): It is a series of chemical reactions used by all aerobic organisms to generate energy through the oxidization of acetate derived from carbohydrates, fats and proteins into carbondioxide. It occurs in the matrix of the mitochondrion.

Electron Transport System : It is the process of electron transfer between an electron donor and an electron acceptor, The resulting electrochemical proton gradient is used to generate chemical energy in the form of ATP. It occurs in the cristae of mitochondria, where ATP is synthesized

Anaerobic respiration (Fermentation) : In the absence of molecular oxygen, pyruvic acid undergones anaerobic respiration or fermentation. Two types of anaerobic respiration are alcoholic fermentation and lactic acid fermentation.

Account of ATP molecules in aerobic respiration : In glycolysis consumption is 2 ATP, Gain in respiration of one glucose molecule is 10 ATP and net gain is 8 ATP (2ATP, 2NADH) and Krebs cycle net gain is 30 ATP. Total net gain is 38 ATP.

Respiratory quotient

The ratio of the volume of CO_2 released to the volume of O_2 taken in respiration is called as respiratory quotient and is denoted as RQ

$$RQ = \frac{\text{Volume of } CO_2}{\text{Volume of } O_2}$$

6.3. GROWTH AND PLANT GROWTH REGULATORS

Growth : It is an irreversible increase in size. It may be evaluated by measurements of mass, length or height, surface area or volume. Growth is restricted only to living cells and is accomplished by metabolic processes involving synthesize of macromolecules, such as nucleic acids, proteins, lipids and polysaccharides at the expense of metabolic energy.

Growth regions : Typical growth regions in plants are the apices of shoot and root. Such growing regions are know as apical meristems, primary meristems or regions of primary growth. These apical meristems are responsible for the increase in length, differentiation of various appendages and formation of plant tissues.

Phases of growth : It occurs in meristematic regions where before completion of this process, a meristematic cell has to pass through the following three phases.

1. Cell formation phase (cell division)
2. Cell elongation phase
3. Cell differentiation (cell maturation)

The cell formation phase is represented by meristematic zone and cell enlargement phase by cell elongation zone.In the last phase or cell maturation, secondary walls are laid down and cell matures and gets differentiated into permanent tissue.

Growth analysis

Leaf Area Index (LAI) $= \dfrac{\text{Leaf area per plant}}{\text{Space occupied by the plant (spacing)}}$

Leaf area duration (LAD) $= \dfrac{\text{LAI (1) + LAI (2)}}{2} \quad \text{x } t_2 - t_1 2$

LAI (1) & 2 = Leaf area index at first and second stage

$t_2 - t_1$ = time interval between the two stages (days)

Leaf Area Ratio(LAR) $= \dfrac{\text{Leaf area per plant}}{\text{Whole plant dry weight}}$

Specific leaf area (SLA) in cm^2/g $= \dfrac{\text{Leaf area}}{\text{Leaf dry weight}}$

Specific leaf weight (SLW) in mg/cm^2 $= \dfrac{\text{Leaf dry weight}}{\text{Leaf area}}$

Net Assimilation Rate (NAR) in $mg/cm^2/$ day $= \dfrac{\text{Log } L_2 - \text{Log } L_1}{L_2 - L_1} \text{ x } \dfrac{W_2 - W_1}{t_2 - t_1}$

Relative growth (RGR) mg/g/ day) $= \dfrac{\text{Log } W_2 - \text{Log } W_1}{t_2 - t_1}$

Crop growth rate (CGR) in $mg/m^2/day$) $= \dfrac{W_2 - W_1}{P\,(t_2 - t_1)}$
P = Spacing of the crop

PHOTOPERIODISM

In plants the flowering requires a certain day length i.e the relative length of day and night which is called as photoperiod.The response of plants to the photoperiod expressed in the form of flowering is called as photoperodism and discovered by Garner and Allard. Depending upon the duration of photoperiod, they classified plants into three categories.

1. Short day plants: Floweirng induced by dark periods longer than the critical length. These plants regime a relatively short day light period (usually 8-10 hours) and a continuous dark period of about 14-16 hours for subsequent flowering. eg. Soyabean, chrysanthemum, tobacco, strawberry

2. Long day plants: Flowering is induced by dark period shorter than the critical length.These plants regime longer day light period (usually 14-16 hours) in a 24 hours cycle for subsequent flowering. These plants are also called as short night plants. eg. Spinach, Sugar beet, Spring wheat, Spring barley and Cabbage

3. Day neutral plants: Flowering is independant of photoperiod. These plants flower in all photoperiod ranging from 5 hours to 24 hours continuous exposure. eg.Tomato, Cotton, Sunflower, Cucumber, Peas varieties and Maize

Vernalization: Some plants require a low temperature treatment during their early stages of the life for subsequent flowering in the later stages.This low temperature treatment was termed as vernalization by Lyseko.

Devernalization: The positive effect of the low temperature treatment on the vernalization of the plant can be contacted by subsequent high temperature. This is called devernalization. The devernalized plant can again be vernalized by subsequent low temperature treatment.

Impotance of vernalization

i. Vernalization shortens the vegetative period of the plant

ii. Vernalization increases cold resistance of the plants

iii. Vernalization increases the resistance of plants to fungal diseases.

Comparison of C_3, C_4 and CAM plants

Particulars	**C_3 Plant**	**C_4 Plant (Hatch slack pathway)**	**CAM (Crassulacean Acid Metabolism)**
Product	C_3 Carbon	C_4 Carbon	
First product	Phospho glyceric acid (PGA)	Malate and Oxaloacidic acid (OAA)	Malic acid
Enzyme involved in carboxylation	Ribulosodiphosphat carboxylase (RuBisco)	Phosphoenol pyruvate carboxylase (PET case), RuBisco	-
Photo respiration	High	Low or negligible (PET case)	-
Photosynthetic efficiencies	Low	High	Lowest
Leaf anatomy	One type of chloroplast	Kranz anatomy (Bundle sheath and mesophyll cell)	
Water use efficiency	Low	High	Highest
Examples	Rice, Wheat, Soybean, Groundnut, Cotton, Sunflower, Barley, Sweet Potato	Maize, Sorghum, Sugarcane, Miner millet	Sisal, Pine apple

Plant growth regulators

Plant growth regulators or phytohormones or natural growth hormones of plants are organic substances produced naturally in higher plants, controlling growth or other physiological functions at a site remote from its place of production and active in minute amounts.

Classifications : Auxins, Gibberellins, Cytokinins, Abscisic acid and Ethylene. Auxins were the hormones first discovered in plants. Later Gibberellins and cytokins were also discovered.

i. Auxins : IAA is the only naturally occurring auxin in plants. The others such as IBA, NAA, 2, 4-D and 2, 4, 5 –T are synthetic auxins.

Physiological effects of auxin : Cell elongation, apical dominance, root initation, prevention of abscission, parthenocarpy, respiration and callus formation

ii. Gibberellins : Gibberellins (GAs) are a group of diterpenoid acids that function as plant growth regulators influncing a range of developmental processes in higher plants

Physiological effects of Gibberellins : Seed germination, dormancy of buds, root growth, elongation of internodes, bolting and flowering, parthenocarpy and synthesis of the enzyme amylase

iii. Cytokinins / Kinetin : Some of the very important and commonly known naturally occurring cytokins are coconut milk factor and Zeatin

Physiological effects of Cytokinins : Cell division, cell enlargement, concentration of apical dominance, dormancy of seeds and delay of senescence

iv. Abscisic acid : Abscisic acid is a naturally occurring growth inhibitor.

*Physiological effects Abscisic acid :*The two main physiological effects are geotropism in roots and stomatal closing, besides other effects are including bud dormancy and seed dormancy, Includes tubrisation, induces senescence of leaves, fruit ripening, abscission of leaves, flowers and fruits and increasing the resistance of temperate zone plants to frost injury.

v. Ethylene: Ethylene is the only natural plant growth hormone exists in gaseous form.

Physiological effects ethylene: It hastens the ripening of fleshy fruits eg. banana, apples, pears, tomatoes, citrus etc., stimulates senescence and abscission of leaves, induce flowering in pine apple, inhibition of root growth, stimulates the formation of adventitious roots, stimulates fading of flowers and stimulates epinastly of leaves.

vi. Growth retardants: There are number of synthesis compounds which prevent the gibberellins from exhibiting their usual responses in plants such as cell enlargement or stem elongation. So they are called as antigibberellins or growth retardants. They are Cycocel (2- chloroethyl trimethyl ammonium chloride (CCC), Phosphon D – (2, 4 – dichlorobenzyl – tributyl phosphonium chloride), AMO – 1618, Morphactins and Maleic hydrazide

PGRs	Use
Glyphosine, Ethepon	Sugarcane ripener
Abscisic acid	Cotton defoliant
NAA	Increasing number of flowers
Ethylene	Regulating ripening of fruits

Global warming: Accumulation of heat or gases causes the warming of earth surface and leads to global warming. Global warming will lead to following effects

1. Rise in temperature
2. Average rise in the level of sea (about 6 cm/decade) due to melting of polar ice.
3. CO_2 concentration steady increasing (CO_2 enrichment).

CO_2 enrichment and crop productivity

1. CO_2 enrichment leads to increased photosynthesis and productivity
2. CO_2 enrichment also decreases stomatal conductance by closing the stomata, their by decrease the transpiration / unit area of the leaf.
3. In C_3 plant the efficiency of ribulose biphosphate carboxylase enzyme is increased
4. Increased CO_2 concentration inhibits photorespiration in plants
5. CO_2 enrichment increased the yield and yield components.

Points to remember

- The cell organelle which is semiautonomous nature is *Mitochondrion*
- The organelle Lysosome is called as suicidal bags
- Chloroplast containing cells in leave is *Parenchyma*
- The metal ion involved in stomatal movement is K^+
- Water potential of pure water is *Zero*
- In plant cell water content is more in *vacuole*
- Cell oxidative phosphorylation takes place in *Mitochondria*
- *Oleosomes* are surrounded by half- unit- membrane
- *Sucrose* is an osmotically active solutes in guard cells
- The pressure difference needed to lift water to the top of a 100 meter tree is *3.0 MPa*
- Which one of the colors of light is effective for stomatal opening? Blue
- Reaction centre of pigment system II is *P-680*
- Lipids are made of *Glycerol and fatty acids*
- Photoreceptor which absorbs light in the UV -Blue region is *Cryptochrome*
- *Transpiration* is necessary evil
- Quantosome is photosynthetic unit and 1 Quantosome contain 200 – 300 chlorophyll molecules
- *Phycophilline* is a red or blue pigments present in Blue green and red algae
- Light reaction is occurs in *grana* and also called Hill reaction
- Dark reaction occurs in stroma of chloroplast – C_3, C_4 and CAM
- *C3 cycle* also called as Calvin cycle/Blackman reaction/ reductive pentose phosphate pathway
- 1 ATP is equal to 7.6 K calories
- 36 ATP is equal to 686 Kcalories
- 1 $NADH_2$ is equal to 3 ATP molecules
- 1$FADH_2$ is equal to 2 ATP molecules
- Energy currency of the cell is ATP
- Both starch and sucrose are synthesized from *Triose Phosphate*
- *GTP* is important for Cellulose biosynthesis
- The element involved in tertiary structure of a protein is Sulphur
- The enzyme related to elemental N fixation in Rhizobium is Nitrogenase
- The process of transfer of energy from carotenoid to chlorophyll is called as Inductive resonance
- Nitrogen assimilation is through GOCAT pathway
- The enzyme accumulated in anaerobic condition is dehydrogenase

- Number of ATP molecule required to fix one molecule of nitrogen by Nitrogenase enzyme of rhizobium is 12
- Symbiotic bacteria rhizobium consumes 2 per cent of the host photosynthates for N fixation land plants allocate 50 per cent of leaf nitrogen to Rubisco
- **FAD** Which transfers electrons during synthesis of ATP in cellular respiration?
- Co-factor required for transamination reaction is **Pyridoxyl phosphate**
- The enzyme **Rubisco** is having very low catalytic power
- The legume crop contains large amount of protein is **Soybean**
- **^{12}C** is the stable isotope
- **Sucrose** is the carbohydrate translocated from leaves to grains and fruits
- **Cellulose** is the single most abundant organic molecule in the biosphere
- Unsaturated fatty acid content is high in **Sunflower oil**
- The relationship between LAI and CGR is **Quadratic** in function
- In which crop senescence of phloem parenchyma limit the assimilate transport? **Maize**
- Crop having high WUE is **Maize**
- **Abiotic Stress responses in plants:** It is defined as the negative impact of non-living factors on the living organisms in a specific environment.
- **Proline** is an amino acid related to moisture stress
- Isozymes are having **Same structure with different activities**
- Mention the kind of stress which induces the synthesis of hsp **High temperature stress**
- Film forming type of Antitranspirent is **Wax**
- Plants tolerance to freezing injury is a consequence of alteration in **Lipid** composition of plasma membrane
- Mechanism in which plants will never allow any strain to moisture stress are called as **Escape**
- Sugar yielding crop which is highly tolerant to soil salinity is **Sugar beet**
- An example of polyhydric alcohol is **Mesoinositol**
- **Early seedling** stage of rice growth is very sensitive to salt stress
- Ozone present in the **Stratosphere** absorbs cosmic rays and protects the living organisms from short wavelength radiation of the sun
- **Tetrazolium** is used for seed viability test
- Give an example for epigeal type germination **Mungbean**
- Seeds have **hydrophilic** colloidal substances
- Chemical dormancy is due to the accumulation of **Phenolics** substances
- Physiologically active phytochrome is **Pfr**
- **Chrysanthemum** is an example for qualitative short day plant
- Pfr form of phytochrome inhibit flowering in **Short Duration Plant** species
- The auxin pumps are activated by a compound called as **Calmodulin**
- Optimum temperature for potato storage is **7.0 – 9.0 ° C**
- The scientist worked on senescence physiology is **Thimann**
- **Enzymes** are called as "agents of life"
- Non-protein part of the enzyme is called as **Co-factor**
- Catalysis occurs at **Active site**
- **Competitive** is the simplest and most common reversible inhibition

AGRICULTURAL EXTENSION 7

L. Muralikrishanan and P. Mooventhan

- Extension is an "**Out – of – School**" system of education for rural people.
- The term "**Extension**" have originated in **Latin**. It means **stretching out**.
- The term "**University Extension"** was first used in 1866.
- The person considered as **Father of extension** and the concept of **Extension Education Process** was given by **Paul Leagans.**
- The person considered as father of extension in India is **K.N.Singh**
- The term "**Extension Education**" was first coined in **United Kingdom**
- **Extension education** is an applied behavioral science and brings about behavioral changes in human beings.
- The philosophy of extension is not "What to Think?." It is **"How to think?"**
- Extension education is a **Non-Formal Education**.
- **Teaching:** Process of arranging / providing situation in which the solving
- **Learning:** Through his own activity to attains behavioral changes in better socialization process.
- Extension education is flexible in nature.
- The **Learners** should occupy the most important **central position in the extension educational process.**

Formal Education	Extension Education (Non-Formal Education)
Teaching is largely confined to the premises of the institution	It is largely outside the four walls of the institution
Learners are homogenous with common goals.	Learners are heterogeneous and have diverse goals.
There is a fixed curriculum, students are examined and degrees are awarded	No fixed curriculum, it is flexible depending on the needs of the learners. No examinations are conducted and no degrees are awarded.
Knowledge flows from teacher to the learners (vertical)	The extension worker also learns from those who he teaches (horizontal). He teaches through local leaders.

- The extension service in USA is called – **Co-operative extension service**
- Co-operative extension service is a consequence of **Smith Liver act -1914**
- In USA –Experimental stations come into existence as part of – **Hatch act- 1903**
- In USA – Land Grand colleges comes into existence as part of – **Moril act - 1862**
- Extension can be effective through **Voluntary change**

- Participation of people in extension programme is **Voluntary.**
- The fundamental objective of extension work is **Development of individual.**
- The basic unit of extension work is **Family**
- Most useful leader in Agricultural Extension is **Voluntary leader**.
- The cone of experience was developed by **Egder Dale**
- **Seaman A Knapp** conducted result demonstration in **Maize**
- The most preferred approach in extension work is **Democratic approach**

Extension Education: Extension is education and its purpose is to change the attitude and practices of people with whom the work is done. The term extension has its origin in the Latin word, *tension* meaning is stretching and *ex,* meaning out. The literal meaning of extension is "***stretching out***". The common use of the term, university extension, was first recorded in the 1840s, in England. Cambridge University, England formally adopted the system in 1873, and was followed by London University in 1876 and Oxford University in 1878.

7.1. RURAL DEVELOPMENT PROGRAMMES

Programme	Initiator	Year	Remark
1. Famine Commission	GOI	1880	To overcome the famine
2. Famine commission (Bengal & Orissa)	GOI	1901	Agricultural Research Institute (1905) and college in PUSA at Bihar
3. Scheme for rural reconstruction - Sandak (Bengal)	Sir Daniel Hamilton	1903	Creating model village Cooperative area
4. Economic conference of mysore (Karnataka)	Dr. Visweswaraya	1914-18	Agri hub based
5. Gurgon project (G.D. Punjab)	F.L.Brayne	1920	Abolition of purdha clean village guides
6. Sriniketan project (Bolepur, Calcutta)	Rabindranath Tagore/Elmher	1921	Village shouts (brati balika)
7. Sellagram (Wanda dist, MP)	Gandhiji	1921	Decentralization "Nei Talim"
8. Marthadam project (Marthandam thavancore)	Spencer harch (YMCA)	1921	5 folded programme (S,M,P,E,S) Demonstration
9. The royal commission report		1928	ICAR
10. Govt of Indian Act (UP)		1935	Village Level Worker (VLW)
11. Indian Village Service (IVS) (UP)	Arthur.T.Mosher	1945	IVS college
12. Firka Development scheme, Madras (34 firka than 50)	Shri Prakasam	1946	Gram sevaks
13. Grow More Food Campaign (GMFC)	Staffs of Dist & Taluk	1947	To agri production
14. Etawah pilot project , Mahewa (UP)	Albert Mayar (USA)	1948	Self helpForever of CDP

15. Nilokheri project (Nilokheri)	S.K.Dey	1948	Mazdoor Manzil
16. Sarvodaya movement (Bombay)	Vinobara Bavey	1948-49	Gandian concept
17. GMF - Enquiry report by V.F. Krishnamoorthy	Government	1952	Reorganized staffing pater booi, TO-4, VIW-12
18. CDP - Community Development Project (55 projects)	Indo-American TEch. Co-op. Agreement - 1952	1952	Nad-obj-res
19. National Extension Service (NES) CDP Community projects	Block Level	1953	Agricultural productionPaul Hoffman visit (Ford fdn.) 1951-15 pilot project
20. Community Development Block (CDB)	BDO, EO,VLWs	1954	People participation 40-50m in sq. mile area
21. Panchayat Raj"Democratic Decentralization" - 1st in Rajasthan	"Balwant Raj Metha" committee	1957	3 tier systemZP, PS, GP
22. Nalagarh committee report (Agri. Admn. Committee)		1958	
23. Intensive Agricultural Dist. programmme - IADP, "Package programme"	I - 7 districts II - 9 districts (1963-64)	1960	30% increase agricultural production
24. Intensive Agricultural Area Programme (IAAP) Revised version of IADP	150 districts (blocks)	1964	20-25% agricultural production
25. The agricultural administrative team report		1967	
26. Rural credit inquiry committee	Headed by S. Venkatapaiah	1969	

First Line Extension System

The main 4 transfer of technology project of ICAR (section of Ex-education-1971)

1. All India Coordinated Project on National Demonstration (AICPND) - 1964
2. Operational Research Project (ORP) - 1974-75
3. Krishi Vigyan Kendra/Farm Science Centre (KVKs) – 1974 (First in Pondicherry, TN)
4. Lab to Land Programme (LLP) – 1979 - (In ICAR golden jubilee celebration)

Front Line Demonstration : The field demonstration conducted under the close supervision of the scientist of National Agricultural Research System (NARS).

Community Development (1950's)

1. Community Development Programme - 1952
2. National Extension Service - 1953
3. Community Development Block - 1954
4. Panchayat Raj - 1957.

Technological Development (1960's)

1. Intensive Agricultural District programme - 1960
2. Intensive Agricultural Area programme - 1964
3. Intensive Cattle Development Programme - 1964 (KVP-Key village programme - 1952)
4. High Yielding Variety Programme - 1966 (AIR)

Development of Social Justice (1970's)

1. Small Farmer Development Agency (SFDA)- 1970-71
2. Drought Prone Area Programme (DPAP)- 1970-71
3. Pilot Project for Tribal Development - 1972-73
4. Training and Visit (T & V) system - 1974
5. Integrated Rural Development Programme (IRDP)- 1978-79
6. Training of Rural Youth for Self-Employment (TRYSEM) - 1979
7. National Rural Educational Programme- NREP - 1980
8. Development of Women and Children in Rural Areas (DWCRA) - 1982

Recently implemented schemes

- Make in India - 25 September 2014
- Swachh Bharat Abhiyan-2 October 2014
- Digital India- 1 July 2015
- Skill India- 16 July 2015
- Beti Bachao, Beti Padhao Yojana- 22 January 2015
- Deen Dayal Upadhyaya Gram Jyoti Yojana (DDUGJY)- 25 July 2015
- Deen Dayal Upadhyaya Grameen Kaushalya Yojana (DDUGKY)- 25 July 2015
- The National Institution for Transforming India (NITI AAYOG)- 01 January 2015
- Shyama Prasad Mukherji Rurban Mission- 21 February 2016
- Startup India, Standup India- 16 January 2016
- Pradhan Mantri Fasal Bima Yojana (PMFBY)- 11 October 2014
- NICRA- National Initiatives for climate resilient Agriculture
- FFS – Farmers field school
- TPDS- Targeted Public Distribution System launched in 1997 for BPL – Below Poverty Line Peoples.
- National food security mission -2007

Recent Programmes

1. **Rashtriya Krishi Vikas Yojana (RKVY):** Launched in 2007. It aims at achieving 4% annual growth in the agriculture sector during the XI Plan period, by ensuring a holistic development of agriculture and allied sectors.

2. **Mahatma Gandhi National Rural Employment Guarantee Scheme (MGNREGA):** Launched in 2006. It aims at enhancing the livehood security of people in rural areas by guaranteeing hundred days of wage-employment in a financial year to a rural household whose adult members volunteer to do unskilled manual work.

3. **Gramin Bhandaran Yojna:** Launched in 2007. The main objectives of the scheme include creation of scientific storage capacity with allied facilities in rural areas to meet the requirements of farmers for storing farm produce, processed farm produce and agricultural inputs. Also, promotion of grading, standardization and quality control of agricultural produce to improve their marketability. The scheme also aims at prevention of distress sale immediately after harvest by providing the facility of pledge financing and marketing credit by strengthening agricultural marketing infrastructure in the country.

4. **Bharat Nirman (2005):** Bharat Nirman, a programme to build rural infrastructure, was launched by the Government of India in 2005.

5. **Pradhan Mantri Adarsh Gram Yojana(PMAGY) :** It is a new scheme launched in 2009-10 by the Union Government on a pilot basis for integrated development of 1,000 villages, each having more than 50% SC population.

6. **National Rural Health Mission (NRHM):** It is launched in 2006 as the central government flagship project that would dramatically change the health care system in rural India.

7. **Sarva Siksha Abhiyan(SSA)** : It is launched in 2001 as a flagship programme of the Government of India pioneered by Atal Bihari Vajpayee for achievement of universalization of elementary education in a time bound manner.

8. **Sampoorna Grameen Rozgar Yojana (SGRY):** It is launched in 2001 with an aims to provide addition wage employment, improved food and nutritional level. It was implemented by Panchayat Raj Institution. Central Government provided the grains free of cost.

9. **Rural Housing - Indira Awaas Yojana (1996):** Ministry of rural development give full support to the Housing and Urban Development Corporation Limited (HUDCO) to construct house for poor people at free of cost. It is sub-scheme of FLEGP

10. **Pradhan Mantri Gramodaya Yojana (2000-2001) :** Allocation of additional central assistance (ACA) to States and UTs to improve the primary health, education, shelter and drinking water facilities.

11. **Rural Employment Generation Programme (1995) :** Creating self-employment opportunities

12. **Prime Minister Rozgar Yojana (1993) :** Self employment for rural educated unemployment youth.

13. **Pradhan Mantri Gram Sanskar Yojana (2000) :** Improve the rural connectivity under central road funded by National Rural Road Development Agency (NRRDA)

14. **Antyodaya Anna Yojana (2000) :** To provide food grains at Rs.2/kg (wheat) and Rs.3/kg (rice)

15. **Valmiki Ambedkar Awas Yojana (2001) :** For construction and up gradation of dwelling units for the slum dwellers.

16. **Kisan Credit Card Scheme (1998-1999) :** Providing crop loans to farmers in a flexible and cost effective manner.

Training and Visit (T&V) System: This through system was introduced in 1974 with the World Bank assistance. It was presumed that transfer of technology trough the 'contact farmers' shall benefit all farmers. This system was introduced in Rajasthan Canal area in Rajasthan and Chambal Command area in Madhya Pradesh. It originated in Turkey and the concept was given by Daniel Benor.

Administrative set up of T & V system.

- 8 Village Level Workers (VLWs) was supervised by AEO
- 8 Agricultural Extension Officer (AEO) was supervised by Sub-division extension officer (SDEO)/ district extension officer (DEO)
- AEO assisted by 3 to 5 number of subject matter specialist (SMS)
- Each VLW cover 800-1000 farm families
- The farmers divided into 8 groups (equal size)
- Contact farmer visited by VLW by every two ways on a fixed day
- Every two weeks, VLW, receive training by SMS
- Every month, SMS contact with research person

Key features of T&V system

1. Professionalism
2. Single line of command
3. Concentration of effort
4. Time-bounded work
5. Field and Farmer orientation
6. Regular and continuous training
7. Linkage with research

7.2 COMMUNICATION

Communication derived from Latin word. It is the discriminatory response (Stevens - 1942) or Communication is the mutual interchanging of ideas (Thayer - 1968). Sharing of ideas, facts, opinions, information and understanding are referred as communication.

- ✍ The instrument use by extension to bring change is **communication**
- ✍ The most basic function of communication is **Influence**
- ✍ The essence of communication is **sharing and understanding**
- ✍ The art of communication is **listening**
- ✍ To get first-hand knowledge on a farm situation and most effective method of transfer of technology in a community is possible through **Farm & Home visit.**
- ✍ Result demonstration is conducted by – **farmer in Farmer's field under the direct supervision of the extension worker.**
- ✍ Media forum is classified under – **Group contact method**
- ✍ Result demonstration works on the **principal of seeing is believing**
- ✍ **Front line demonstration:** Conducted by research worker farmers field
- ✍ Method demonstration works on the **principal of learning by doing**
- ✍ **Method demonstration:** To show the technique of doing things

- ✍ The method which build the confidence of the researcher, extension worker and the farmer – **Minikit trial**
- ✍ A symbolic display of models, specimens, charts in a sequence around a theme is called **Exhibition**
- ✍ The major drawback of mass media in context of extension work is **Lack if free feedback**
- ✍ Trust worthy ness and competency are elements of – **Credibility**
- ✍ The most important element in the communication process is **Receiver**
- ✍ Acceptance and rejection are the one of the kind of **Feedback**
- ✍ The process through which we arrive at expectations and anticipations of internal psychological state of mind –**Empathy**
- ✍ **Communication gap** refers to the difference between what was communicated by the extension agent and what has actually been received by the audience.
- ✍ Communication has four basic functions of **Information, instructive, persuasive and integrative** functions.
- ✍ Mass media is generally identified with modern mass media like **Television and Radio**
- ✍ Poster/ Mass communication devices are most effective at **awareness stage.**
- ✍ A three dimensional recognizable imitation of an object is **Model**
- ✍ The communication model suited for extension education is **Leagans model**
- ✍ The SMCR Model of communication is given by **Berlo**
- ✍ LCD – Liquid Crystal Display
- ✍ The ABC of Journalism is related to **Accuracy, Brevity and Clarity**
- ✍ Comprehensive information about the topic is **Bulletin**
- ✍ The first farm magazine in India is **Kethi**
- ✍ The first newspaper started in India was – **Bengal Gazette**

Communication models

1. Aristotle's model of communication - Persuasion model
2. Lassusells communication sequence - Rudimentary model
3. Stimulus - response model by New Comb
4. Schram's model - Field of experience
5. Shannon and Weaver's model - Mathematical model for electronic media
6. Berlo Model - S-M-E-M-C-D-R
7. Leagan's model - Model for extension communication
8. Hovland precession model - For attitudinal change
9. Westley and Macleans - Mass Media Model
10. Liley and Liley Model - Mass communication as one social system.
11. Gerbner's Model - perception
12. Theories of mass media effect by Defleur (B)
13. The medium is the message - Mc Luhan

Four levels of communication by Thayer (1968)

1. Intra-personal : Within one's self
2. Inter-personal : With others
3. Intra-organizational : Within organisation
4. Inter-organizational : With other oganisation

   Systems : Operation, regularity and maintenance and development.

- Change agent is Extension agent
- Message as the 'physical product" of the source and encoder
- Receiver is the most important element in communication process

Nature of communication

1. Communication is the process
2. Communication employs many means
3. Communication involves interdependence

Flow of communication

1. Down-ward
2. Up-ward
3. Horizontal
4. Diagonal

Communication effectiveness:

- Communication encounter: Achieving "commonality" between sender and receiver by comprehension (clear)
- Communication performance: Measure the success of any communication in term of subsequent behavior or receiver.

Effectives: Comprehension, Validity and Utility (E = C+V+U)

Validity: Consonance, credibility and congruity

Channel Capacity: $C = W\log_2 \frac{P+N}{N}$ bits/sec

P → Average signal power
N → Average noise power
W → Band width
C → Maximum channel capacity in cycle/sec.

Dimensions of message - By Berlo (1960)

1. Message code - Symbols/structure, used in military
2. Message content - Material selected by source
3. Message treatment - Selecting and arranging both code and content for comprehension to receiver.

Classification of communication channel

1. Inter personal : Face-to-face type
2. Mass Media : Radio, TV (Impersonal cosmopolite)
3. Localite : Personal localite (opinion / local leader)
4. Cosmo polite : Personal cosmopolite (extension agent)

- **Inter personal** : Face-to-face, two-way, "Attitude formation effect"
- **Mass Media:** Interposed, one-way and "knowledge change effect"

Extension teaching/ Communication methods

1. Individual contact

a) Farm and home visit: Direct, face-to-face contact by extension agent with farmer in field / home.

b) Farmer's call: Call made by farmer

c) Personal letter: Written by extension group to particular farmer.

e) Farm clinics: Facility developed and extended to the farmers for diagnosis and treatment of farm problems and provides advice to farmers

f) Flag method: Latin American countries (Kenya)

2) Group Contact

a) Result demonstration: Method of motivating people for adoption of new technologies by show in its distinctly superior result. Conducted by farmers participating own home with the superficial guidance of extension agent. It is "Trust worthiness" oriented.

b) Method demonstration: Is given before group of people to show how to carry out an entirely new practice / old practice in a better way. Conducted by extension agent and it is skill oriented.

c) Group Meetings: Democratic way of arriving decisions by a group of people

d) Small group training: Technique of imparting specific skills to a group of people under user based.

e) Field day / Farmer's day: Method of motivating people to adopt a new technology by showing what has been actually achieved by applying the practice under field condition.

f) Study tour: Group of interested person accompanied with extension agent moves out of neighborhood to study and learn significant improvement in farm / outside Motivating people by showing what others have been able to achieve and Group meeting and group discussion is the oldest and important method of extension teaching

g) Peripatetic team visit

- This method frequently used PAV and other farmer training centre with the help of 3-5 SMS.
- Visit village with pre-drawn schedule, presentation on "some timely topic"
- Question asking to farmer. "on-the-spot" advice given to the farmers
- More emphasis on answering question and organizing debates.
- It is different from group meeting and group discussion.

Mass Communication Methods

1. Farm publication: Prepared by extension agent.

   a) Leaf let - Single printed sheet

   b) Folder - Single printed sheet (big) folded once/twice

   c) Bulleting - Booklet

   d) Journal, Magazine

2. Mass Meeting: To communicate interesting and useful information to a large audience at a time

3. **Campaign:** Intensive "educational activity" for motivating and mobilizing a community to action to solve a problem urgently felt by it.

4. **Exhibition :** Systematic display of models, specimens, chart, photo, picture poster etc in a sequence around a theme to create awareness and interest in the community.

5. **Newspaper :** Bunch of printed paper properly folded which contain news.

6. **Radio :** Rural and farm broadcasting, field recording, delivery talk, discussion, rural radio forum, farm school on All India Radio (AIR)

7. **Television :** Farm telecast.

8. **Farmer's fairs / melas :** Part of outculture.

9. **Circular letter:** Intimate letter by ex-agent to large farmer distributed periodically.

Media Forums (1956): (Combining "mass media channels" with inter personal channels)

- Media forums are organized small groups of people (12-15) who met regularly receive a mass media programme and discuss its contents (Radio, radiophonics, TV, Vide Cassettes, Audio tape, etc.)
- Attitude change and group decisions are made
- First in "Canada", In India " Charcha Mandal.

Farm Telecast

- The service was started by AIR, Delhi and programmes were telecast twice a week for a duration of one hour each a day. Pilot project of agricultural communication initiated by Dr. Vikram Sarabhai (1967). Program titled Krishi Dharshan
- Aug, 15 1962 considered as a landmark in history of TV in India, TV networks linked with satellite INSAT-1A and colour TV also introduced by this year. Government set up Doordarshan (April 1, 1976) **94% by seeing and hearing**
- Panjab Agricultural University (PAU) gave first pre-coded audio cassette in 1985 during "Kisan Diwas"

Audio-Visual (AV) aids

- It is the multi-sensory approach
- Most of the information learned by seeing and hearing
- According to Hass and Packer (1964)

  1. Sight - 87.01%
  2. Hearing - 7.0%
  3. Smelling - 3.5%
  4. Touching - 1.5%
  5. Tasting - 1.0%

Classification of AV-aids by evaluation:

A) I generation media - Hand made chart, graph, model, hand-written etc.

B) II generation media-Printed text illustration graphs

C) III generation media- Photo, slides, film-strip, film, radio

D) IV generation media-TV, Computer, programmed instruction language laboratory

Classification according to sensory involved

A) Audio aids

1. Tape recorder
2. Radio
3. Public address system (PAS)
4. Telephone

B) Visual Aids (Projected)

1. Slides
2. Film strip
3. Opaque projection
4. Over head projection

C) Visual Aids (Non - projected)

1. Chalk board, translide
2. Bulletin Board
3. Picture, photo, poster
4. Diagram, map, chart, graph
5. Specimen, model drama

D) Audio Visual (Projected)

1. Motion picture
2. Video

E) Audio-visual (Non-project)

1. Drama
2. Puppet
3. Talking doll

Audio Aids

1. **Tape Recorder:** Equipment for recording sound on magnetic tape by electro-magnetic process
2. **Public address system:** Is a set of equipment to amplify sound so that it is audible to large audience over a distance.
3. **Telephone:** People can communicate both ways to distance places.

Visual aids

- Specimen - Is a sample which represent the whole
- Model - Is a miniature replica of an object
- Diorama - Is a scenic representation of a original with specimen, model and picture.
- Graphic Aids (Drawings + text + picture)
- Flash cards: These are the series of cards which when present before the audience in proper sequence, tell a complete story, size of cards is 10" x 12", Range of group is 20-30, and 30-45 cm (ICAR)
- Flannel Graph: This is a board with a surface of felt, flannel (or) khadda cloth which is used for fixing picture on the back of which there is piece of sand paper to stick. Presenting detailed concept. Board provide action like effect. Most adoptable and flexible.

Charts: Letter size for chart is 2.2" for title, 1.5-2" for sub-title, 1"-2" for body

- **Flow chart**: Cycle of plant growth
- **Outline chart**: The organization of content into key points and sub-points.
- **Tabular chart**: Ability to show time relationship.
- **Organization chart**: Eg. Vice chancellor to assistant professor
- **Letter size for chart**: 21.5" Title, 1.5-2" - Sub title, 1"-2" for Body

Graphs : Visual representation of numerical data.

- **Line Graph**: Precise and accurate of all graphs and it shows 'trend" and relationship between two data.
- **Bar graph** is a simplest of all graphs to read
- **Pie graph** is in circle/pie and divided into sectors
- **Pictorial graph** is representational figure, population, wheat prodution
- **Area and solid figure graph** is represented bycomparison of area
- **Bar graph** followed by line graphs best for farmers

Poster

- Modern poster born in 1867, Paris (France)
- First great modern poster artist, Jules Cheret
- Used for creating **"awareness"**
- It has characteristic of brevity (not more than 5 words) simplicity, idea, layout, colour
- Size 20" x 30" not more less than 50cm x 75cm

Type of projection

1. **Direct Projection :** Slide projectors, film-strip LCD projectors and Film projectors
2. **Indirect projection:** Overhead projector - OHP projectors
3. **Reflective projection** - Opaque projector.

Mock-up

- 3D Imation of a thing, it is only for learning purpose
- It may not be similar in appearance.
  Eg. Clock in schools showing movement of hands / mock-up of driver's seat in training school.

Note

- Elements of learning situation given by Paul Leagans
- Cone of Experience given by Edgar Dale's
- Extension Education Process given by Paul Leagans - 1967
- Steps in extension teaching given by Wilson & Gallup
- The spread of technology in a social system is **Diffusion**
- The desire to make best course of action is **Adoption**
- The idea which is perceived as new is termed as **Innovation**

- **Time** is a very important consideration in the process of Diffusion
- **Frequency of adopters** when plotted over time follows a Normal curve
- **The rate of adoption** of an innovation in a social system over time follows S shape curve
- The degree to which two or more interacting individuals are **similar** in certain attributes is **Homophily**
- The degree to which two or more interacting individuals are **dissimilar** in certain attributes is **Heterophily**
- **Adopter categories** are the classification of the members of the social system on the basis of **Innovativeness**
- The adopter category considered as deviant by other members of the society is **Innovators/ Venturesome (2.5%)**
- **Opinion leaders/ Legitimizers** belong to the category of **Early adopters (13.5%)**
- **Last stage** in the adopter categories are **Laggards/ Traditional (16%)**

7.3. RURAL SOCIOLOGY AND IMPORTANT CONCEPT

Rural sociology

It is a field of sociology associated with the study of social life in non-metropolitan areas. It is the scientific study of social arrangements and behaviour amongst people distanced from points of concentrated population or economic activity.

The word *sociology* (or *"sociologie"*) is derived from both Latin and Greek origins. The Latin word: *socius*, "companion"; *-ology*, "the study of", and in Greek *lógos*, "word", "knowledge". It was first coined in 1780 by the French essayist Emmanuel-Joseph Sieyès (1748–1836)

- Father of rural sociology - August Comet
- August comet coined "sociology" (1842)
- Sociology refer "theory of companion"
- Sociology is a "science" notes in social philosophy
- Sociology is the **science of Human relationship**
- Sociology is the scientific study of man's behaviour in relation to other groups and individual, with as whom he interacts.
- Human relationship and human interaction are crux of the sociology
- Rural sociology is a specialized field of sociology.
- Positive philosophy given by Auguste Comet (1798-1857)
- Principles of sociology by Herbert spencer
- Dynamic sociology given by Lester ward
- Crime and urbanism given by Fredinand Tonnies
- Transportation, race relation and social growth (Cooley Williams)

✍ **Formal groups** are organized, have definite roles, rules of operation of its members

✍ **Informal groups** are **not** formally organized, have definite roles, rules of operation of its members

✍ **Community** refers to a group of mutually dependent people, living more in compact continuous geographical area.

- **Social system** is a set of inter related units that are engaged in joint problem solving to accomplish a common goal.
- Family, Religion, Government, Economy and education are the **social institutions**.
- **Culture** is learned through the process of socialization.
- **Ethnocentrism** is the tendency of the people to consider their own culture of high value and superior to all.
- **Norms** are the established behavior patterns for the members of a social system.
- **Mores** are socially acceptable ways of behavior
- **Attitude** as a mental state of readiness
- **Personality** is the unique, integrated and organized system of all behavior of a person
- **Stereotypes** are fixed images formed in one's mind about people practices.

Management and Programme planning

- A blue print for action is **Plan**
- The plan of work when arranged chronologically is **Calendar of work**
- The specification of work to be done in order to accomplish particular objective – **Project**
- **Logical framework technique** is greatly used in Project planning
- **"POSCORB"** as management function was developed by **L. Gulick**
- The term "**entrepreneur**" is coined by **Richard Cantinlon**
- Opinion leader is the degree to which an individual is able to **influence other individual's attitudes.**
- **Entrepreneur is an innovator** who invariably links the innovation with the market to earn profit.

Demographic change

- **Population change** $P_2 = P_1 + Ni + Nm$
  - Immigrants - Enter
  - Emigrants - Leave (Difference of two is not migration)
- **Rate of Natural increase** = $\frac{\text{Birth year - death year}}{\text{Total population}} \times 100$
- **Birth Rate** = $\frac{\text{No. of live birth /year}}{\text{Total population}} \times 100$
- **Death Rate** = $\frac{\text{No. of death /year}}{\text{Total population}} \times 100$
- **Age and sex ratio** = SR = $\frac{\text{Males}}{\text{Females}} \times 100$
- **Life expectancy** = Average life span

Case of system

- Derived from "Portuguese" word casta meaning linege/race
- A case is a social category whose numbers are assigned a permanent status within a given social hierarchy and whose contracts are restricted accordingly.
- Most rigid and clearly graded types of social stratification.
- "Caste" referred as a "Extreme form of closed class system".

IMPORTANT CONCEPTS

Innovative extension Approaches

- The person considered as father of PRA (Participatory Rural Appraisal) – **Robert Chambers**
- The term PRA was first used in **Kenya**.
- The collection of data in PRA is based on **Multi-Disciplinary team**
- The PRA technique that shows **cause effect relationship – Flow diagrams**
- The PRA technique that represent the degree of importance of individual and institutions in decision making - **Ven diagram**
- The technology flow in the **farmer first model** is basically from farmer to farmer
- Participatory development approach/ Farming System Approach is otherwise called as a **Bottom up approach**
- **The participatory extension approach recognizes** the importance of Learning from farmers
- The most important factor in Farming System Research/ Extension is **Farmer**
- The general micro unit of an agro ecosystem – **Village**

Innovative Extension approaches in India

- Institute Village Linkage Programme (IVLP) uses the concept of **Agro-eco system analysis**
- ATMA expands as – **Agricultural Technology and Management Agency**
- ATMA is a **registered society** operates at **District level.**
- **Farmers information and advisory centers under** ATMA operates at Block level
- The apex training institute at state level to give training support to ATMA – **SAMETI**
- SREP stands for **Strategic Research and Extension Plan**- it is important component in the functioning of ATMA
- ATIC expands to – **Agricultural Technology Information Centre**
- The use of computers in technology transfer – **Cypher extension**
- **The e- Chaupal** concept in extension delivery is promoted by – **ITC**
- The network that connect agricultural produce wholesale markets and rural markets - **AGMARKNET**
- The toll free telephone number on Kissan call centre is **-1800-180-1551 or 1551**
- **"Rice Doctor"** an interactive pest diagnostic expert system is developed by – IRRI
- GIS- **Geographical Information System**
- **E – Governance** is the use of information communication Technology (ICT) by the government to **engage citizens through feedbacks and governance**
- **Boomi project** - Land records in Karnataka
- DTP – **Desktop Publishing**

Critical path method (CPM)

The critical path method (CPM) is an algorithm for scheduling a set of project activities. It is an important tool for effective project management. CPM developed by Morgan R. Walker of DuPont and James E. Kelley, Jr. of Remington Rand.

1. For solving schedule problems in an industrial setting.

2. It is more concern with costs of project scheduling and how to minimise them.
3. It is the project management technique useful in the basic management function of planning, scheduling and control.
4. It not concerned with certain job times.
5. It deals with time-cost trade-offs
6. Mostly used in project construction
7. Obtaining the trade-off between cost and completion date for large projects
8. It is deterministic rather than probability
9. The time-cost trade-off for an individual activity is linear, and its activity-oriented.
10. 10.CPM specifies a method of finding out this optimum point, the least cost schedule

PERT

The Program (or Project) Evaluation and Review Technique (PERT), is a statistical tool, used in project management, that is designed to analyze and represent the tasks involved in completing a given project. First developed by Booz Allen Hamilton and the United States Navy in the 1950s, it is commonly used in conjunction with the critical path method (CPM).

1. Mostly used in research and development project
2. Find out uncertainties with the use of Three time estimates (optimistic, pessimistic and more likely)
3. Find out specific activities and their network relationship need for planning development of new technologies.
4. PERT calculate the expected value of an activity duration as a weighted average of three time estimates.
5. PERT also calculated the variability in the completion of an activity on the basis of SD.
6. PERT is event oriented.

Factors favoring for privatisation of Agricultural Extension Services

1. Difficult in getting fund from government to extension machinery.
2. Reduced donor contribution to Agri. Term and extension
3. Poor quality of service provided by govt. agencies.
4. Global competition after GATT agreement
5. Transformation of Agriculture from subsistence to profit oriented one
6. Department of Private Research and Extension set-up with permits technology.

Agricultural Technology Management Agency (ATMA) - 2005

- It is one of the innovations technology disseminations components of (National Agricultural Technology Project) NATP.
- District level service focus on interpreting research and extension activities to help in decentralism the management of agricultural technology transfer
- Members are Department of Agriculture, Animal Husbandry, Horticulture and Fisheries

Objectives

1. To strengthen "Research - extension -Farmer' linkage

2. To develop new partnership with the private institutions including NGO.
3. To provide an effective mechanism for coordination and management. Activities of different agencies involve in technology adoption and discrimination at distinct level.
4. To increase the quality and type of technologies being determinate
5. To move towards shared ownership of the agricultural technology system by share holders.

Farming system's reserve/extension FSK/E evolved in 1970's

Character

1. Farmer based
2. Problem solving
3. Comprehensive
4. Inter disciplinary
5. Complementary
6. Interactive
7. Dynamic
8. Responsible approach to society

Participatory Technology Development

- PTD is a process of bringing together the local communities with their own knowledge and capacity for research on one plant form with the researchers and other professionals.
- PTO process is aimed at strengthening the "existing experimental capacity of farmers' in search of sustainable innovation-generation system.
- ITK gives farmers equal status as on expert steps in PTD:

Steps

1. Build rapport with farmers
2. Identification of ITK
3. Joint collaboration
4. Sharing results
5. Sustainability and consolidating the system

The Farmer Field School (FFS) is a group-based learning process that has been used by a number of governments, NGOs and international agencies to promote Integrated Pest Management (IPM). The first FFS were designed and managed by the UN Food and Agriculture Organisation in **Indonesia** in 1989 since then more than two million farmers across Asia have participated in this type of learning.

Kisan Call Centres: (Toll free number - (1800-180-1551) The Department of Agriculture & Cooperation (DAC), Ministry of Agriculture & Farmers Welfare, Govt. of India launched Kisan Call Centres on January 21, 2004 across the country to deliver extension services to the farming community. The purpose of these call centres is to respond to issues raised by farmers, instantly, in the local language. There are call centres for every state which are expected to handle traffic from any part of the country. Queries related to agriculture and allied sectors are being addressed through these call centres.

Village Knowledge Centre (VKC) serves as information dissemination centre providing instant access to farmers to latest information/ knowledge available in the field of agriculture, starting from crop production to marketing.

A "VKC In-charge" who looks after the operations of the VKC mans every VKC. It was initiated by M.S.Swaminathan Research Foundation in 1998 at Pondicherry in South India.

Points to remember

- TV introduced in India – 1959
- Fiscal commission by GOI in 1949
- Minikit programme for rice - 1971
- Minimum need programme - 1972
- National Demonstration Project - 1965-66
- Doordarshan established in 1976
- SITE established in 1975-76
- Krishi Darshan initiated by Vikram Sharabai in 1967
- INSAT-1A + TV network linked in 1982
- Farm and home unit of (Akasaani) - 1966
- Pre-coded audio cassette released by PAU in 1985
- Media Forum established in 1956
- National Planning Commission in 1950
- National Development Council in 1952
- First central agricultural university in 1933 at Manipur (Imphal)
- Manage set up in 1987 at Hyderabad
- ICAR first involve in Extension activities in 1966 with National Demonstration Programme.
- National Commission on Agriculture - 1976
- National Commission on farmers - 2004
- National Food for Work Programme - 2004
- The Technology Mission on Oilseeds -1986
- The Technology Mission on pulses -1990-91
- The Technology Mission on Oilpalm – 1992
- The Technology Mission on Maize - 1995-96
- Accelerated irrigation benefits programme -1996-97
- National commission on farmers -2004
- ICAR established in July 16, 1929. The president of ICAR is Union Agricultural minister
- ICAR is a registered cooperative
- ICAR was established based on the recomendation of the Royal commission -1928
- DARE was established in 1973
- MANAGE expands to National Institute of Agricultural Extension Management – established in 1986.
- The first agricultural university was established - G.B. Pant university at Pantnagar -1960

- The National Demonstration Scheme (NDS) by Kalwar and Subramanyam
- The number of SAUs in India – 74
- The first line system of extension in India is controlled by – ICAR institutes / SAUs
- The first KVK was established in Pondichery in 1974
- The funding of KVK (Farm Science Centre) is done by ICAR
- Mandates of KVK are Front line demonstration, On farm Testing and Vocational trainings
- The integration of TOT programmes of ICAR with KVK happened in 1992
- NARS- National Agricultural Research System
- TARP- Technology Assessment and refinement Project
- NATP was funded by – World Bank
- TRIPS- Trade Related Intellectual Property Rights
- The Article under WTO that deals with agricultural sector trade reforms – Article 20
- All domestic support measures considered to distort production and trade falls in – Amber box
- The people under BPL in India -28%
- India's Position in the UNDP's Human Development Report for 2011 is -131th Rank
- NABARD- National Bank for Agriculture and Rural Development established in 1982
- The chairman of the district rural development agency is District Collector
- The right to information act was passed in the year -2005
- International women day –March 8
- Agriculture was made a state subject through Government of India Act (1919)
- The T&V system/ Single line commend of extension was introduced in India - Danial Benor
- Operation flood was launched in 1971
- Total number of KVKs in India – 634

Category : Collection of items that have at least "one common Characteristic"
Eg. General, OBC, SC, ST, 2004 - B.Sc. (Ag) Batch students

Aggregation : Collection of items in Physical proximity of one another
Eg. Cinema Theatre

Customs : Socially prescribed form of behaviour
Folkways : Approved form of behaviour/Informal rules, no moral standards
Society : Lived together, sharing common values, general principles
Group : Reciprocal-communication and Interaction.
Community : Mutually dependent, Geographic area, norms, values, common interest.
Social system : Interrelated units, joint problem solving
Social stratification : Hierarchically arranged strain in a community
Cooperation : Working together, common objectives, goals
Accommodation : Group/person accept each other
Assimilation : Fusing/blending, complete merging

Competition	:	Compete with each other to reach/survival of fittest, monocracies
Conflict	:	Struggle, try to eliminate, abolish, wipe out
Values	:	Expression of preference
Beliefs	:	Mental convictions
Mores	:	Acceptable way of behaviour/ involve moral standards
Norms	:	Established behaviour pattern/acceptable or unacceptable way of behaviour
Laws	:	Grow out of mores/ more complex
Organization	:	Group with special concern
Conventions	:	Customs regularly more significant behaviour
Social control	:	Pattern of influence to maintain orders
Culture	:	Pattern of learned behaviour
Attitudes	:	Mental state of readiness
Emotion	:	State of being moved
Prejudice	:	Pre-judgment/judgment before due examination
Stereotypes	:	Fixed image without any direct experience.
Crowd	:	Temporary collection of people reacting to stimuli
Contravention	:	Strive to prevent each other
Codes	:	Is a statement of policies, principles/rules that guides the behaviour
Policies	:	They are guides to decision making
Procedures	:	That is guided to action
Rules	:	It spell out specific required action
Objectives	:	Status end results
Strategies	:	It indicate the direction
Policies	:	It indicates the discretion (Freedom to decide).
Society	:	Is a group of people who have lived together, sharing common values and general interests, long enough
Attitude	:	Positive/negative effect associated with some psychological objects

AGRICULTURAL ECONOMICS 8

L. Muralikrishnan and S. Menaka

Economics is the science that deals with production, exchange and consumption of various commodities in economic systems. The word economics was derived from two Greek words, oikos (a house) and nemein (to manage) which would mean 'managing an household' using the limited funds available, in the most satisfactory manner possible.

8.1.CONCEPTS IN ECONOMICS

Economics definition

Adam smith (1776) defined economics as the science of wealth in his book An Inquiry into Nature and Causes of Wealth of Nations.

According to Marshall Economics is primarily a study of mankind in the ordinary business of life.

Agricultural Economics : An applied social science that deals with the production, distribution, and consumption of agricultural or farming goods and services.(or) an applied phase of social science of economics in which attention is given to all aspects of problems related to agriculture.

Subject matter of Economics : It can be studied through a) traditional approach and (b) modern approach.

a) Traditional Approach : Economics is studied under five major divisions namely consumption, production, exchange, distribution and public finance.

Consumption : The satisfaction of human wants through the use of goods and services is called consumption.

Production : Goods that satisfy human wants are viewed as "bundles of utility".

Exchange : Goods are produced not only for self-consumption, but also for sales. They are sold to buyers in markets. The process of buying and selling constitutes exchange.

Distribution : The production of any agricultural commodity requires four factors, viz., land, labour, capital and organization. The land owner gets rent, the labourer earns wage, the capitalist is given with interest and the entrepreneur is rewarded with profit. The process of determining rent, wage, interest and profit is called distribution.

Public finance : It studies how the government gets money and how it spends it.

b) Modern Approach : The study of economics is divided into: i) Micro-economics and ii) Macro-economics.

Micro-economics : The subject matter of economics dealing with individual (or) tiny aggregates of population is called microeconomics. Microeconomics is otherwise called as price theory.

Macro-economics : The subject matter of economics dealing with large aggregates (or) economy as a whole is called Macroeconomics. Macroeconomics is otherwise called as Income theory.

Keynes - Father of modern economics

There are two approaches in economics viz., i) cardinal and ii) ordinal.

The law of diminishing marginal utility, law of equimarginal utility, consumer surplus etc., falls under cardinal approach. According to cardinal approach, utility can be measured in terms of unit called *utils*. Indifference curve analysis falls under ordinal approach. According to ordinal approach, utility cannot be measured.

Economic holdings : The size of holding which provides full employment to the farm family and provides a reasonable standard of life is termed as economic holding

According to the Central Land Reforms Committee the economic holding size for a family (having 5 members) is

- 10 acres i.e. 4 ha. For irrigated land
- 27 acres i.e. 10.9 ha. For partially irrigated land
- 54 acres i.e. 21.85 ha. For non-irrigated land

On the basis of ownership of agriculture land holding, there are three types of common farmers in India:

- **Marginal Farmers**: Such farmers who have less than one hectare unirrigated land are called marginal farmers. Marginal farmers constitute 51% of the total farm holdings
- **Small Farmers**: Such farmers has one to two (1-2) hectare unirrigated land and constitute 19% of the total farm holdings in India
- **Agricultural Labour**: Such farmers are landless and having no any landed property

Holdings of farmers

Farmers	Land Holding (ha)
Marginal	Less than 1
Small	1 – 4
Medium	4 – 10
Large	Above 10

Agricultural credit

Loans	Purposes
Short term	For purchasing seeds, fertilizers and for meeting out family requirements
Medium term	To purchase agricultural equipments, animal and for land improvements
Long term	To purchase land and expensive agricultural equipments and for repayment of old loans

Finance according to period

Loan	Term (Year)
Short term	1 to 1.5
Medium term	1 to 5
Long term	5 to 30

Factors affecting shift in demand (Shift factors)

Increase in demand	Decrease in demand
Rise in income	Fall in income
Rise in price of substitute	Fall in price of the substitute
Change in taste favourable to commodity	Change in taste against the commodity
Increase in population (size of market)	Decrease in population (size of market)
Favourable redistribution of income	Unfavourable redistribution of income for the people who consume the commodity

Components of Assets and Liabilities

Liabilities	Assets
Current Liabilities • Creditors • Bills payable • Bank borrowings • Provisions for rent • Wages payable • Income received in advance • Installments of long-term loans • Deposits from public	**Current assets** • Cash • Marketable securities • Accounts receivable • Bills receivable • Inventory • Prepaid expenses • Loans and advances given
Intermediate liabilities • Livestock loan • Machinary loan • Unsecured loans	**Intermediate assets** • Dairy cattle • Bullocks • Poultry birds • Machinery and equipment • Tractor
Long-term liabilities • Deposits from public • Debentures • Secured loans • Equity • Share capital • Reserves and surplus	**Fixed Assets** • Tangible fixed assets • Intangible fixed assets • Long-term investments • Other non-current assets like deferred charges

8.2. AGRICULTURAL MARKETING

Marketing: It is a social and managerial process by which individuals and groups obtain what they need and want through creating, offering and exchanging products of value with others.

Agricultural marketing: It is the study of all the activities, agencies and policies involved in the procurement of farm inputs by the farmers and the movement of agricultural products from the farms to the consumers(Acharya and Agarwal).

Market structure

Market structure	Number of firms or producers or sellers	Degree of product differentiation	Firm's degree of control over price	Part of economy where prevalent
Perfect competition	Many sellers	Homogeneous product	No control over price	Farm commodities
Imperfect competition				
a. Monopolistic competition	Many sellers	Differentiated product	Some control over price	Retail trade
b. Pure oligopoly	Few sellers product	Homogeneous over price	Some control	Steel, chemicals etc.
c. Differentiated oligopoly	Few sellers product	Differentiated	Some control over price	Automobiles, computers, etc.
d. Monopoly	One seller close substitutes	Product with no control over price	Considerable electricity, etc.	Railways, posts,

- **Monopsony** is a market structure in which there is only one buyer instead of one seller.
- **Duopsony** is a market situation in which there are two buyers for the commodity.
- **Oligopsony** is a market structure in which there are only few buyers.
- **Bilateral monopoly** is a market structure in which a single seller faces a single buyer.

Conditions for the perfect market

- Large number of buyers and sellers
- Product produced by all sellers is homogenous
- Both sellers and buyers have perfect knowledge about price
- No legal or institutional restrictions on buyers and sellers
- A competition that is not pure is called Imperfect Competition. It exists in two forms

Monopolistic competition : Many producers but each produces and sells differentiated products. Each one has a monopoly on its own product.

Oligopoly : A few sellers, dealing in homogenous or differentiated product is small, complete interdependence among the sellers with regard to their price output policies.

There is extreme case also i.e. **Monopoly** means practically one seller or producer.

Consumer's Surplus (CS): The consumer's surplus was first referred by French Economist **Dupuit** and was further developed by **Marshall**. Consumer surplus is the difference between the amount of money the consumer actually pays and the amount he is willing to pay for that commodity.

- **Economic surplus (ES)** = Producer's surplus (PS) + Consumer's Surplus (CS)
- **Producer's Surplus (PS)** = Market price – Minimum variable cost
- **Marketable Surplus:** The portion of the total produce in stock which the farmer is willing to sell
- **Marketed Surplus:** The portion which is actually brought to the market at a particular time for sale

Export promotion agencies in India

a) Ministry of Food Processing Industry

b) Foreign Investment Promotion Board

c) Indian Investment Centre

d) Confederation of Indian Industry(CII)

e) Federation of Indian Export Organisation (FIEO)

f) Confederation Of Indian Food Trade and Industry(CIFTI)

g) National Dairy Development Board (NDDB)

h) The Agricultural and Processed Food Products Export Development Authority (APEDA)

i) Marine Products Export Development Authority (MPEDA)

8.3.COST CONCEPTS

The cost is the outlay of funds for productive services. There are many cost concepts used in agricultural economics and farm management i.e. Cost A_1, Cost A_2, Cost B, Cost C; working or operational cost and fixed costs; explicit and implicit cost, opportunity cost, labour costs etc.,

Cost A_1	☞ Value of hired human labour (Permanent and Casual) ☞ Value of owned bullock labour ☞ Value of hired bullock labour ☞ Value of owned machinery ☞ Hired machinery charges ☞ Value of fertilizers ☞ Value of manure (Produced on farm and purchased) ☞ Value of seed (both farm-produced and purchased) ☞ Value of insecticides and fungicides ☞ Irrigation charges (owned and hired tube-wells, pump sets etc.,) ☞ Canal-water charges ☞ Land revenue, cesses and other taxes ☞ Depreciation on farm implements (both bullock drawn and worked with human labour)

		☞ Depreciation on farm buildings, farm machinery and irrigation structures ☞ Interest on working capital ☞ Miscellaneous expenses (wages of artisans, cost of ropes and repairs to small farm implements)
Cost A_2	=	**Cost A_1** + Rent paid for **Leased Land**
Cost B	=	**Cost A_2** + Imputed Value of **Owned Land** (Less land revenue paid there upon) + Imputed interest on **Owned Fixed Capital** (Excluding land)
Cost C	=	**Cost B** + Imputed Value of **Family Labour** where cost C is the cost of cultivation (**Cost A_1** + Rent Paid for Leased land + Imputed value of owned land + Imputed interest on owned fixed capital + Imputed value of family labour)
Therefore,		
Cost A_1	=	**Cost C** - Rent Paid for Leased land - Imputed value of owned land - Imputed interest on owned fixed capital - Imputed value of family labour
Cost B	=	**Cost C** - Imputed value of family labour
Cost A_2	=	**Cost B** - Imputed value of owned land - Imputed interest on owned fixed Capital
Net Income	=	**Gross Return – Cost C**

Fixed Cost (or) Overhead Cost : Such cost does not change in magnitude as the amount of output of the production process changes and are incurred even when production is not undertaken.

Variable Cost (or) Prime Cost : It is the cost of using the variable inputs. Such variable inputs are directly related to the production.

Farming Expenses = f (farm output)

It means higher the production; higher will be the variable cost. Variable cost is also known as Prime Cost (or) Special Cost (or) Direct Cost

Implicit and Explicit Cost : Economic Cost = Accounting (Explicit) Cost + Implicit Cost

Opportunity Cost (or) Social cost (or) Alternative Cost : The farm resources have normally a number of alternative uses. For example a farmer raises paddy on his farm instead of maize, it means the farmer utilizes the other opportunity by giving up the first alternative. Hence, the Social Cost of raising paddy will be the amount of maize sacrificed in the process. Therefore in modern economics the real cost is presented as opportunity cost (or) alternative cost

Cost Elements : The cost has different elements but broadly two categories, viz.,

Fixed Cost and **Variable Cost :** The cost elements are derived from these two costs

Production Function Curve

I Stage

II Stage

OUTPUT

Inflexion point

INPUT

Properties of the TP, MP and AP under each of the Three Stages

Sl.No.	Stage I	Stage II	Stage III
1.	Starts from the origin and ends where MPP = APP	Starts from where APP is maximum and ends where MPP is zero.	Starts from where MPP is zero or TPP is maximum.
2.	TPP increases at increasing rate upto the point of inflection	TPP increases at decreasing rate	TPP decreases at increasing rate
3.	APP is increasing throughout this stage.	APP decreases	APP decreases
4.	MPP increases upto the point of inflection	MPP decreases	MPP becomes negative.
5.	MPP > APP	MPP < APP	-
6.	$E_p > 1$	$E_p < 1$	$E_p < 0$
7.	Technical efficiency of variable and fixed resources increases	Technical efficiency of variable resource decreases but of the fixed resources increases	Technical efficiency of variable and fixed resources decreases

8.	Fixed resources are abundant and variable resource is scarce	Variable resource is abundant and fixed resources are scarce	Variable resource is in excess capacity
9.	Sub-optimal (irrational) stage of production	Optimal (rational) stage of production	Supra-optimal (irrational) stage of production
10.	Scope for the reorganization of resources	No scope for the reorganization	Scope for the reorganization
11.	MVP >MIC	MVP = MIC	MVP < MIC
12.	MR > MC	MR = MC	MR < MC

Rational and Irrational Stages of Production Function Curve

- **Stage I (Irrational Stages):** In this stage, AP and MP are increasing, but MP is always greater than AP. Hence, more units of input increase production in greater proportion
- **Stage III (Irrational Stages):** In this stage, total production is decreasing and MP is less than zero. Hence, one gets less and less product for increasing amounts of input
- **Stage II (Rational Stage):** It is the relevant stage of production, because the MP, though decreasing is not negative.

Relationship between MP, TP & AP

What happens to MP, when TP is maximum	MP is equal to zero
When TP is zero, what happens to AP	AP is zero
When MP is increasing, how does AP behave	AP also increase
When MP is at its maximum, is TP at the point of inflation?	Yes
When AP is at its maximum, is the ratio of output to input maximum?	Yes, because MP = AP
When AP is increasing, how does MP behave?	MP is greater than AP
When AP is decreasing, what is the position of MP?	MP is lesser than AP
How does TP behave, when	
• MP is rising	• TP increases at a increasing rate
• AP is increasing	• TP is increasing at a decreasing rate after MP is at its maximum

Difference Between

Production Economics	Farm Management
Macro approach	Micro approach
Inter-farm science	Intra-farm science
Decisions are made for aggregate resources of nation	Decisions are taken for individual farm

Cloud Seeding materials

Cold clouds: Dry ice and Silver Iodide

Warm clouds: Water drops and Common salt

Monsoon : "Monsoon" is derived from Arabic word "MAUSIM" means season

South west monsoon : Contribute more than 70% of the total rainfall of the country SW monsoon is "Grand period of rainfall" in India

North East Monsoon : This is also called as Retreating monsoon (or) Returning monsoon.

The depression in Bay of Bengal is common and it gives rainfall in Andhrapradesh, TN and also Karnataka

Season: Season is a period in a year comprising few month during which the prevailing climate does not vary widely

Seasons of India (IMD classification)

1.South west monsoon (June – September) or Monsoon period or Kharif or Rainy season
2.North east monsoon (October – December) or Post Monsoon or Post Rainy season
3.Winter (Dec/Jan- Feb.) or Cold weather period
4.Summer or Pre monsoon (March / April –May/June) or Hot weather period

Seasons for Agricultural Crops

1.Hot weather (Mid February –Mid June).
2.Kharif or rainy season (Mid June –Mid October).
3.Rabi (Mid October –Mid February).
4.Spring falls between January and March.

Rainfall distribution in Tamilnadu in different seasons		
Rainfall season	**Quantity (mm)**	**Percentage share**
South West monsoon	311.7	32
North East Monsoon	457.8	47
Cold weather period	48.7	5
Hot weather period	155.9	16

DROUGHT

Aridity: It is a permanent climatic feature and is the culmination of a number of long term processes.

Drought: It is a temporary condition that occurs for a short period due to deficient precipitation for vegetation, river flow, water supply and human consumption.

1. **Permanent drought**: This is characteristic of the desert climate where sparse vegetation growing is adapted to drought and agriculture is possible only by irrigation during entire crop season.
2. **Seasonal drought:** This is found in climates with well defined rainy and dry seasons. Most of the arid and semiarid zones fall in this category.
3. **Contingent drought**: This involves an abnormal failure of rainfall.
4. **Invisible drought:** This can occur even when there is frequent rain in an area. When rainfall is inadequate to meet the evapo-transpiration losses, the result is borderline water deficiency in soil resulting in less than optimum yield.

8.4. FARM EFFICIENCY MEASURES AND ECONOMIC RATIOS

$$\text{Production Efficiency} = \frac{\text{Yield per acre}}{\text{Yield of locality}} \times 100$$

$$\text{Intensity of cropping} = \frac{\text{Total Cropped Area}}{\text{Total cultivated area}} \times 100$$

$$\text{Productive man work unit perman (PMWU)} = \frac{\text{Total PMWU}}{\text{Man equivalent}}$$

$$\text{Machinery cost per acre} = \frac{\text{Total machinery cost}}{\text{Total cropped area}}$$

$$\text{Operating cost ratio} = \frac{\text{Total operation cost}}{\text{Total profit}}$$

$$\text{Overhead charges ratio} = \frac{\text{Total fixed cost per year}}{\text{Gross income}}$$

$$\text{Ratio of capital turnover} = \frac{\text{Gross income}}{\text{Total farm income}} \times 100$$

Net cash income	=	Total cash receipt - Total cash operating cost
Net farm income	=	Net cash income ± change in inventory
Farm earnings	=	Net farm income + value of farm product
Family labour earnings	=	Farm earning - Interest on farm capitals
Percentage returns to capital average capital investment	=	Ratio of farm earning - imputed value of family labour to
Returns to management	=	Family labour earning - imputed value of family labour

$$\text{Marginal Utility} = \frac{\text{Change in Total Utility}}{\text{Change in Quantity Consumed}}$$

$$\text{Cross Elasticity of Demand Ec} = \frac{\text{Percentage Change in Quantity Demanded of X}}{\text{Percentage Change In Price of Y}}$$

$$\text{Pay back period} = \frac{\text{Cash outlay (or) Original cost of asset}}{\text{Annual cash inflows}}$$

$$\text{Post pay back profitability index} = \frac{\text{Post pay back profits}}{\text{Investments}} \times 100$$

$$\text{Average Rate of Return (ARR)} = \frac{\text{Average annual profits}}{\text{Net investment in the project}} \times 100$$

Net Present Value (NPV)

$$NPV = \sum_{t=1}^{N} \frac{Bt}{(1+i)t} - \sum_{t=1}^{n} \frac{Ct}{(1+i)t}$$

Bt = Benefit in year 't'

Ct = Cost in year 't'

Benefit Cost Ratio (BCR) : This is the ratio of present worth of benefit stream to present worth of cost stream.

$$BCR = \frac{\sum_{t=1}^{n} Bt/(1+i)^t}{\sum_{t=1}^{n} Ct/(1+i)\,t}$$

Internal Rate of Return (IRR)

Internal rate of return is that discount rate which makes the net present worth of incremental net benefit equal to zero.

$$\sum_{t=1}^{N} - \frac{B_t}{(1+i)^t} - \sum_{t=1}^{n} - \frac{C_t}{(1+i)^t} = 0$$

$$IRR = LDR + \text{Difference between the two discount rates} \times \left| \frac{\text{Present worth of incremental net benefit stream (cash flow) at lower discount rate}}{\text{Sum of the absolute values of present of cash flow at two discount rates}} \right|$$

IRR = Internal rate of return

LDR = Lower discount rate

Time Value of Money

Compounding is the process of finding future value of present amount

Future value = PV x (1 + interest rate per conversion period) $^{\text{no. of conversion periods}}$

$V_n = V_n (1+i)^n$

PV = Present Value

$(1+i)^n$ = is the compounding factor

n = number of conversion period (years)

The term $(1+i)^n$ can always be computed. However, the procedure becomes tedious for higher values of n. fortunately, numerical results of such equations have been tabulated for widely ranging value of i and n.

Discounting is the process of finding the present value of future amount

$$\text{Present value } (V_o) = \frac{\sum \text{Future amount}}{(1 + \text{interest rate})^n}$$

$$V_o = \sum \frac{V^n}{(1+i)^n}$$

$\frac{1}{(1+i)^n}$ = discount factor, i = interest rate, n = no. of conversion periods

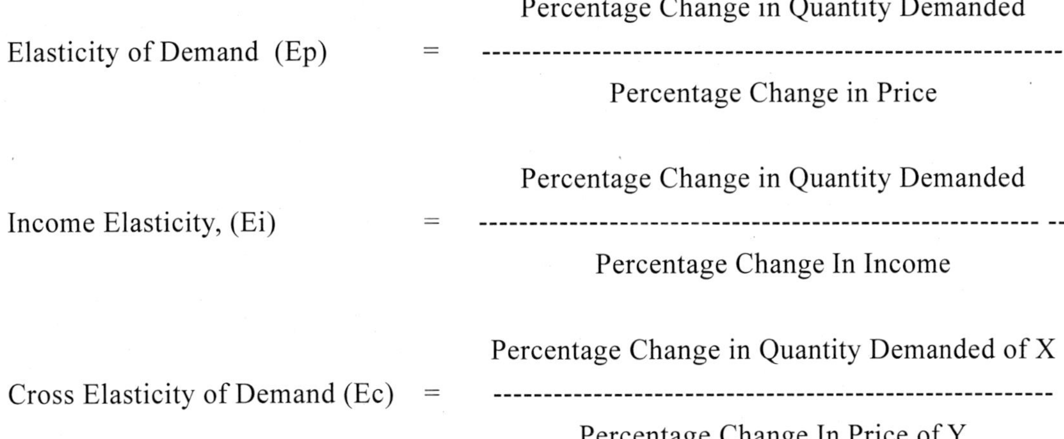

$$\text{Elasticity of Demand (Ep)} = \frac{\text{Percentage Change in Quantity Demanded}}{\text{Percentage Change in Price}}$$

$$\text{Income Elasticity, (Ei)} = \frac{\text{Percentage Change in Quantity Demanded}}{\text{Percentage Change In Income}}$$

$$\text{Cross Elasticity of Demand (Ec)} = \frac{\text{Percentage Change in Quantity Demanded of X}}{\text{Percentage Change In Price of Y}}$$

$$\text{Cost of equity} = \frac{\text{Current dividend per share}}{\text{Market price per share}} + \text{Expected annual rate of growth in dividends}$$

$$\text{Cost of preference} = \frac{\text{Annual preference dividend}}{\text{Market price of preference stock}}$$

$$\text{Cost of debt} = \frac{\text{Annual interest (1 - tax rate)}}{\text{Market value of debt}}$$

Net Working Capital = Current Assets - Current Liabilities

$$\text{Price Elasticity (Ep)} = \frac{\text{Relataive Change in Quantity}}{\text{Relative Change in Price}}$$

- **Completely Inelastic Demand :** When demand completely unresponsive to the price change i.e. **Ep = 0**
- **Relatively Inelastic Demand :** When the increase in quantity sold proportionately is smaller than the decrease in price i.e. **0< Ep<1**
- **Unit Elastic :** When proportionate change in the quantity demanded is just equal to the proportionate change in its price i.e. **Ep = 1**
- **Elastic Demand :** When proportionate change in quantity demanded is more than proportionate price change i.e. **Ep > 1**

$$\text{Substitution ratio} = \frac{\text{Quantity of input replaced}}{\text{Quantity of input added}}$$

$$\text{Price ratio} = \frac{\text{Price of input being added}}{\text{Price of input being replaced}}$$

$$\text{Price Elasticity (Ep)} = \frac{\text{Relataive Change in Quantity}}{\text{Relative Change in Price}}$$

$$\text{Current ratio} = \frac{\text{Total current assets}}{\text{Total current liabilities}}$$

$$\text{Intermediate ratio (or) Working ratio} = \frac{\text{Total current assets + Total intermediate assets}}{\text{Total current liabilities + Total intermediate liabilities}}$$

$$\text{Current liability ratio} = \frac{\text{Current liabilities}}{\text{Owners equity}}$$

$$\text{Debt - equity ratio} = \frac{\text{Total debts}}{\text{Owners equity}}$$

$$\text{Equity value ratio} = \frac{\text{Owners equity}}{\text{Value of assets}}$$

$$\text{Operating ratio} = \frac{\text{Total operating expenses}}{\text{Gross income}}$$

$$\text{Gross ratio} = \frac{\text{Total expenses}}{\text{Gross income}}$$

Theories developed by different Economists

Theory	Economist
Theory of profit	A. Walker
Modern theory of interest	Hicks. Hansen
Wage fund theory	J. S. Mill
Population theory	Malthus
Theory of multiplier	Keynes
Theory of disguised unemployment	Nurkses
Theory of unlimited supplies of labour	Lewis
Low level equilibrium trap	Nelson
Theory of undeveloped countries	Keynes
The stages of economic growth and classical theory of economic development	W. W. Rostow
Theory of das capital	Karl Max
Theory of economic growth	John Fei & Gustav Ranis
Theory of push	N. Rosenstein Rodan
Theory of social dualism	J. K. Boeke
Theory of circular causation	Myrdal
Requirement of steady growth	Harrod Damar
The model of distribution	Kaldor
The model of profit and growth	Psinette
The model of capital accumulation	Joan Robinson
The model of economic growth	R.E Meade
The model of growth	Kaldor
The single sector model	Mahalanobis
Theory of rent	Ricardo
Theory of time preference	Fisher
Theory of employment	Keynes
Liquidity preference theory	Keynes
Monetory theory	R.G. Hawtrey

Theory of consumption	Keynes
Modern theory of wages and employment	Keynes
Sunspot theory	Stanley Jevon
Under consumption theory	J.A.Hobbson
Quantity theory of money	Milton Friedman
Theory of inflation	A.P. Lerner
Theory of absolute advantage	Adam Smith
Modern theory of international trade	Bertil Ohlin

Points to rember

- New entry in the balance sheet is ***net-worth***
- The financial statement of a farm business can be provided by ***net-worth statement***
- The solvency ratio which is highly useful in measuring overall financial position of a farm business is ***net capital ratio***
- The statement which provides repayment capacity of the farmer is ***fund flow statement***
- The statement which measures the credit worthiness of the farm business is ***balance sheet***
- Expense to income ratio is ***efficiency ratio***
- The point at which total cost and total revenue is equal is ***break even point***
- Farm buildings can be valued by ***cost-depreciation method***
- Production function is ***physical relationship between inputs and output***
- A statement which gives a complete list of purchase and sales of assets is ***cash flow statement***
- The word marginal in "economic" term means ***additional***
- The type of relationship in which all other inputs are fixed except one to produce a product is known as ***factor product***
- When MP = 0, ***TPP is at maximum***
- The stage which optimize the resources, maximize the production is ***II Stage***
- Slope of iso-cost line is ***downward***
- A line that connects the points of equal slope of successive product indifference curve is ***isocline***
- A technique which is used to evaluate profitability of a particular type of equipment is ***partial budgeting / enterprise budgeting***
- National income in India is calculated by ***CSO (Central Statistical Organisation)***
- In India, National Income is calculated by ***Product Method and Income Method***
- Inflation means ***persistent rise in general price level***
- Production possibility curve is concave downwards because ***law of increasing opportunity cost***
- The production function which is highly useful in fertilizer related studies is ***quadratic production function***
- Theory of rent is based on ***law of diminishing marginal returns***

- All possible combination of two products with a given level of input is ***production possibility frontier***
- The purpose of regulated market is to ***eliminate healthy malpractices***
- Linkages of co-operative marketing societies with PACS was recommended by ***commission on agricultural credit***
- Apex body of AGMARK is ***STC***
- Advisory of central government regarding agricultural commodity prices is ***CACP (Centre for Agricultural Cost and Prices)***
- Minimum support price and procurement price are announced ***before sowing and before harvesting of the produce***
- When a firm performs more than one activity, it is called ***Vertical integration***
- Rejection of null hypothesis when it is false is known as ***no error***
- Debt-equity ration is ***Deferred liabilities/Net worth***
- The theory of demand is explained by ***marginal product curves***
- Demand function of a commodity is influenced by ***income***
- The firm is in equilibrium when ***MC=MR***
- A banker to government of India is ***Reserve Bank of India***
- Disposal Income is ***personal income - personal tax***
- The present value of future returns is calculated by the formula ***PV=R/(1+r)***
- Optimum factor combination is achieved ***at the point of least cost combination***
- Economic area of production can be located in ***stage II***
- After the inflation point, the TPP curve increases at ***decreasing rate***
- Markets are classified as forward and spot markets based on ***nature of transaction***

AGRICULTURAL STATISTICS

9

P. Kannan and P. Arunachlam

Statistic deals with the methods of collecting numerical observation and analyzing the data and interpreting them.

Data : The data are distinguished as two types namely qualitative and quantitative.

i. The *qualitative data* are distinguished by some quality or attribute, in this case the data can be observed but not measured and expressed as nominal scale or categorical data. eg. flower colour, shape of fruit, smell, taste and gender.

ii. In *quantitative characters* are the individuals distinguished by a measurement. This case the characters can be measured and expressed in numbers or numerical data. The variation in case may be discrete or continuous. eg. Plant height, fruit weight and temperature.

9.1. STATISTICAL METHODS

Measures of central tendency : Three measure of the central tendency of a frequency distribution are mean, median and mode.

i. Mean / average / arithmetic mean : It is sum of all of the data values divided by the number of data values and it affected by extreme values.

ii. Median : It is the middle value of the data set when it has been arranged in ascending or descending order. This is not at all affected by extreme values.

iii. Mode : It is the value that occurs most often in a distribution. Not affected by extreme values. It used to find the ideal size of different age group. eg: size of readymade garments, shoe size

Harmonic mean : Reciprocal of the arithmetic mean of the reciprocal of a given value. It deals with rate, price and speed.

Measures of dispersion : It is the degree to which the data tend to spread about the mean value is called dispersion.

i. Range : The range of a distribution with a discrete random variable is the difference between the maximum value and the minimum value.

ii. Standard Deviation (SD) : It is the square root of the arithmetic mean of the squares of the deviations measures for the mean or square root of the variance.

iii. Variance : It is the average of the squared deviation from the mean or the square of the standard deviation

iv. Standard Error (SE) : It is the measure of the mean difference between sample estimate of mean (X) and the population parameter (μ).

v. Coefficient of Variation (CV): It is defined as the ratio of the standard deviation to the mean. This is expressed as percentage.

Correlation Coefficient(r) : Karl Pearson studied the association between two or more variables. It refers to degree and direction of association between the two or more than two variables

Regression Coefficient (b) : Sir Francis Galton stated, it is the measure of degree of dependence of one variable on the other variable(s). In this case one variable is dependent and other as independent variable.

Difference between correlation and regression

Correlation	Regression
It measure the mutual (degree and direction) relationship between two or more variables	It measure the functional (degree of dependence) relationship between two or more variables
It does not involve cause and effect relationship	It indicates cause and effect relationship
It is confined to linear relationship	It is for both linear and non-linear relationship
The value lies between - 1 to + 1	It ranges from - α to +α (it takes any number)

Sampling methods : It is the selection of a part of the population to represent the whole population is known as the sampling and the part selected is known as the sample. The sampling methods are

1. **Probability Sampling Methods**
   - i. A simple random sample
   - ii. A stratified sample
   - iii. A cluster sample

2. **Non Probability Sampling Methods**
   - i. Convenience Sampling
   - ii. Quota Sampling
   - iii. The judgement sample

Probability : It is a measure of the expectation that an event will occur and ranges from 0 to 1

$$\text{Probability} = \frac{\text{Number of favourable cases}}{\text{Total number of equally likely cases or exhaustive number of cases}}$$

1. Normal distribution (ND) : It is a continuous probability distribution. Continuous distributions are represented by a curve and the area under the curve is equal to1. The graph of a normal distribution is called the normal curve. The properties of normal distribution are

i. The mean, median and mode are equal.

ii. The normal curve is bell-shaped and symmetric about the mean.

iii. The total area under the curve is equal to one.

Skewness : It is a measure of symmetry or more precisely called the lack of symmetry. It is zero for a normal distribution and any symmetric data should have a skewness near zero. Negative values for the skewness indicate data that are skewed left and positive values for the skewness indicate data that are skewed right.

Kurtosis : It is a measure of whether the data are peaked or flat relative to a normal distribution

2. Binomial distribution (BD) : Random variables of binomial distribution are discrete one. Ex. The tossing coin considering head as success and tail as failure, quality check of products considering non-defective as success and defective as failure.

The four requirements for a binomial distribution must are

i. Number of tests or observation is fixed

ii. Each observation must be independent

iii. Each observation represents one of two outcomes (Success or Failure)

iv. No test or observation has any impact on any other test

3. Poisson distribution (PD) : A Poisson random variable is the number of successes that result from a Poisson experiment. The probability distribution of a Poisson random variable is called a Poisson distribution. Ex. printing errors in a book, number of deaths in a district in a given period. The properties of Poisson distribution are

i. The mean of the distribution is equal to μ (μ is mean number of successes that occur in a specified region).

ii. The variance is also equal to μ

iii. The mean and variance are equal

Types of hypothesis : It is used to decide whether the differences under the study are significant or non significant

- **Null hypothesis :** A hypothesis of no significant difference between two parameters
- **Alternative hypothesis :** A hypothesis of significant difference between two parameters

Errors

- Type I error: Hypothesis is true but test is rejected
- Type II error: Hypothesis false but test is accepted

Comparision of two means

1. t-test : It is used when the sample size is small (< 30). It is of two types

i. Student'-t test (Sir William Gossel): It is used when the sample size is small and the observations are paired.

ii. Fischer's-t test: It is used when the sample size is small and the observations are not paired

2. Z test : It is used when the sample size is large (>30 samples)

Comparision of several means : It is used to test the significant difference among the several means and it is tested by F test. F values are based on the analysis of variance (ANOVA).

9.2. EXPERIMENTAL DESIGNS

Experimental designs are used to test a set of treatments to draw valid conclusion about a particular problem. An experiment has to be designed by taking into consideration of aspects, *viz.*, objective of the study, experimental material availability, cost, time, etc.

Experimental unit : A subject (or) a group of object (or) the total material to which a treatment is applied in a trial in a single replication is known as an experimental unit.

Treatment : A treatment is a substance or known factor that is administered or allocated to one or more experimental units to estimate the effect pertaining to certain characteristics or for comparing it with others.

Principles experimental design : The basic principles of any design of experiment are replication, randomization and local control of error.

1. **Replication:** The repetition of a treatment is known as replication. The advantages of replication are:

i. Replications are essential to obtain a valid estimate of the experimental error variance.

ii. Replications are necessarily required to attach a probability statement with estimate treatment differences

iii. The larger number of replications reduces the standard error of the treatment means. But one cannot take more than certain number of replications because of the management problems and limitation of resources

2. **Randomization :** Allocation of the treatments to the different plots by a random process is known as randomization of treatments. It gives equal chance to all the treatments being allocated to different experimental unit it is known as randomization. The advantages of randomization are to get an unbiased estimate of experimental error and to get valid conclusions

3. **Local control :** It is a device to maintain greater homogeneity of experimental units a block of an experiment (or) as a whole. The advantages of local control are it reduces the experimental error, to make designs more efficient, it makes any test of significance more sensitive and powerful and to reduce the experimental error and consequently helps the investigator to detect the small real difference between treatment means.

Experimental error : Variation due to uncontrolled or environmental factors

Correction factor : It is the square of grand total divided by the number of observations.

Degrees of Freedom (df) : The number of degrees of freedom is the number of values in the final calculation of a statistic that are free to vary (or) independently allotted.

Types of Experimental Design

The choice of the experimental design depends on the factors like number and nature of treatment, objective of the experiment and resources available.

1. Completely Randomized Design (CRD)

Completely randomized design is the basic single-factor experiment. This design is used when the experimental materials is limited and homogenous. In laboratory experiments and greenhouse studies it is easy to achieve homogeneity of experimental materials.

ANOVA for the CRD: 8 treatments with 3 replications

Sources of variation	df	SS	MSS	*F ratio*
Treatments	t-1 =7			
Error	(t-1)x(r-1)= 14			
Total	rt-1=23			

Merits of a CRD

- Its layout is very easy
- There is complete flexibility in this design i.e. any number of treatments and replications for each treatment can be tried
- Whole experimental material can be utilized in this design
- This design yields maximum degrees of freedom for experimental error
- The analysis of data is simplest as compared to any other design

Demerits of a CRD

- ✣ The design is suitable for a small number of treatments
- ✣ It is difficult to find homogeneous experimental units in all respect.
- ✣ CRD is not suitable for field experiments as compared to other experimental designs

2. Randomized blocks design (RBD)

This design is used when the fertility gradient runs in one direction. It is the most frequently used design in agricultural experiments.

ANOVA for the RBD : 8 treatments with 3 replications

Sources of variation	*df*	SS	MSS	*F ratio*
Replication/Blocks	r-1=2			
Treatment	t-1=7			
Error	(t-1)x(r-1)=14			
Total	rt-1=23			

Advantages of RBD

- ✣ It increases the precision of the experiment. This is due to the reduction of experiment error by adoption of local control
- ✣ The amount of information got in RBD is more as compared to CRD
- ✣ Any number of replications can be included in this design

Disadvantages of RBD

When the number of treatments is increased, the block size will increase. If the block size is large it may be difficult to maintain homogeneity within blocks. Consequently the experimental error will be large and hence RBD may not be suitable for large number of treatments.

Design for CRD in pot culture and RBD in field for 5 treatments with 3 replications

CRD		
A	A	A
B	B	B
C	C	C
D	D	D
E	E	E

Where A, B, C, D and E represents different treatments in the pot

RBD		
RI(block1)	**RII(block2)**	**RIII(block3)**
B	C	D
A	D	E
E	A	C
D	E	B
C	B	A

High ⟶ Low

Direction of the fertility gradient

3. Latin Square Design (LSD)

When the fertility gradient runs in both directions in the field, then the blocking has to be done in both directions. In such situations a Latin Square Design can be adopted. A LATIN Square is a balanced two-way classification, because of the two ways blocking reduce the experimental error considerably. This design is an improvement over RBD. In this design the number of rows, columns and treatments are all equal, each treatment occurs only once in each row and in each column, each row and each column is a complete block.

ANOVA for the LSD with 5 treatments

Sources of variation	*df*	SS	MSS	*F ratio*
Rows	r-1=4			
Columns	c-1=4			
Treatments	t-1=4			
Error	(t-1)(t-2)=12			
Total	rc-1=24			

Advantages of LSD

- ✣ LSD is able to control two way variability that is not possible in CRD or RBD.
- ✣ The analysis of data does not entails any complications
- ✣ The estimation and analysis in case of missing data is simple.

Demerits of LSD

- ✣ LSD can accommodate a limited number of treatments
- ✣ LSD for less than five treatments does not provide adequate degrees of freedom for experimental error.
- ✣ LSD of order 2, 3 and 4 may not be as efficient as a CRD or RBD with same number of replications.

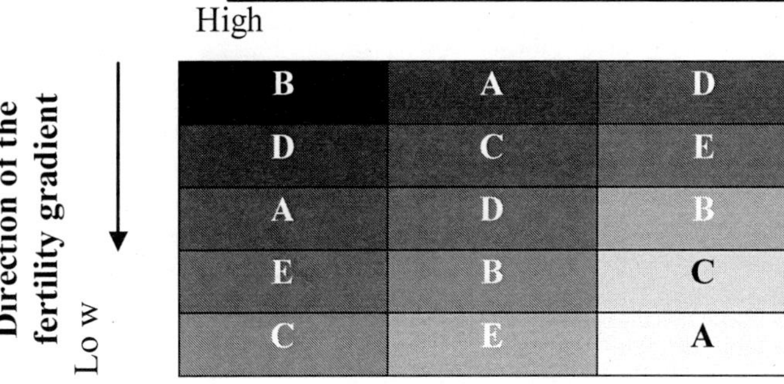

4. Factorial Experiments

In practice, the response of biological organisms to the factor of interest is expected to differ under different levels of other factors. For example, the yields of paddy varieties may differ under different rates of fertilizer application, spacing and irrigation schedules. When several factors are investigated simultaneously in a single experiment, such experiments are known as factorial experiments.

4a. Split-plot design

In this design large plots are taken for the factors that require less precision. Next, each of the large plots divided to smaller plots in order to accommodate the other factor. The different types of treatments are allotted at random to their respective plots. Such arrangement is called split- plot design, the larger plots are called main plots and the smaller plots within the larger plots are called sub-plots. The factor levels allotted to the sub-plots are called sub-plot treatments. It may be noted that a spilt-plot design is a two factor factorial experiment with a very particular randomized assignment of treatments.

Advantages

- ✤ It provides increased precision in the estimates of sub-plot treatment and interaction compared to factorial RBD with two factors.
- ✤ It is a confounded design in which the main plot treatments are unfounded with whole plots.
- ✤ Inter block information can be utilized in the analysis of data.
- ✤ Increased precision attained on the sub plot treatments and interactions of sub plot x main plot treatments as compares to the main plot treatment alone.
- ✤ It is quite feasible to use two, three (or) more split plot design known as double split, triple split (or) multiple split-plot design.

Disadvantages

- ✤ Effects of main plot treatments are estimated with less precision than they are in factorial RBD with two factors.
- ✤ Analysis becomes more complex when missing data occur.
- ✤ The difference in the precision of the estimates of the main plot treatments and sub-plot treatments is due to the sizes of the main plot and sub-plot.

4b. Split-split plot design

The split-split plot design is an extension of split-plot design. This design is used when there are three factors each requiring a different level of precision.

The layout of split-split plot design is similar to split- plot design. Each replication is dividing into number of main plots. The main plots are divided into sub-plots. The sub-plots are further divided into sub-sub-plots (ultimate plot). The less important factor is allocated at random to the main plots; somewhat more important factor to the sub-plots and the most important factor to the sub-plot.

ANOVA for the Split Split-plot design : The design for the three factors with different levels is given below

Main plot : 3 levels of irrigation (I1, I2, I3); Sub plot: Spacing 4 levels (S1, S2, S3, S4) and Ultimate plot: 4 levels of nitrogen (N1, N2, N3, N4) with 3 replications (r)

Sources of variation	*df*	SS	MSS	*F ratio*
Replications(r)	r-1= 2			
Nitrogen (N)	N-1= 3			
Error (a)	(N-1)x(r-1)=6			
Total	**(Nxr)-1=11**			
Spacing (S)	S-1=3			
Interaction between N x S	(N-1)x(S-1)=9			
Error (b)	{(S-1)+[(N-1)x(S-1)]} x (r-1)=24			
Total	**(NxSxr)-1=47**			
Irrigation (I)	I-1=2			
Interaction between I x S	(I-1)x(S-1)=6			
Interaction between Ix N	(I-1)x(N-1)=6			

Interaction between I x S x N Error (c)	(I-1)x (N-1)x(S-1)=18 {(I-1)+[(N-1)x(S-1) x(I-1)]}x (r-1)=40			
Total	**(NxSxIxr)-1=143**			

5. Strip plot design

Another variant of split plot design is the strip-plot design. It is also known as split block design. If two factors are involved and if both the factors require large plot sizes, it is difficult to carry out the experiment in a split plot design. In some other situations a higher precision may be required for the interaction than the precision for the two factors. The strip-plot design is suitable for such experiments. In strip plot design each block is divided into number of vertical and horizontal strips depending on the levels of the respective factors. The vertical strip treatments are laid out either in randomized blocks or in latin square. The intersection plots provide information on the interaction of two factors.

5a. Strip-split design

The strip- split design is an extension of strip plot design which accommodates a third factor. When the intersection plot design is divided into plots it results in strip-split plot design. The randomization procedure is similar to the strip-plot design as for as strips are concerned. The third factor is randomly allocated within the intersection plots.

Chi square test (Karl Pearson) : It is a statistical test commonly used to compare observed data with data we would expect to obtain according to a specific hypothesis. Chi-Square test requirements are

i. Independent observations
ii. Adequate sample size, at least 10
iii. Simple random sample
iv. Data in frequency form
v. The data to be strictly categories (one or more)
vi. All observations must be used for analysis

There are three test of Chi-square

i. *Chi-square test for independence:* The test is applied when you have two categorical variables from a single population.

ii. *Chi-Square Goodness of Fit Test :* This test is commonly used to test association of variables in two-way tables. Where the assumed model of independence is evaluated against the observed data. Chi-square is the sum of the squared difference between observed (O) and the expected (E) data divided by the expected data in all possible categories. The formula for calculating chi-square (X^2) is = $(O-E)^2/E$.

iii. *Chi-square test of homogeneity :* The test is applied to a single categorical variable from two different populations. It is used to determine whether frequency counts are distributed identically across different populations.

NON-PARAMETRIC TEST : In some situation like agricultural extension, pathology research works observations are taken in terms of score or rank. In such condition the observation are no longer consider as distributed in normal. In such cases non parametrical test are used.

i. Sign test
ii. Paired sample
iii. Two independent samples
iv. Rank correlation

HORTICULTURE

10

Dr. P. Muthukumar, S. Arul and Dr. P. Ganeshamoorthi

1. Area, Production and Productivity of Horticultural Crops (2013-2014)

Particulars	Area (M ha)	Area share (%)	Production (Mt)	Production share (%)	Productivity (Mt/ha)	India Position	World scenario
Fruits (including nuts)	7.21	29.80	88.97	32.07	12.30	2nd (13.6%)	1st China (20.9%)
Vegetables	9.39	38.81	162.89	58.73	17.30	2nd (14%)	1st China (49.5%)
Flowers (Loose)	0.25	1.03	1.75	0.82	6.90	-	-
Spices	3.16	13.06	5.90	2.12	1.90	-	-
Aromatic and medicinal crops	0.49	2.02	0.89	0.32	1.8	-	-
Plantation crops	3.67	16.30	16.30	5.87	4.40	-	-
Total	**24.19**	-	**277.35**	-	**11.5**	-	-

2. Leading Horticultural Crops in India during (2013-14)

Particulars	Area	Production	Productivity
Fruits	Mango > Citrus> Banana	Banana > Mango > Citrus	Papaya > Banana > Apple
Vegetables	Potato >Tomato > Onion	Potato > Onion > Tomato	Tapioca > Cabbage > Potato
Plantation crops	Coconut > Cashew > Arecanut	Coconut > Cashew > Arecanut	Coconut > Cashewnut
Spices	Chilli > Garlic > Turmeric	Chilli > Garlic > Turmeric	Garlic > Turmeric

3. State-Wise leading Area, Production and Productivity of in Horticultural crops during (2013-14)

Crops	Area	Production	Productivity
Fruits	Maharashtra > Andhra Pradesh > Gujarat	Maharashtra > Andhra Pradesh > Gujarat	Madhya Pradesh
Vegetables	West Bengal > UP> Bihar	West Bengal> UP > Bihar	Tamil Nadu
Plantation crops	Kerala	Kerala	Kerala
Spices	Rajasthan	Andhra Pradesh	Arunachal Pradesh
Cut flowers	West Bengal	West Bengal>Karnataka > Odisha	Bihar
Loose flowers	Tamil Nadu	Tamil Nadu>Karnataka > Madhya Pradesh	-

- Horticulture is derived from the Latin word hortus means garden and cultura means cultivation
- Horticulture plays an important role in India's economy accounting for about 30.7% of India's agricultural GDP from 13.7% of cropped area
- Horticulture crops occupy only 14.0% of the total cropped area
- Highest number of released varieties in horticultural crops: Vegetables (57%), plantation and spices (22%), fruits (22%)

Fruits:

- India: 2nd largest producer of fruits after China
- India accounts 13.59% of global total fruit production
- In India, Fruits occupy 29.82% of area and 32.07% of total horticultural area and production respectively, with average productivity of 12.3 Mt/ha.
- States leading fruit area: Maharashtra
- States leading fruit production: Maharashtra (15%)
- States leading fruit productivity: Madhya Pradesh (28 MT/ha)
- Among fruit crops, production share of Major fruit crops: Banana (33.4%), Mango (20.7%), Citrus (12.5%)
- Among fruit crops, production share of states in Major fruit crops: Maharashtra (15%), Andhra Pradesh (12%), Gujarat (9%)

Vegetables:

- India is the 2nd largest producer of vegetables after China
- India shares about 14.04 % of the total vegetable production of the world
- Vegetables occupy 38.9% of area and 61% of total horticultural area and production respectively
- Vegetables shares 2% of the total cropped area
- Annual per capita availability of vegetables: 120 kg/person

Floriculture:

- The area under floriculture production in India was 0.25 Mha with a production of 2.29 Mt of loose flowers and 0.47 thousand tonnes cut flowers (2013-14)
- In India 98.5% of flowers grown under open cultivation. Only 1.5% flowers were grown under green house.
- In India about 90% area is under traditional flower that is mostly loose flowers
- In India more than 70% of the floricultural exports from dry flowers
- India's total export of floriculture was Rs. 479.42 crores (2015-16)
- There are more than 300 export-oriented units in India. More than 50% of the floriculture units located in Karnataka, Andhra Pradesh and Tamil Nadu
- Leading global floricultural export market: Netherlands (58%), Columbia (14%)

10.1. POMOLOGY

- Study of fruit crops is Pomology
- Pomology word derived from Latin word pomum means fruits and the Greek word *logy* means study
- Father of systematic pomology: De Candolle
- Leading top three fruit producing countries in the world
- China (20.9%)
- India (13.6%)
- Brazil

Classification of fruit crops based on climate:

- **Tropical fruits:** Mango, banana, citrus, papaya, sapota, guava, grapes, pineapple, jack
- **Subtropical fruits:** Mangosteen, avocado, litchi, Rambutan, loquat, durian, persimmon, passion fruit
- **Arid and semi-arid fruits:** Ber, Aonla, Annonaceous Fruits, Pomegranate, Date Palm, Fig, Bael, Phalsa and Jamun
- **Temperate fruits:** Apple, Pear, Quince, Peach, Plum, Pear, Almond, Apricot, Cherry, Walnut and Pecanut

Common names of important Fruits

- Mango-National fruit of India, King of fruit
- Banana- Apple of paradise, Tree of paradise, Kalpataru
- Jamun - Indian black berry
- Mangosteen, Litchi: Queen of fruit
- Guava: Apple of tropics, Poor man's apple
- Apple: King of temperate fruits
- Walnut: King of nuts, Peacanut- Queen of nuts
- Ber- Poor man's fruit, King of arid fruit

System of planting:

Square system:

- Simplest and ubiquitous system of planting, most convenient system
- Most commonly followed system (easy layout)

Rectangular system:

- Used in high density planting (HDP) e.g. Mango, Aonla
- Easy cultural operation and easy mechanical operations

Hexagonal system or equilateral triangle or septule system:

- Accommodates 15% more no. of plants than square system
- Very difficult to layout

Quincunx system or diagonal system:

- Accommodates1.5 -2 times double the number of plants (89%) more than square system
- Filler plants are generally used in quincunx system
- This system accommodates 10% more plants than the square system
- Suitable filler plants: Guava, Kinnow, Phalsa, Plum, Peach, Papaya, Banana, Pomegranate

Triangular system:

- ✤ Based on the principle of isolateral triangle
- ✤ Accommodates 11% lesser no of plants than square system
- ✤ Mostly used for high density planting (HDP)
- ✤ Problem in triangular system: Intercultural operation

Contour system

- ✤ Commonly used in hilly regions and undulated topography (slope exceeds 10%)
- ✤ Double hedge row contour planting system accommodates 22% higher plants than the single hedge row contour planting
- ✤ Main advantage: reduce the soil erosion and conserve the soil moisture
  - « Double hedge row planting followed in Guava and Pineapple
  - « HDP- apple, guava, pineapple, mango (Amarapali)

Training system:

- ✤ Training is a new practice in which tree growth is directed into a desired shape and form
- ✤ It includes tying, staking or supporting young plants
- ✤ Main purpose: Shaping of young fruit trees

Commercial training systems in fruit crops

Training systems	**Examples**
Central leader	Walnut, pecan nut, apple
Open center or vase system	Peach, Japanese plum, nectarine
Modified leader	European plum, sweet cherry, pear
Bower system	Grapes
Espalier system	Apple
Cordons	Peach
Single stem system	Citrus, fig, Annona
Multiple stem system	Pomegranate
Two arm kniffin system	Passion fruit

Pruning system:

- ✦ Pruning is the removal of a portion of a tree to correct or maintain tree structure

Main objectives:

- ♣ Regulation of shape and growth of tree
- ♣ Enhance the production and quality of fruits
- ✦ Pruning is done into 2 ways

  1. Thinning out: Removal of undesirable shoots or branches without leaving any stub
- ♣ Encourage the tree growth *e.g.* Mango, loquat, quince, olive

  2. Heading back: Removal of terminal portion of the shoots, branches or limb leaving its basal portion
- ♣ Reduce the tree size by topping and hedging

Propagation methods in fruit crops:

- Stratification: (Moist chilling) seeds are exposed to low temperature (2-5°C) e.g. Apple
- Potassium nitrate (0.2%) concentration will improve seed germination e.g. Guava
- The plants or shoots composed with normal and mutated cells are called Chimeras e.g. Citrus, apple, ornamental plants
- Recalcitrant seeds: Mango, Mangosteen, Jackfruit, Citrus, Cocoa, Cinnamon
- Seed propagation- Acid lime, Papaya, Passion fruit, Phalsa, Mangosteen
- Simple or serpentine Layering: Guava, Pomegranate
- Tip layering : Black berry
- Mound layering : Guava, Goose berry, apple root stocks
- Trench layering or Etiolation method e.g. Cherry
- Air layering or Gootee or marcottage : Litchi
- T budding (Shield budding) – Citrus, Pome fruits, Rose, Ber
- Patch budding: Bael
- T budding: Citrus
- Veneer grafting- Mango (commercial method)
- Epicotyl grafting- Mango (coastal regions)
- Bridge grafting or Repair grafting- Rejuvenation of old orchards e.g. Temperate fruits
- Softwood grafting- Cashew, Mango
- Inarching or approach grafting is most popular and extensively used propagation of mango and sapota in Tamil Nadu
- Double working- Pear
- Rhizomes (Modified stem): Banana
- Suckers, Crowns, slips: Pineapple
- Tuberous roots- Sweet Potato, Dahlia (massive enlargement of secondary roots)
- Runner (Specialized stem)- Strawberry
- Sword suckers: Banana
- Offsets (Lateral shoot or branch): Date Palm
- Asexually propagated crops are true-to-type
- Polyembryony: occurrence of more than one embryo in a seed. Germination of many seedlings in a single seed e.g. Mango, Citrus, Jamun
- Polyembryony – Mango (Olour, Bappakkai, Kurukkan, Chandrakaran, Goa, Bellary)
- Apomixis: Development of embryo without fertilization, Apomictic seedlings are identical to mother plants (true-to-type)

Type of apomixis

- Recurrent apomixis- e.g. Allium, Malus
- Adventitious or Nucellar embryony- e.g. Citrus, jamun
- Non recurrent apomixis- e.g. *Solanum nigrum* and Allium spp
- Vegetative apomixis-e.g. Agave and grass species

Common root stocks used in fruit crops:

- ★ Kurukkan- salt tolerant polyembryonic root stock of mango
- ★ Pusa Srijan (Aneuploid-82), (2n=24): Potential dwarfing rootstock for HDP

- ★ M-9: Dwarf rootstock of apple
- ★ M-27: Ultra dwarf rootstock of apple
- ★ *Manilkhara hexandra* (Pala) is a common root stock of Sapota
- ★ Troyer' Citrange: Most promising rootstock for Nagpur mandarin and Kinnow mandarin in India
- ★ Trifoliate orange: Cold hardy, tolerant to Phytophthora, Tristeza, RKN and dwarfing
- ★ Ultra dwarf root stock of citrus- *Severinia bouxifolia*
- ★ Grapes nematode resistant root stock- Dogridge, Salt creek
- ★ Salt tolerant grapes rootstock: Salt creek (Ramsey)
- ★ Best grapes rootstock for drought and salinity: Dogridge (used for all conditions)
- ★ Grapes saline tolerant : Solanis, 1616
- ★ Resistant Rootstock for apple woolly aphid- Northern Spy
- ★ Quince C: Ultra dwarf for pear, suitable for HDP

Fruit types:

Fruit type	Crops
Berry	Banana, Papaya, Grapes, Sapota, Arecanut, Avocado
Modified berry	**Balausta**- Pomegranate, **Amphisarica**- Wood apple, Bael, **Pome**, Apple, Pear, Quince, Loquat, **Drupe**- Mango, Peach, Plum, Ber, Coconut, cherry, coffee, cocoa, **Hesperidium**- Citrus, **Capsule**- Amla, Carambola, **Nut**- Cashew, Litchi, Rambutan
Aggregate fruits	Custard apple, Strawberry
Multiple fruits	**Syconus**- Fig, **Sorosis-** Jackfruit, Pine apple, Breadfruit, Mulberry

Edible portion of fruit crops:

Edible part(s)	Fruit crops
Pericarp	Custard apple, Avocado, Ber, Datepalm
Pericarp and placenta	Grape
Mesocarp	Mango, Papaya, Sapota, Passion fruit, Mulberry
Mesocarp and endocarp	Banana, Aonla, Apricot
Mesocarp and epicarp	Peach, Plum, Persimmon, Phalsa, Cherry, Jamun, Karonda
Fleshy peduncle	Cashew apple
Fleshy aril	Litchi
Fleshy receptacle	Fig
Fleshy thalamus	Apple, Pear, Quince, Strawberry, Loquat
Thalamus and pericarp	Guava
Succulent placentae	Bael, Wood apple
Bracts and Perianth	Jackfruit, pineapple
Juicy placental hairs	Citrus
Juicy seed coat	Pomegranate
Cotyledon	Peacanut, pistachio nut, Cashewnut, Almond
Endosperm	Coconut

Major bioactive compounds (pigments) present in fruit crops

Colour	Pigments	Examples
Orange	β-carotene	Mango, Pineapple
Red-purple	Anthocyanins	Grapes, Pomegranate, Blackberries, Raspberries, Blueberries
Orange	Caricaxanthin	Papaya
Red	Lycopene	Red flesh Papaya, red flesh guava
Orange-yellow	Flavonoids	Peach, Papaya, Orange, Tangerine
Yellow-green	Lutein and zeaxanthin	Avocado
Green	Chlorophyll	Guava
Yellow	Xanthophyll	Guava

Important nutrition rich fruit crops

- Dry karonda is richest source of iron
- Cashewnut is richest source of protein
- Litchi is richest source of calcium
- Vitamin A: Mango (4800 IU) > Papaya (2000 IU)
- Vitamin C: West Indian cherry 1500 mg/100g >Amla (600 mg /100g) > Guava (300mg/100g)
- Anola: Very rich in Vitamin-C (600 mg/100g) maximum in mature fruits
- Anola: Preparation of hair dyes and hair oils, Medicinal value: Vitality restorer

Origin of fruit crops:

- Tropical America: Sapota, Guava, Papaya
- Central America: Avocado
- Mango: Indo-Burma (South East Asia)
- India origin crops: Jack fruit, Acid lime, Jamun, Aonla, phalsa, wood apple
- Mexico: Sapota
- Brazil: Pineapple
- Chinese origin- Litchi, kiwi, sweet orange, persimmon (national fruit of Japan), peach
- Iran: Pomegranate

Important terms

Banana

- Propping (giving support to pseudostem)
- Denavelling (Removal of male buds) and Desuckering (Removal of unwanted suckers)
- Mattocking (Cutting of pseudostem after harvest leaving of 0.6m stump)

Special pruning techniques in horticultural crops:

Use of Growth Regulators:

- To overcome the alternate bearing or off-season flower induction in Mango- Paclobutrazol @ 2 ml/tree

Special Techniques	Purpose	Examples
Root pruning: Removal of roots 40cm away from the plant	To make dwarf, to induce flowering, fruitfulness and determining the flowering time	Rose, Mandarin, Guava
Ringing: Removal of complete ring of bark from a branch or a trunk	To increase fruit bud formation	Mango, Grapes
Dehorning: To removal of overcrowding and intermingling of branches	To induce flowering	Mango (Vidharbha Region of Maharashtra)
Notching: Partial ringing of a branches above a dormant lateral bud	To induce the fruiting branches & increase the bearing area of the plant	Poona Fig (Pune Region of Maharashtra)
Nicking: Partial ringing of a branches below a dormant bud	To increase the flowering shoots To induce spurs from buds	Apple and Poona Fig (Pune Region of Maharashtra)
Smudging: Practice of smoking around the trees	To induce the offseason crops	Mango (Philippines)
Bending: Bending of branches or shoots	To increase the lateral branches and fruit production	Guava (Allahabad Region in UP, Deccan region)
Coppicing: Complete removal of trunk leaving 30-35cm stump	To increase the vigorous shoots	Cinnamon, Cinchona, Eucalyptus
Pollarding: Removal of growing point in shade trees	To encourage the side branches	Silver oak
Lopping: Reducing the canopy cover in shade trees	To permit more sunlight	Silver oak
Thinning: Removal of part of flower bud or small fruits from a heavy crop	To increase the fruit size To reduce the alternate bearing tendency	Grapes, Peach, Plum, Quince
Disbudding: Removal of unwanted flower buds in cluster of flowers	To induce the showy good quality flowers	Carnation, Chrysanthemum, Dahlia
Pinching: Removal of terminal growing point in herbaceous plants	To induce the lateral or side shoot production	Carnation, Chrysanthemum
Girdling: Removal of 2-3mm white strip of barks around the them	To increase the berry size	Grapes
Leaf pruning: Removal of old and senescence leaves	-	Date palm
Top working or top-grafting or top budding: Changing the established plants, trees, shrubs or vines with a desirable cultivars	-	Mango, Apple, Nutmeg

- The use P333 or Paclobutrazol or Cultar @ 1.25 to 10 g.a.i/tree. Commercialized method in India to manipulate flowering by post- harvest application to the soil significantly promotes flowering and fruiting.
- Paclobutrazol is a gibberellins bio-synthesis inhibitor. It controls the emergence of new vegetative flush.
- Improvement of earlier flowering and fruiting: KNO_3 1% at the time of flowering
- Commercial hormone used for root formation in cuttings- IBA
- Induction of parthenocarpy in grapes: GA_3
- To elongation of berry clusters in seedless grapes: GA_3 @ 10-20 ppm
- To increase berry size, clusters dipped in GA_3 600 ppm alone or GA_3 @ 30 ppm with BA 10 ppm or 2 ppm CPPU at two times at Bajra grain size and red gram size berries stage
- Suppression of vine growth and increase the fruitfulness of buds in grapes: CCC @ 250-500 ppm
- Bud breaking in North India: Dormex (Hydrogen Cyanamide (HCN) 1.5% applied dormant buds during January
- Hydrogen Cyanamide: Induces uniform and early bud sprout
- For uniform bud break: Swabbing of buds with hydrogen Cyanamide: 1.5-4% within 24-48 hours for forward pruning
- Bud breaking in South India: Thiourea @ 4%
- Uniform flowering in pine apple: Ethephon combination with (Urea 2% +Ca/Na carbonate 0.04%)
- To avoid rainy season crop (production of poor quality fruits)- Spraying at summer season at peak flowering time: Urea 10 to 20 %
- To overcome the chilling requirement in temperate fruits- GA_3
- To control pre-harvest fruit drop in Mango, Citrus: 2, 4-D @ 20 ppm
- To increase the latex flow in rubber- Ethrel
- Deblossoming agent- NAA @ 200 ppm e.g. Mango, Guava
- Fruit thinning in peach: Ethephon @ 100-300 ppm
- Fruit thinning agent in date palm: Ethephon
- Grape guard packaging material contains sodium bisulphite, control the post-harvest fungal diseases
- To reduce the rainy season crop- Deblossoming done through applying NAA @ 600 ppm + Urea @ 1.5%
- Guava thinning agent: NAA @ 80-100 ppm at 10% of anthesis (NAD, 2,4-D)

Important varieties in fruit crops:

Mango varieties/hybrids

Hybrids	Parents	Special features
Ambika	Amrapali × Janardhan Pasand	Late ripening variety
Arunika	Amrapali × Vanraj	
Pusa Peetembar	Amarapali × Lal Sundari	Yellow colour variety
Pusa Arunima	Amarapali × Sensation	Regular bearer
Pusa Lalima	Dashehari ×Sensation	Red colour variety

Pusa Pratibha	Amarapali ×Sensation	Regular bearer
Pusa Shreshth	Amarapali ×Sensation	Regular bearer
Mallika	Neelum × Dashehari	Regular bearer- Mid season
Amrapali	Dashehari ×Neelum	Regular bearer, Desert variety
Ratna	Ratnagiri Alphonso ×Neelum	Regular bearer
Sindhu	Ratna × Alphonso	Seedless
Arka Udaya	Amarapali × Arka Anmol	-
Arka Aruna	Banganapalli × Alphonso	Regular bearer- Suitable for HDP
Arka Puneet	Alphonso× Banganapalli	Regular bearer- Suitable for canning
Arka Neelkiran	Alphonso × Neelum	Regular bearer- Suitable for export
Arka Anmol	Alphonso × Janardhan Pasand	Regular bearer- Suitable for export
PKM-2	Neelum × Mulgoa	-
PKM-1	Chinnaswarnarekha × Neelum	-
Prabha Sankar	Bombay × Kalapady	Regular bearer and late season variety
Sunder Langra	Langra × Sunder Prasad	Free from malformation disorder

Varieties

Pusa Surya	Selection from Eldon	-
Neelum	Kasaladdu	Regular bearer, late season mango
Totapuri	Kallamai or Kilimooku	Suitable for processing industry (pulp)
Rumani	-	Apple shaped variety
Alphonso	Khader or Badami or Gundu	Suitable for export, susceptible to spongy tissue
Dashehari	-	Mid season, Susceptible to mango malformation
Mulgoa	-	Late season and shy bearing
Langra	Turpentine flavour	Most popular in North India, Biennial bearer, Midseason
Paiyur-1	-	Suited for HDP
Niranjan	-	Off-season fruit bearing variety
Pairi	-	Regular bearer, Early maturing
Mankurad	-	Mid season

Specific features:

- ♣ Good mango varieties contain 20% of TSS
- ♣ Dwarfing cultivars: Ambalavi, Kalapady
- ♣ North Indian mangoes Langra and Dashehari are alternate bearers
- ♣ Off season mango (Fruit maturity: January to February) : Kanyakumari district of Tamil Nadu due to microclimate; Cultivars: Neelum, Rumani, Bangalora
- ♣ Regular bearer varieties: Totapuri or Bangalora, Neelum, Ratna, Sindhu

- Most popular varieties in North India: Dashahari, Chausa
- Canning variety: Alphonso
- Off season variety: Niranjan
- Most suitable variety for canning purpose: Alphonso
- Amarapali is suitable for high density planting
- Sindhu is a seedless variety of mango

Banana:

- Dwarf Cavendish (AAA): Leading commercial cultivar in India. It occupy about -58% of the banana production
- Grand nine (AAA): Internationally accepted variety, popular all over India
- Robusta (AAA): Semi-tall sport of Dwarf Cavendish
- Rasthali (AAB) Choicest table variety
- Hill banana (AAB): Virupakshi, Sirumalai Suitable for higher elevation regions
- Red Banana (AAA) Long duration of cropping (16 months), grown only backyard gardens

Papaya:

- Highest papain yielding variety: CO-5
- Suitable for high density planting: Pusa Nanha
- Transgenic papaya varieties: Rainbow, Sun Up
- Red lady, Zinda: Popular private company Gynodioecious hybrids

Grapes:

- Thompson Seedless: Ruling variety in India
- Coloured seedless varieties: Fantasy Seedless, Sharad Seedless and Crimson Seedless
- Green seedless varieties: Thompson Seedless, Tas-A-Ganesh, Sonaka

Guava

- Pink flesh varieties: Arka Kiran, Arka Rashmi
- Lucknow-46: Pear or pyriform fruit shape
- Lucknow-49 (Sardar Guava): National variety, widely cultivated variety
- Thai guava: big fruited types

Pine apple:

- Kew and Giant Kew (Commercially grown in India)
- Amritha (Kew × Ripley Queen):1st hybrid in India Pineapple Research Centre, Mannuthi, KAU.

Anola:

- NA-4, NA-6, NA-8: Selection from Chakaiya
- NA-5, NA-9, NA-10: Selection from Banarasi
- Resistant to fruit necrosis: NA-6, NA-7, Chakaiya
- Goma Aishwarya: Clonal selection of NA-7
- NA-7 (Neelum): commercially cultivated variety in India

Apple:

- ♣ Ambri: longest shelf life and indigenous variety of India
- ♣ Red delicious is most popular variety in India
- ♣ Hazratbali (Benoni) earliest variety in Kashmir valley
- ♣ Winter banana is a variety of apple

Pear

- ★ Bartlett or Williams or William Bartlett is the most popular variety all over the world
- ★ Kieffer: Well adapted widely grown pear in India
- ★ Varieties of Pine apple: Giant Kew, Queen, Mauritius
- ★ Peach varieties: Florida Sun, Shane-e-Punjab
- ★ Artificial interspecific hybrid: Arka Sahan (Island Gem (Atemoya) × Mammoth (Custard apple)
- ★ Goma Kirti is a ber variety
- ★ Kaveri (Purple × Golden): F_1 hybrid: Resistant to collar rot, wilt, brown leaf spot and nematodes
- ★ NA 7, BSR 1 varieties of Amla
- ★ Most popular seedless variety of grape in Tamil Nadu- Thompson Seedless
- ★ Ganesh, Jyoti, Ruby is a variety of Pomegranate
- ★ Bhagwa (Kesar): Popular pomegranate Indian variety, soft seeded variety
- ★ Early ripening variety: Halawy (Most popular variety in india)

Important varieties released from TNAU

	Breeding Methods	Special Features
Sapota		
CO-1	Cricket Ball × Oval	
CO-2	Clonal selection from Baramasi	
CO-3	Cricket Ball × Vavilavalasa	Suitable for High density planting (HDP)
PKM-1	Clonal selection- Guthi	Production of two types of fruit shape- Oval & Round
PKM-2	Guthi × Kirthi Barthi	
PKM-3	Guthi × Cricket Ball	Suitable for high density planting (HDP), Cluster bearing
PKM-4	Clonal Selection-PKM-1	Spindle shaped fruit (Pinkish flesh)
PKM-5	Clonal Selection- Virdhunagar	Offseason variety
Annona		
	APK(Ca)-1: Clonal selection from Courtallam in Tamil Nadu	Drought tolerant and suitable for arid tracts of Tamil Nadu
Papaya		
CO-1	Selection- Ranchi type	

CO-2	Selection- Local type	Suitable for papain extraction
CO-3	CO-2 × Sunrise Solo	Suitable for home gardening
CO-4	CO-1 × Washington	
CO-5	Selection- Washington	Suitable for papain extraction
CO-6	Selection- Giant	Dual purpose (both table and latex)
CO-7	CP75 (Pusa Delicious × Co-3 Coorg) × Honey Dew	Gynodioecious variety
CO-8	derived through improvement of CO.2	Dioecious, Red pulped
Mango	PKM-1, PKM-2, Paiyur, Banganapalli, Bangalora, Neelum, Rumani, Mulgoa, Alphonso, Senthura, Kalepad, Imam Pasand.	

Important pests and diseases in fruits:

- Mango Nut weevil: *Cryptorrynchus mangifera*
- Mango mealy bug*: Drosicha mangiferae*
- Fruit fly: *Daccus dorsalis*: Major problem in export of mango fruits
- Anthracnose (*Colletotrichum gloeosporioides*): Most serious storage disease of mango
- Vapour heat treatment (VHT) is done to control mango fruit fly and stone weevil
- Rhizome or Corm weevil: *Cosomopolitus sordidus*: Major pest
- Pseudo stem weevil: *Odiphorous longicollis*
- Panama wilt in banana is caused by *Fusarium oxysporium* f sp. *cubense*
- Yellow sigatoka leaf spot (*Mycosphaerela musicola*): Most serious disease in subtropical regions
- Guava wilt is caused by *Fusarium* spp.
- Guava fruit fly (*Daucus dorsalis*): More dangerous pest during Rainy season fruit crop
- Citrus canker caused by bacteria *Xanthomonas campestris*
- Guava wilt is a serious disease most common in alkaline soil
- Heart rot or Stump rot (*Phytophthora parasitica*): Emits foul smell-Predominant in alkaline soil- Severe in dry and wet regions
- Apple scab- *Venturia inequalis:* Fungal disease
- Woolly apple aphid (*Eriosoma langierum*) is the most devastating pest in the world
- San Jose Scale (*Quadraspidiotus perniciosus*) is the most serious pest of apple
- Apple scab is caused by fungus *Venturia inaequalis*: Most serious disease in the world
- Viral diseases- Tristeza, Exocortis in citrus
- Tea mosquito bug important pest of guava and cashew
- Ber Fruit fly (*Carpomiya vesuviana*) is the major pest of ber. Mainly infest at pea stage
- Pomegranate (Anar) butterfly or fruit borer (*Virachola isocrates*) Serious pest all over India
- Pomegranate Bacterial blight (*Xanthomonas oxonopodis* pv. *punicae*)

Important diseases of fruit crops and their vectors:

Diseases	Vectors
Bunchy top of banana	Banana aphid (Pentalonia nigronervosa)
Banana Bract Mosaic Virus (BBMV)	Aphids
Banana streak virus (BSV)	Mealy bug
Greening (MLOs)	Citrus psylla
Papaya ring spot virus	Aphids
Papaya leaf curl	White fly
Citrus Tristeza	Aphids (Toxoptera citricida)
Pine apple wilt (virus)	Mealy Bug

Physiological disorders:

Crops	Disorders	Causes and Remedies
Mango	Spongy tissue	Alphonso susceptible
	Mango malformation (serious in North India)	Susceptible: Bombay green Resistant varieties: Illaichi, Bahaduran, Manjeera, Aliff • Deblossoming, • Spraying NAA 200 ppm in October
	Black tip(chimney disease)	Brick kiln+ Bo deficiency
Banana	Improper bunch filling	K deficiency
	Kottai vazhai	Susceptible variety: Poovan
	Neervazhai	Susceptible variety: Nendran
Citrus	Granulation	Sweet orange susceptible
	Leaf mottling/frenching	Zn deficiency
	Exanthema/dieback	Cu deficiency
Grapes	Hen and chicken	Boron deficiency
	Millerandae or Shot berries	Boron deficiency
Guava	Bronzing	Zn deficiency
Aonla	Fruit necrosis	Boron deficiency
Pomegranate	Fruit cracking	Boron deficiency
Apple	Bitter bit	Ca deficiency
	Hardy corky tissue, fruit cracking, Blossom blast	Boron deficiency
Pear	Fruit cracking	Boron deficiency
	Black end, Cork spot	Calcium deficiency
Pine apple	Multiple Crown/Fasciation	Genetical/nutritional factors

Miscellaneous

- Papain is used for digests proteins and used as a meat tenderizer
- Climacteric fruits: Mango, Banana, Apple, Sapota, Guava
- Non-climacteric fruits: Citrus fruits, Pineapple, Pomegranate
- Micro grafting: transferring small shoot apices on to rootstocks. It is done through in vitro or in vivo used for recovery of citrus clones free from virus diseases
- Shoot tip grafting (STG): done for free from virus/virus like organisms in Citrus spp.
- Kinnow mandarin: King (*C. nobilis*)× Willow leaf (*C. deliciosa*)
- Kinnow mandarin Interspecific hybrid developed by Dr. H.B. Frost at Citrus Experiment Station, California, USA in 1915. Most popular in Punjab and Haryana.
- Bud forecasting-Grapes
- Nipping-Grapes
- Cultivation of grapes is known as Viticulture
- Science of wine making: Enology
- Red colour in grapes due to anthocyanins (Malvin 40-60%)
- For uniform bud break: Swabbing of buds with hydrogen Cyanamide: 1.5-4% within 24-48hrs for forward pruning
- Bud breaking in South India: Thiourea @ 4%
- Ber is a summer deciduous fruit tree
- Nectarines is a smooth skinned peaches

Parthenocarpy

- Vegetative parthenocarpy: Banana, fig, pine apple
- Stimulative parthenocarpy: Litchi, Bread fruit, Black Corianth varieties of grapes
- Stenospermocarpy: Grapes, Sindhu (Seedless variety of mango)

10.2. OLERICULTURE

☆ Olericulture word derived from Latin word oleris means potherb and the English word culture means culture of vegetables

Crops	Origin	Economic Parts Used	Fruit Type
Tomato	South America (Peru)	Fruit	Berry
Potato	South America (Peru)	Tuber (Modified stem)	Berry
Onion	Central Asia	Bulbs (Modified leaves)	
Brinjal	India	Immature fruit	Berry
Chilli	Tropical America	Fruit	Berry
Okra	South Africa or India	Tender pods	Capsule

Cole Crops			
Cabbage	Mediterranean region	Head	Siliqua
Cauliflower	Mediterranean region	Curd (Pre-floral apical stem)	Siliqua
Knol-Khol	Mediterranean region	Knob (Modified stem)	Siliqua
Carrot	Afghanistan and Iran	Root	Schizocarp
Cucurbits			
Cucumber	India	Immature fruit	Pepo
Musk melon	Tropical Africa	Fruit	Pepo
Pumpkin	South America	Ripened fruit	Pepo
Watermelon	South Africa	Fruit	Pepo

Crops	**Varieties**	**Special Characters**
Tomato	Arka Samrat Arka Rakshak	Resistance to ToLCV, bacterial wilt and early blight
	Arka Ananya	Resistance to ToLCV and bacterial wilt
	Arka Abhijit	Resistant to bacterial wilt
	Arka Abhay, Arka Alok	Bacterial wilt resistant
	Arka Vardan, Arka Vishal	Resistant to root knot nematode
Potato	Kufri Swarna	Resistant to golden cyst nematode
	Kufri Jyoti	Resistant to late blight
	Kufri Surya	Early maturing, heat tolerant (20°C) and hopper-burn resistant potato variety°
	Kufri Chipsona-1, Kufri Chipsona-2, Kufri Himsona Kufri Surya	Suitable for processing (French fries, chips)
	Kufri Swarna and Kufri Dewa	Resistant to golden cyst nematode
Chilli	Arka Sweta, Arka Meghana Arka Harita, Arka Kyathi	New CMS based hybrids
	CH-1, CH-3 (GMS based)-PAU, Punjab	First F_1 hybrid from public sector
	Arka Lohit, Arka Abhir	Suitable for colour extraction
Okra	Pusa Sawani	First yellow vein mosaic virus resistant
	Arka Anamika, Arka Abhay	Yellow vein mosaic virus resistant
	Punjab Padmini	
	Parbhani Kranti	
Onion	Arka Vishwas	Suitable for growing in Rabi season and suitable for export

	Arka Swadista	White colour onion
	Arka Sona	Suitable for growing in Rabi season and suitable for export
	Arka Bindu	Pickling type variety
	N-53	Suitable for kharif (rainy) season
	Bhima Shwata, Bhima ShubraBhima Safed	White colour onion varieties
Aggregatum Onion	CO-2	Propagation done by seed
Sweet potato	Sree Kanaka, Sree Rethna	â-carotene rich variety
Cassava	Sree Padmanabha	CMD resistant cassava variety
	Sree Visakam	â-carotene rich variety
Amaranthus	Badi Chaulai (*A. tricolor*)	Suitable for summer and rainy season
	Chhoti Chaulai (*A. blitum*)	Suitable for spring-summer season
	Pusa Kiran	Rainy season variety
	Pusa Kirti	Summer season variety
	Pusa Lal Chaulai	Stem deep red colour, red dye extraction Suitable for kitchen gardening and ornamental variety
Palak	Pusa Jyoti	Selection - colchicine induced polyploid (2%)-
	Pusa Harit (Sugar beet × Local Palk) for hills	Late bolting type, tolerant to alkaline soil Suitable
Cucumber	Pusa Uday, Pusa Barkha	New varieties
Water melon	Arka Muthu, Arka Akash, Arka Manik	New varieties
	Arka Jyoti, Arka Madhura, Arka Aiswarya	Hybrids
Pumpkin	Arka Chandan	Suitable for home gardening

Important vegetable varieties released from TNAU

Crop	**TNAU**	**Breeding Methods**	**Special Features**
Tomato	CO-1	Selection from Kalyanpur	Semi-determinate
	CO-2	Introduction from Russia	Indeterminate
	CO-3 (Marutham)	Mutant of CO-1	Determinate
	PKM-1	Induced Mutant of Annanji	Determinate type, Green flesh, Suitable for long distance transport
	Paiyur-1	Pusa Ruby × CO-3	Rainfed variety

Tomato Hybrids: COTH1, COTH2, TNAU Tomato Hybrid CO 3			
Chilli	CO-3	-	Dual purpose
Brinjal		-	
	MDU-1	-	Purple colour
	Annamalai	-	Aphid resistant
	KKM-1	-	White colour,Suitable for diabetics patient
Okra	MDU-1	Mutant of Pusa Sawani	
Bittergourd	MDU-1	Induced Mutant	-
Snake Gourd	MDU-1	F1 Hybrid	-
	CO-2	Short Fruit Type	-
Moringa	PKM-1	Propagation by Seed	Annual type
	PKM-2	Propagation by Seed	Annual type, long fruited type, Suitable for Kitchen garden
Sweet Potato	CO-3		Orange flesh-High â-carotene variety
Cassava	CO-2		Suit for starch industry
	CO-3		Cassava mosaic virus (CMV) resistant
Aggregatum onion	CO-(On)-5	Propagation by seed	-
Amaranthus	CO-1	*A. dubius*	-
	CO-2	*A. tricolor*	-
	CO-3	*A. tristis*	Clipping type
	CO-4	*A. hypochondriacus*	Suitable green cum grain type
	CO-5	*A. blitum*	Induced tetraploid variety (Sirukeerai)
Cluster bean	MDU-1	-	Long stem type

Origin, Economic Parts and Fruit Types of Important Vegetable Crops

Important vegetable varieties released from TNAU

Important nutritional aspects in vegetable crops:

★ Micronutrient malnutrition is a lack of essential vitamins (Vitamin A, B, and C) and minerals (Zn, Fe, Ca)

★ Vitamin A is essential for vision, immune response, epithelial cell growth and repair, bone growth, reproduction and maintenance of the surface linings of the eyes

★ VAD. Vitamin A deficiency: Night blindness, Conjunctival xerosis

★ Xanthophylls (lutein and zeaxanthin): plant pigments that selectively accumulate in the macula of the retina of the human eye.

- ★ Lutein and zeaxanthin are found in green, certain yellow/orange vegetables and squash
- ★ Tomato red colour is due to lycopene
- ★ Red colour in Chilli is due to presence of capsanthin
- ★ Pungent principle in chilli is capsaicin
- ★ Heat units is the maturity index of garden pea
- ★ ICMR recommendation for Balance Diet:
  - 300 g of Vegetables/Day, 125 g Green Leafy vegetables, 100 g Root and tuber vegetables and 75 g Other Vegetables
- ★ Quercetin- Anti-cancer activity e.g. Onion and Garlic
- ★ Bitter gourd: Anti-diabetic (Hypoglycemic) Cheratin
- ★ Cabbage: Anti-cancer property (Indole-3-Carbinol)
- ★ Broccoli: Anti-cancer property (Sulphoraphane)
- ★ Red colour in Beetroot is due to â-cyanin whereas Yellow colour is due to â-xanthins

Health promoting compounds present in vegetable crops

Compounds	**Source of vegetables**	**Biological activity**
1.Carotenoids		
â carotenes	Carrot, kale, pumpkin, spinach	Provitamin A activity
Lycopene	Tomato, water melon	Antioxidant and anticancer activity
Lutein	Kale, spinach degeneration (AMD)	Prevention of age-related macular
Zeaxanthin	Lettuce, parsley degeneration (AMD)	Prevention of age-related macular
2. Polyphenols		
Anthocyanidins	Red cabbage, purple cauliflower	
Cyanidin	Purple carrot, red onion	Antioxidant and anticancer activity
Malvidin	Rhubarb	Reduction cardio vascular disease (CVD)
Nasunin	Purple brinjal	Antioxidant activity
Flavonols		
Quercetin	Yellow onion, broccoli, curly kale	Anti-obesity activity
Myricetin	Garlic Broccoli, cherry tomato	Anti-diabetes activity
Kaempferol	Leek	Antioxidant and anticancer activity
Flavones		
Apigenin	Celery, parsley	Antioxidant and anticancer activity
Luteolin	Capsicum	Anti-obesity activity
Isoflavones	Garden pea	Anticancer and CVD activity

Phenolic acids		
Chlorogenic acid	Black carrot	Anti-obesity activity
3. Glucosinolates		
Glucoraphanin	Green and purple broccoli Purple cauliflower, red cabbage	Anticancer activity
Glucoiberin	Cauliflower, Cabbage, Brussels sprouts, kale, Purple cauliflower	Anticancer activity
Sinigrin	Cauliflower, cabbage, kale Brussels sprouts	Anticancer activity
Glucoraphasatin	Radish	Prevention of Cholesterol gallstones
4. Sulphides/thiols		
Methiin	Onion	Antibacterial and antifungal
Alliin	Garlic	Antibacterial and antifungal
5. Phenolic compounds		
Chlorogenic acid	Carrot (purple), potato Sweet potato, red colour lettuce	Antichronic activity anti-obesity
Betalains		
Betacyanins		Antioxidant and anticancer activity
Betaxanthins		Antioxidant and anticancer activity
Folate	Spinach broccoli, Brussels sprout depression disorders	Anti-neural tube defects (NTDs) and
Vitamin C	Green chilli, Tomato, Cauliflower Cabbage, Turnip leaves, Brussels sprout	Prevention of scurvy

Toxic substances present in vegetable crops:

- ♣ Phytic acid: Beans and Peas
- ♣ Oxalic acids and Oxalate: Leafy vegetables
- ♣ Calcium oxalate: Colocasia and Elephant foot yam (Acridity)
- ♣ Hydrogen cyanic acid (HCN): Cassava
- ♣ Solanine: Potato, Tomatine, Tomato
- ♣ Glucosinolates: Cole Crops
- ♣ Solasodine: Brinjal
- ♣ Serotonin: Watermelon

Temperature requirement for vegetable crops:

- ☆ Cauliflower thermosensitive crop
- ☆ Average temperature for lycopene formation in tomato: 21-24°C

- ☆ Lycopene production drops off rapidly above 27°C
- ☆ Potato do not form tuber above 21°C
- ☆ Ideal temperature for onion bulb development: 15.5-25°C
- ▪ Seed to seed method (in situ): Practiced for production of foundation and certified seed
- ▪ Head to seed method: Only for nucleus and breeder seed production
- ☆ Pinching of male flowers on female plant is commercially practiced for hybrid seed production
- ☆ Seed shattering is a major problem for okra seed production
- ☆ Agra Petha is a famous sweet prepared from ash gourd

Important pest and diseases of vegetable crops:

- ☆ Charcoal rot is a most serious storage disease of potato
- ☆ Common Scab of potato is most serious disease in alkaline soil
- ☆ Phomopsis blight (Seed borne) is a major problem in seed production of Brinjal
- ☆ Anthracnose or Fruit Rot or Bird Eye's Spot (seed borne) is a serious diseases in chilli
- ☆ Potato Wart (Fungi) is endemic to Darjeeling region
- ☆ Wart and Bacterial wilt is tuber borne disease of potato
- ☆ Little leaf of brinjal is transmitted by Jassids/Leaf hoppers
- ☆ Tomato Leaf Curl Virus (TLCV) is transmitted by whitefly
- ☆ Tomato Spotted Wilt Virus (TSWV) is transmitted by thrips
- ☆ Yellow Vein Mosaic Virus (YVMV) in okra is transmitted by white fly
- ☆ Fruit fly is very serious pest of most of the cucurbitaceous vegetables
- ☆ Cucumber mosaic is transmitted by aphids
- ☆ Gummy stem blight: *Didymella bryoniae*-major problem in hot and humid conditions (tropical, subtropical regions and greenhouses).
- ☆ Musk melon Powdery mildew: *Podosphaera xanthii*
- ☆ Big vein of Lettuce is caused by Virus, which is transmitted by *Olpidium brassicae* fungus
- ☆ Shoot and fruit borer (*Earias vittella*) is the most common occurring pest in okra
- ☆ Most devastating disease of okra: Yellow Vein Mosaic Virus (YVMV): Vector: Whitefly (*Bemisia tabaci*)
- ☆ Potato golden cyst nematode (*Globodera rostochinensis*) is major problem in Southern Hills
- ☆ Cucumber green mottle mosaic virus (CGMMV): Transmitted by seed and mechanical
- ☆ Most serious pest in sweet potato: Sweet Potato weevil (*Cylas formicarius*) severe damage in summer crop
- ☆ Cassava mosaic virus (CMV) is most serious transmitted by whitefly (*Bemisia tabaci*)

Important physiological disorder in vegetable crops:

Crops	Disorders	Causes
Tomato	Fruit Cracking	Boron deficiency
	Blossom End Rot (BER)	Calcium deficiency
	Puffiness	Poor pollination
	Cat Face	Abnormal growing condition
	Blotchy Ripening	Potassium deficiency
Cauliflower	Riceyness	Unfavourable temperature
	Leafyness	High temperature
	Browning or Red rot	Boron deficiency
	Whip tail	Molybdenum
Potato	Black heart	Poor ventilation
	Collar rot	Excessive N_2 application
Radish	Pithiness	excess NPK and soil moisture stress
Carrot	Carrot splitting	Excessive N, Boron deficiency, Change in soil moisture
	Cavity spot	Excessive N, Ca deficiency
Garlic	Bulb sprouting	excessive N2 and soil moisture

FLORICULTURE

- ★ **Hedges**: Shrub is planted on boundary for fencing, used for ornamental and protective purpose e.g. Lantana, Inga
- ★ **Edges** (20-30cm): Low growing perennial plants are grown on the border of plots or beds e.g. *Irisine, Alternanthera*
- ★ **Topiary:** Art of training plants into different shapes like birds, animals, domes and Umbrellas e.g. *Clerodendraon, Duranta, Thuja*
- ★ **Herbaceous Border:** Planting of annuals in the border of plots
- ★ **Hanging Baskets**: Trailing of plants in container which is suitable for indoor and outdoor conditions e.g. *Verbena, Petunia, Zebrina*
- ★ **Carpet Bedding:** Covering on area of bed or series of beds with dense low growing herbaceous plants e.g. *Alternanthera, Coleus, Portuluca*
- ★ *Cassia fistula* is known as Indian Laburnum or Gold shower tree (Yellow flower colour)
- ★ Formal style garden (Symmetrical axis): Persian Gardens and Mughal Gardens
- ★ Informal style garden (Asymmetrical axis): Japanese Gardens
- ★ Free style garden: Rose Garden of Ludhiana, English Garden, Lal Bagh
- ★ **Pinching**: Removal of apical shoots leaving about 5-6 nodes on the plants. It is done at 30-35days after planting. It improves the quality of flower
- ★ Pinching practice is done in carnation and chrysanthemum

- ★ Disbudding is followed in carnation and chrysanthemum
- ★ **Disbudding**: To remove the unwanted buds. It improves the size of flower
- ★ Pot-pourri is known as adding fragrance to dry flower arrangement
- ★ Lawn is known as "**Heart of Garden**"
- ★ Bermuda Grass or Hariyali (*Cyanodon dactylon*) is the most commonly used Lawn grass in Playground, Cricket ground
- ★ Quickest method of lawn planting: Turfing
- ★ Dibbling is the most common and cheapest method of lawn planting
- ★ Japanese flower arrangement is known as "**Ikebana**"
- ★ **Bonsai**: Art of growing miniature trees and shrubs by extreme dwarfing in pots
- ★ Bonsai is originated in China but famous in Japan
- ★ Japanese garden is also known as "**Nature in Miniature**"
- ★ Crossandra wilt (*Rhizoctonia* and *Fusarium*) is transmitted by Root lesion nematode (RLN) (*Pratylenchus sp.*)
- ★ Bud opening Solution: High concentration of sucrose + biocides e.g. Rose
- ★ Vase solution: Low concentration of sucrose + biocides
- ★ Rose - Queen of Flowers
- ★ Dahlia - King of Flowers
- ★ Largest producer of cut flower in the world: Netherlands
- ★ Largest producer of loose flower in India- Tamil Nadu
- ★ Largest producer of cut flower in India- West Bengal
- ★ Anthurium ornamental part: Spathe
- ★ Bougainvillea ornamental part: Bracts
- ★ Netting is common cultural practice in Carnation
- ★ Highest oil content in *Jasminum sp.*: *Jasminum auriculatum* (0.36%)
- ★ Delhi Crossandra is propagated by stem cuttings
- ★ Blue colour in rose is due to Delphinidin
- ★ Blue colour rose 1st developed in Japan
- ★ Commercial root stock for Rose: *Rosa indica* var. *odorata*
- ★ Gladiator is very popular variety among rose growers of Nasik and Pune (Open condition)
- ★ Chocolate brown colour patented rose variety: Mohini (Aneuploidy)
- ★ Highly fragrant rose variety: Rose Sherbet
- ★ Chrysanthemum variety: Pusa Anmol is photo and thermo- insensitive and therefore flowers thrice in a year (Oct-Nov, Feb-Mar and May-Jun).
- ★ Cymbidium (Temperate orchid) is now among the top 10 cut flowers of the world market
- ★ Most widely cultivated tropical orchid: Dendrobium
- ★ Type of inflorescence in Anthurium: Spadix
- ★ Gladiolus commercial propagation: Corms
- ★ Gladiolus inflorescence: spikes

Gardens and Floricultural Research institutes:

- ★ Bryant Park is located at Kodaikanal, Tamil Nadu
- ★ Botanical Garden is located at Ooty, Tamil Nadu
- ★ International Flower Auction Center (IFAC) is located at Bangalore, Karnataka
- ★ Largest flower market is located at Alsmeer, Netherlands

Propagation methods in flower crops:

Crops	Propagation
Rose	T Budding
Gladiolus	Corms
Carnation	Terminal Cuttings
Marigold	Seeds
Chrysanthemum	Cuttings, Suckers
Jasmine	Ground Layering
Tuberose	Bulbs
Crossandra	Seeds

- ♣ Leaf cutting- Begonia, Bryophyllum, Sansevieria
- ♣ Leaf bud cuttings: Hydrangea
- ♣ Hardwood cuttings: Fig, Grape, Rose
- ♣ Semi hard wood cuttings: Coleus, Geranium
- ♣ Softwood cuttings: Jasmine
- ♣ Herbaceous cuttings – Coleus, Geranium, Sweet potato
- ♣ Root cuttings – Red raspberry, Bread fruit

Important varieties in flower crops:

Crops	Varieties/features
Rose	
Hybrid tea	Poornima, Priyadarshini, Abhisarika, Abhaya, Pusa Sonia
Floribundas	Pusa Barahamasi, Pusa Pitamber, Pusa Virangana
Mullai- *Jasminum auriculatum*	CO-1 -Resistant to Gall mites
	CO-2
	Pari Mullai
Pitchi- *Jasminum grandiflorum*	CO-1-Pitchi
	CO-2-Pitchi
	Arka Surabhi

Tuberose	
Single flowered	Rajat Rekha, Prajwal, Srinagar
Double flowered	Suvarna Rekha, Vaibhav, Suwasini
Crossandra	Arka Kanaka (bright orange colour)
	Arka Ambara (Biggest corolla size, orange red colour)
	Co-1 (yellowish orange), Soundarya (Pink colour)
French marigold	Pusa Deep: Maroon colour Pusa Arpita: Orange colour
African marigold	Pusa Narangi Gainda: Orange colourPusa Basanti Gainda: Sulphur yellow colour Pusa Bahar: Yellow colour
Carnation	Arka Flame: Tolerant to Fusarium wilt and nematodesArka Tejas
Hibiscus	Thilagam, Punnagai and CO-3

Important varieties in flower crops:

Physiological disorders

- ♣ Rose Bent neck: Too early maturity
- ♣ Rose: Blackening of petals is due to low temperature or high temperature
- ♣ Carnation: Calyx splitting is very common in double carnation, Factors: Nutritional, environmental and genetical factors
- ♣ Sleepiness disorder is due to production of high ethylene/high stress
- ♣ Curly tip disorder is due to low temperature, low light and N2 deficiency
- ♣ Slab side is type of malformation during cool periods

10.3. SPICES

- ★ The term 'Spice' was derived from the Latin "species aromatacea", means 'Fruits of the earth'
- ★ Most of the spices are native of India (Land of Spices)
- ★ Spices not only add flavour and taste to our food but also enhance keeping quality and medicinal values of food
- ★ Spices are high value and low volume commodities of commerce in the world market.
- ★ India being the world leader in seed spice production, consumption and export
- ★ Home of spices/Spices bowel of India- Kerala

Black Pepper- *Piper nigrum*- Piperaceae **- King of Spices (2n = 52), Origin-Western ghats**

- o Perennial climbing vine
- o Commercially propagated by runner shoots
- o Piperine is responsible for pungency, average content of Piperine- 1.7–7.4%
- o Bush pepper means by planting lateral branches of pepper, the vine can be grown as a bush.,

- o Bush pepper is propagated by plagiotrophic shoots
- o Inflorescence type- pendent spike (Catkin)
- o Pepper is naturally self-pollinated crop due to presence of geitonogamy
- o Single seeded berry, Edible portion- Fleshy pericarp and hard endocarp
- o Black pepper is generally harvested at fully mature and nearly ripe stage, whereas white pepper at fully ripe stage
- o Volatile oil is responsible for the aroma and flavour
- o Drying recovery ranges from 28–38%
- o Drying recovery:
- ♣ Black pepper: 33%
- ♣ White pepper: 25%
- o Oleoresin is produced by solvent extraction of pepper powder
- o Pepper yellows (or slow decline) complex disease caused by the Burrowing nematode (*Radopholus similis*) and *P. capsici*

Cardamom- *Elettaria cardamomum, Zingiberaceae* (2n = 48), Origin- Western ghats

- o Small cardamom popularly known as **'Queen of Spices'**
- o Shade loving plant, propagated by Rhizomes
- o Generally cured cardamom have 12% moisture, curing temperature 45–50ºC @ 10–12hrs
- o Large cardamom or Nepal cardamom: *Amomum subulatum,* Origin- Eastern Himalayas, Propagated by suckers
- o Katte viral disease is transmitted by aphids (*Pentalonia nigronervosa* f sp. *cardamomi*)
- o Azhukal or capsule rot is caused by *Phytophthora spp.*

Ginger-*Zingiber officinale,* Zingiberaceae (2n=22), Origin-South East Asia

- Major pungent principle of ginger: Gingerol
- Commercially propagated by Rhizomes
- Unbleached ginger : Peeled rhizomes washed and sun drying
- Bleached ginger: Peeled rhizomes soaked in 2% lime H_2O for 6 hrs
- Rio-de-Janeiro is most popular variety
- Recovery of dry ginger**: 16-18%**
- Major pungent principle of ginger: Gingerol
- **Varieties**: IISR-Varada, Suprabha, Suruchi, Suravi, Himagiri, IISR-Mahima, IISR-Rejatha
- Soft rot or rhizome rot: *Phythium aphanidermatum,* soil borne disease

Turmeric-*Curcuma longa*, Zingiberaceae (2n =3x = 63), Origin- South East Asia

- It is used as a condiment, dye, drug and cosmetic in addition to its use in religious ceremonies
- Curcumin is the golden-yellow pigment present in turmeric. It is used at levels of 5–20 ppm.
- India is a leading producer and exporter of turmeric in the world

- Curcumin (4-7%) is the principle colouring pigment in turmeric
- Curcumin is used as therapeutic effects and anti-cancer property
- Varieties: Prabha, Prathibha, Sudarsana, Suguna, Co-1, Sugandham, BSR-1
- Curcumin (4-7%) is the principle colouring pigment in turmeric
- Processing 3 steps: Curing => Polishing => Colouring

10.4. SEED SPICES

Coriander: *Coriandrum sativaum*: Apiacae: Origin: Mediterranean region

- ♣ Fruit- Schizocarp
- ♣ Coriander has been used as an antispasmodic, carminative, stimulant, and stomachic
- ♣ Black cotton soil is more suitable for coriander rainfed cultivation
- ♣ Direct sown crop: Done by broadcasting
- ♣ Seed rate:
- Irrigated condition:10-16 kg/ha
- Rainfed condition: 15-20 kg/ha
- ♣ **Varieties:**
- CO-1- Pure line selection-Suitable for green and grains
- CO-2- Reselection from culture of P2 of Gujarat- Dual purpose
- CO-3- Reselection from Acc. No. 695- Dual purpose
- ♣ Coriander wilt is caused by *Fusarium oxysporium* (Dropping of leaves, Epinasty)
- ♣ Powdery mildew: (*Erysiphe polygoni*) serious disease in rainfed crop

Fenugreek: *Trigonella foenum-graecum*: Fabaceae: 2n=16, Origin: Mediterranean origin

- Seeds contain 'Diosgenin' used in preparation of contraceptives
- Common methi: *Trigonella foenum-graecum*- White flower, Straight pods
- Kasuri methi: *Trigonella corniculata* (Multiple cutting)- Rosette leaves, Long vegetative growth period, Bright orange to yellow, curved or sickle-shaped pods

Vanilla-*Vanillia planifolia*- Orchidaceae **-Princes of Spices**

- ♣ Shade loving plant, climbing vine
- ♣ Vanillin (2-2.5%) the main flavouring compound of vanilla
- ♣ Propagated by Stem cuttings
- ♣ Vanillin is developed as a result of â-glycosidase action on the glucosides in vanilla beans during the process of curing
- ♣ Vanillin is extracted by Hydro-alcoholic extraction method
- ♣ Main pollination problem: Vanilla flower contains complex flower parts, due to presence of *rostellum* makes natural pollination is impossible

- ♣ Artificial pollination is necessary for fruit setting
- ♣ Vanilla fruits is harvested at fully mature and ripe stage

Tree Spices:

- ★ **Clove:** Economic part- Dried unopened flower bud- Dioecious tree
- ★ **Problem in Clove**: Alternate bearing habit
- ★ Propagated by seeds or mother cloves
- ★ **Nutmeg**- Economic part- Red aril and Mace, Dioecious Tree
- ★ **Problem in Nutmeg**: Dioecious tree, Top working is important operation for unproductive male plants converted into productive female trees
- ★ **Cinnamon**- Economic part- Bark
- ★ All Spice (*Pimento dioica*)- Economic part- Dried immature fruits
- ★ Its flavour is a mixing of Cinnamon + Clove + Nutmeg
- ★ Tamarind (*Tamarindus indicus* L.), Origin: Tropical Africa
- ★ Tamarind: Contains tartaric acid (8%), **Varieties:** PKM-1(Cluster bearing), Urigam

Other Spices and Condiments:

- ♣ Garlic-True odour- Allicin (Anti-bacterial properties)
- ♣ Garlic Long day plant, commercially grown in Ooty, Kodaikanal
- ♣ Garlic propagated by cloves (500 kg/ha)
- ♣ Saffron-economic part- Dried stigma
- ♣ Responsible for the colouring agents of saffron: Cicrocrocin
- ♣ Asafoetida- Economic part- Oleo gum
- ♣ Condiment paprika is a variant of Capsicum
- ♣ The colour in paprika is due to carotenoids, namely capsanthin and capsorubin, comprising 60% of total carotenoids

10.5. PLANTATION CROPS

Tea- *Camellia sinensis*-Theaceae-Queen of Beverage Crop, Origin- China, Calcifuge crop

- ✦ India is the largest producer, consumer and exporter in world tea industry
- ✦ Responsible for briskness, brightness and colour of the liquor: Aflavins and arubigins
- ✦ Responsible for colour of black tea/major polyphenols: Theaflavins and Thearubigens
- ✦ Major phenolics in green tea: Catechins
- ✦ Variety UPASI-9 (Athrey)-most suitable for higher pH soils
- ✦ UPASI-10 (Pandiyan)- Resistant to drought and winds, suitable for higher elevations
- ✦ Propagated by Single node cuttings
- ✦ In South India, Silver oak (*Grevillea robusta*) is used as a permanent shade tree

- Skiffing is the lightest form of pruning
- Collar pruning is the severe most pruning for rejuvenation of tea garden
- Tipping is the practice of removal of terminal portion of shoot
- Best plucking or harvesting in tea is two leaves and a bud
- Yellowing of tea is due to sulphur deficiency

Processing of Tea:

- CTC (Crush, Tear, Curl): Withering=>Rolling=> Fermentation =>Drying=> Grading

Pest and disease

- Red coffee borer (*Zeuzera coffeae*) is the major pest of tea
- Mites is a serious problem; Major mite: Pink mite (*Acaphylla theae*)
- Blister blight of tea is caused by *Exobasidium vexans*- Translucent spot on tender leaves and stem

Coffee-*Coffea arabica* – Rubiaceae -(2n=22), Origin- Ethiopia, Fruit- Drupe

Important species:

- Arabica coffee: *Coffea arabica*- 2n=4X=44, Self-pollinated, self-fertile, allotetraploid, high elevation, shrub, Origin: Ethiopia
- Robusta coffee: *C. canephora*- 2n=2X=22, Cross pollinated, self-sterile, lower elevation, tree

Origin: Central Africa

- Tree coffee: *Coffea liberica*
- **Temporary shade tree**:
- Dadap (*Erythrina indica)*-Most commonly used in India and Silver oak *(Grevillea robusta)*
- **Permanent shade tree**:
- *Albizzia lebbeck, Cedrella tonna, Dalbergia latifolia, Ficus glomerata, Maesopsis eminii*
- Permanent shade for coffee is (*Grevillea robusta*) and temporary shade is *Indigofera teysanonii*
- Pruning system in coffee is single stem open centre system
- Topping is a cultural practice in wet coffee processing
- Central Coffee Research Institute (CCRI) is located at Balahanur, Karnataka.
- Cauvery (Caturra × Hybrido-de-Timor) is popular hybrid derivative in India
- Coffee leaf rust is caused by *Hamelia vestatrix* introduced from Sri Lanka
- Permanent shade for coffee is (*Grevillea robusta*) and temporary shade is *Indigofera teysanonii*
- San Roman is a mutant variety
- Coffee leaf rust (Hamelia vestatrix) introduced from Sri Lanka
- Coffee berry borer (*Hypothenemus hampei*): Most serious pest in the world

Cocoa-*Theobroma cocoa*- Malvaceae- (2n=20) - **Food of God**- Beverage crop

- Native to Amazon valley of South America, Propagation: seed, Fruit type: Drupe
- Rich source of Fat (37%) and Protein (7%)
- Largely grown as a homestead crop (Mixed Crop) in Kerala in the coconut and arecanut gardens

- ☆ Cocoa is considered as a functional food because rich source of polyphenols and antioxidant properties
- ☆ Centering (Cultural operation) is a process of removal of lateral branches
- ☆ Two major varieties (Naturally evolved):
- ✦ Criollo-Dark red pod colour
- ✦ Forestero-Yellow pod colour
- ☆ Forstero is a recommended variety by CPCRI, Kerala
- ☆ Bearing habit- Cauliflorous, Self-incompatibility, Natural pollination by Insects,
- ☆ Major disease: Black pod or stem canker disease (*Phytophthora palmivora*)- Pods turn chocolate brown to black
- ☆ Serious pest: Cocoa mealy bug (*Planococcus lilacinus*)

Coconut-*Cocos nucifera-Araceae,* Kalpavriksha (2n=32) - **King of Species-Tree of Heaven**

- ✦ India is the 3rd largest producer after Indonesia, Philippines
- ✦ Monoecious palm
- ✦ Coconut oil contains lauric acid
- ✦ Suitable for homestead system of farming , mixed cropping and intercropping
- ✦ Inflorescence type- Spadix, Pollinators- Honey bees
- ✦ Milling copra is most popular in Southern India (Kerala: 60-65%)
- ✦ Cup copra is most popular in Northern India
- ✦ Coir is obtained from mesocarp
- ✦ Multitier cropping is followed in coconut gardens
- ✦ Tapping of coconut is done for toddy
- ✦ Coconut Research Station, Veppankulam, Pattukottai.
- ✦ Tall cultivars: Kalpa Mitra, Kalpa Pratibha, Kalpa Thenu, Kalpa Tharu, Kalpa Haritha, Chandra Kalpa
- ✦ Dwarf cultivars: Kalpa Raksha, Chowghat Orange Dwarf (COD), Kalpa Sree (oil purpose), Kalpa Jyothi (yellow), Kalpa Surya (orange)

Pest and disease

- ✦ Rhinocerous beetle (*Orycetes rhinoceros*): Characteristic geometric cut (inverted 'V' shape)
- ✦ Red palm weevil (*Rhynchophorus ferrugineus*): Oozing out of a viscous brown fluid and longitudinal splitting of leaf bases
- ✦ Eriophyid mite (*Aceria guerreronis*) is serious pest in coconut
- ✦ Stem bleeding (*Thielaviopsis paradoxa*)- Exudation of reddish brown liquid on trunk
- ✦ Thanjaur wilt (*Ganoderma lucidum*)- Bleeding patches on the stem- Root feeding
- ✦ Bud rot (*Phytophthora palmivora*): Yellowing of two young leaves and emitting foul odour
- ✦ Root wilt caused by Phytoplasma, Vector: Lace bug (*Stephanitis typica*) and Leaf hopper- (Proutista moesta) Abnormal bending of ribbing of the leaflet and marginal necrosis of leaflet-Prevalent throughout Kerala and Tamil Nadu

Physiological disorders

- Rosette or Little leaf is due to zinc deficiency
- Crown choking or Button shedding: boron deficiency
- Button shedding (Immature nut fall) is due to lack of pollination and fertilization, defects in the flowers, physiological disorders, genetic nature of the variety, pest and disease and unfavourable environment

Arecanut- *Areca catechu*, Araceae- (2n=32) - Origin: Malayan Archipelago-

- Mixed stand as a backyard crop
- Polyphenols and tannins responsible for astringent taste of nuts
- India ranks first in both area (49%) and production (50%) of Arecanut
- Mangla, Sumangla, Sreemangla, Mohitnagar is important cultivars
- Inflorescence type- Spadix, Fruit- Single seeded berry
- Chali/Kottapak: Prepared from dried ripened (9 months old) arecanut is popular in Northern India and Western India
- Kalipak is prepared from immature dark greennut (6-7 months), popular in Kerala and Karnataka

Pest and disease:

- Koleroga or Mahali disease is caused by *Phytophthora arecae*
- Anab-e-Roga is caused by *Ganoderma lucidum*
- Polyphenols and tannins responsible for astringent taste of nuts
- Yellow leaf disease (YLD) is caused by MLO (Mycoplasma Like Organism)

Rubber- *Hevea brasiliensis*- Euphorbiaceae- Origin- Brazil-

- Latex is obtained from the bark of the rubber tree by tapping
- Tapping- Process of collection of latex from bark of the tree
- Latex contains 32% dry rubber content
- Propagation: Forket budding
- Rubber Research Institute of India (RRII) is located at Kottayam, Kerala
- Alternative source for natural rubber: Guayule (*Parthenium argentatum*) as a source of high quality latex
- Clonal seeds (Highly heterogenous): Seeds of clones are termed clonal seeds
- Coagulants: Acetic or formic acid
- Anti-coagulants: Formalin, Ammonia, Sodium sulphite
- RRII-105 is very high yielding clone
- Measuring of rubber percentage: Metrolac, Latex meter
- Tapping panel dryness (TPD) or brown blast is the major physiological disorder in rubber
- Exact causes of TPD is unknown, Recommendation: Low frequency tapping
- Most common and widely followed processing: Sheet rubber
- Brown Blast: physiological disorder
- Abnormal Leaf fall is *Phytophthora botryosa* and *P. palmivora*

Oil Palm- *Elaeis guineensis*, Araceae, "**Small Holders Irrigated Crop**"- Origin- West Africa

- ♣ Oil palm: highest edible oil yielding perennial crop
- ♣ Oil palm: high yielding crop (more than 5 tonnes of oil per hectare per year) as compared to other oil yielding crops that are yielding around one tonne of oil per hectare
- ♣ It produces two distinct oils, i.e., palm oil and palm kernel oil, which have commercially used for culinary and industrial purpose
- ♣ Palm oil is derived from the fleshy mesocarp of the fruit, which contains about 45-55 % of oil
- ♣ Three variants/types:
- ▪ Dura (Shell is present)
- ▪ Pisifera (shell absent)
- ▪ Tenera (Dura × Pisifera)- High mesocarp content
- ♣ Commercially propagated by Seed, Cross pollinated crop,
- ♣ Pollinating by Weevil (*Elaeidobius kamerunicus*)
- ♣ Oil Palm is extracted from Mesocarp (Palm Oil) and Kernal (Palm Kernal Oil)
- ♣ Palm oil is rich in Palmitic Acid
- ♣ Leaf pruning is commercially followed in oilpalm
- ♣ Ablation is a removal of male and female flowers in early stage of plantation
- ♣ Stripping is a removal of berries from the bunches after harvest

Cashewnut: ***Anacardium occidentale*-** **Anacardiaceae-** Plough crop, "**Gold mine of waste land**", Dollar earning crop

- ▪ Due to its nutritive value and commerce designated as "Zero cholesterol nut or white gold"
- ▪ India largest processor, exporter, importer and consumer of cashew
- ▪ The cashew was introduced from Brazil to india (16th century) by Portuguese
- ▪ Introduced to India to cover bare hills and for soil conservation
- ▪ Propagated by epicotyl soft wood grafting
- ▪ Cashew apple is fleshy peduncle
- ▪ Type of Fruit: Drupaceous nut
- ▪ Priyanka is a export variety
- ▪ Cashewnut Research Station, Vridhachalam, Tamil Nadu.
- ▪ Major pest: Tea mosquito bug (Helopeltis antonii)

Palmyrah- *Borassus flabellifer* –Araceae- 'Kalpaga thara'-**"State Tree of Tamil Nadu",** Origin-Tropical Africa

- ☆ Tree of Life
- ☆ SVPR-1 is a semi-dwarf variety
- ☆ Tapping- Extraction of sap from the inflorescence
- ☆ Aripanai and Vallupanai-Method of extraction of tapping in male palm

10.6. MEDICINAL PLANTS

S.No	Medicinal Plants	Part Used	Variety	Alkaloids	Uses
1.	Aloe Vera	Leaves	-	Aloin	Laxative Purpose
2.	Asgand (Aswagandha) Indian Gin-geng	Roots	Jawahar Asgandh-20	Withanine Somniferine	Aphrodisiac property Immuno-modulator Anti-stress
3.	Medicinal Yam	Tubers	Arka Upkar, FB(c)-1	Diosgenin	Production of Contraceptive Pills
4.	Fox-Glove	Leaves	-	Digoxin	Heart Disease
5.	Opium	Capsule (Latex)	Telia, Dholia	Codeine	Painkiller
6.	Dill or Sowa	Seeds	Sona, ALFT-2	Carvone	Improve digestion, Control vomiting, Carminative property
7.	Guggel (Kiluvai)	Oleogum Resin	-	Z and E-Gugglusteroids	Anti-inflammatory Hypoglycemic
8.	Henbene	Leaves	-	Hyoscyamine	Anti-asthma Anti-cholicgenic
9.	Isabgol	Husk, seed	Gujarat Isabgol-1,2	Mucilage	Anti-diarrhoea Laxative
10.	Khasi-Kateri	Fruits	Arka Sanjeevani	Solasodine	Production of Contraceptive Pills
11.	Liquorice	Roots	Haryana Mulathi 1	Glycyrrhizin	Anti-inflammatory Spasmolytic activity
12.	Periwinkle	Roots and Leaves	Nirmal, Dhawal	Vinblastine Vincristine Ajmalcine	Tranquilizer Anticancer Anti-neoplastic
13.	Pipali	Unripe Fruit (Female Spike)	Vishwam	Piperine	Improve appetite Laxative
14.	Sarpagandha	Roots	RS-1	Serpentine Ajmalicine Reserpine	Anti-Blood pressure
15.	Senna	Leaves and Pods	KKM-1	Sennosides (A, B,C)	Laxative and Constipation
16.	Kalmegh **(King of Bitter)**	Whole Plants	-	-	Jaundice
17.	Glory Lily **(State Flower of Tamil Nadu)**	Seeds	-	Colchicine	Anti-Gout(Joint Pain)
18.	Cinchona	Bark	-	Quinine	Treatment of Malaria
19.	Datura	Whole plant		Hyoscine, Tropane	Preanasthetic surgery

10.7. AROMATIC PLANTS

S.No	Aromatic Plants	Part Used	Oil Content	Uses
1.	Ambrette/Muskdana	Seeds	Farnesol and Ambreetolide	Musk odour used in incense stickes, Panmasala, Perfumery, Cometics and Scents
2.	Celery	Seeds	Selinene, limonene Purees	Seasoning, Flavouring Sauces,
3.	Chamomile	Flowers	Blue Oil	Pain Relieving balms
4.	Davana	Leaves	Hydrocarbons decoration, bouquets	Delicate fragrance for floral
5.	Indian Basil	Leaves	Methyl Chavicol Eugenol and Linalool	Flavouring foods
6.	Java Citronella	Leaves	Citronellal scented soaps	Mosquito repellants Deodorants,
7.	Kewada	Flowers	Luplin	Kewada water
8.	Lemon Grass	Leaves	Citral	Starting material for Ionones
9.	Mint	Leaves	Menthol, Carvone	Scenting Supari
10.	Palmarosa Grass	Leaves	Geraniol	Soaps, Perfumery industry
11.	Patchouli	Leaves	Patchouliol	Fixative Property
12.	Rose Geranium	Leaves and	Rhodinol Flowers	Perfume industry
13.	Vettiver	Roots	Citronellal	Carminative Property
14.	Indian Basil	Leaves and Inflorescence	Methyl Chavicol	Flavouring foods
15.	Rosemary	Whole herb	Camphene, Cineol	Anti-cancer, anti-oxidant
16.	Oil bearing Rose	Flowers	Citronellol, Geraniol	Rose oil, Ruha Gulab

10.8. POST-HARVEST TECHNOLOGY

- Total post harvest losses in fruits and vegetables: 20-40%
- Post-harvest losses in India are estimated to range from 14-36% in fruits and 10-25% in vegetables
- All fruits are acidic in nature whereas vegetables are alkaline except tomato
- More acidic fruit tissue is generally attacked by fungi while vegetables having pH above 4.5 are more commonly attacked by bacteria
  - Bacterial soft rot of potato by *Erwinia spp.*
  - Dry rot by *Fusarium spp.*,
  - Black rot of sweet potato by *Ceratocystis fimbriata*

- Water soft rot of carrot by *Sclerotinia sclerotiorum*
- Leafy vegetables soft rot by *Erwinia carotovora* and dry rot by *Fusarium* spp.

Classification of fruits based on climacteric pattern:

- Climacteric fruits can be defined as the fruit showing a large increase in carbon dioxide and ethylene production rates coincided with ripening, non-climacteric fruits show no change in the CO_2 and ethylene production rates during ripening

1. Climacteric fruit:

- Increase of respiration and ethylene production during ripening
- Fruits can harvested at unripe stage and can be allowed for ripening during storage after harvest
- Apple, pear, peach, plum, persimmon, apricot, avocado, tomato, watermelon, kiwi, fig, mango, banana, papaya, guava, blueberry, cherimoya, ber, melons, sapota, passion fruit

2. Non-climacteric fruit:

- Steady decline in respiration without ethylene production during ripening
- Fruits harvested only at ripe stage
- Citrus fruits, Grapes, Grape fruit, Cucumber, Pineapple, Strawberry, Tree tomato, Sweet Cherry, Carambola, Cherries, Litchi, Loquat, Olive, Pomegranate, *Rin* and *Nor* Tomato

Post-harvest practices:

- **Curing**: Hardening of epidermal layer (outer tissue) of bulb and root crops by exposing to high RH and temperature *e.g.* Onion, Garlic, Sweet Potato, Tapioca
- **Degreening**:
- Degradation of chlorophyll in mango, banana, tomato, citrus fruits by artificial application of ethylene
- Best degreening temperature @ 27-°C, 85-90% RH
- Most widely used growth regulator for degreening: Ethrel
- Artificial ripening: Ethylene or Ethrel (Mango, banana)
- Synthetic ripening: Calcium carbide produces acetylene gas (CaC_2) (Used earlier, now banned)
- Ethylene absorber or scrubber: $KMnO_4$
- Ethylene ripening champer: fruits exposed to low level of ethylene 100-150 ppm an air tight champer for 24 hours to ripen (temperature 16-25°C with RH 90-95%)
- Banana ripening champer: 100 ppm, 24 hours, temperature 16-18°C with RH 90-95%
- Mango ripening champer: 100 ppm, 24 hours, temperature 20-22°C with RH 90-95%
- Ethylene inhibitors: Silverthiosulphate (STS), Silver nitrate ($AgNo_3$), 1-MCP
- **Class I Preservative**: Sugar, Salt, Spices
- **Class II Preservative**: Sodium benzoate, Potassium metabisulphite
- Sugar and Salt act as a preservative due to their osmotic effect
- **Curing**: Hardening of epidermal layer (outer tissue) of bulb and root crops by exposed high RH and temperature e.g. Curing is done in onion, garlic, potato, sweet potato, cassava and yam
- Maximum safe temperature for onion curing: 37.8°C for 6 days
- Temperature for artificial curing of onion: 40°C for 16 hrs
- Most effective temperature for potato curing: 20°C @ 85%RH
- **Pre-cooling**: Rapid removal of field heat from harvested horticultural produce e.g. Okra, Garden Pea
- **Degreening**: Removal of chlorophyll from Mango, Banana, Tomato, Citrus fruits by artificial application of ethylene

- Best degreening temperature @ 27-°C, 85-90 % RH
- **Blanching**: To inactivation of natural enzymes and microbes in fruits and vegetables
- **Blanching**: As a pre-treatment (Boiling water exposure of for fruits and vegetables) e.g. Okra, Cabbage, Beans
- ♣ is also known as scalding, parboiling or pre-cooking
- ♣ Inactivates natural enzymes and microbes in fruits and vegetables
- ♣ Blanching: Act as a pre-treatment (Boiling water exposure of for vegetables) e.g. Okra, Cabbage, Beans
- ♣ Blanching temperature: 82.2-93.3°C (@ 2-5 min. Average blanching time: 6 minutes
- Browning: Many cut fruits and vegetables have tendency to turn brown when exposure to air is due to polyphenol oxidase (PPO) activity
- Lye peeling: Peach, apricot, sweet orange, mandarin and vegetables (Carrot and sweet potato)
- Lye peeling is done by 1-2% NaOH (Boiling caustic soda)
- **Waxing**: Application of fungicides/growth substance and other chemicals along with edible wax for increase shelf-life e.g Apple, Pomegranate
- Canning by Sterilization process in fruits (Syruping) and vegetables (Brining)
- Pasteurization: Heating of fruit juices @ 85-90°C for 30min
- Irradiation or Cold Sterilization: To increase shelf life of bulbs and delay ripening
- Ultra-Filtration or Cold Process: To removal of microbes from juice
- Salt act as a preservative: 15-25%
- Fruit syrup, jam, jelly, marmalade, preserve, candy, crystallised fruit and glazed fruit are preserved by sugar
- Sugar act as a preservative: 62-65%
- Alcohol acts as a preservative in wines: 14%
- Acetic acid (vinegar), citric acid (lime juice) and lactic acid are commonly used for preservation
- Citric acid is added to many fruit squashes, jams and jellies to increase the acidity and prevent mould growth
- Vinegar contains 5% acetic acid
- For pickles, chutneys, sauces and ketchups-vinegar is recommended

Syruping:

- ♣ A solution of sugar in water is called as syrup
- ♣ Syruping is done only for fruits
- ♣ Sucrose syrup is used for canning
- ♣ Syruping concentration: 20 to 55° C
- ♣ Syrup should be filled at 79 to 82° C, leaving head space of 0.3 to 0.5cm

Brining:

- ♣ A solution of salt in water is called as brine
- ♣ Brining is done only for vegetables
- ♣ Brining concentration:1-3%
- ♣ Brining filled at 79 to 82° C, leaving head space of 0.3 to 0.5cm

Jelly:

- ★ **Jelly** (65% sugar, 0.5-0.75% Acidity): Semi-solid product: Boiling clear strained solution of pectin containing fruit extract, free from pulp after addition of sugar and acid

- ✦ Pectin content in jelly: 0.5-1.0%
- ✦ Acid content in jelly: 0.5-0.75%
- ✦ Total soluble solids (TSS) in jelly: 65%
- ★ Most important constituent of jelly: Pectin
- ★ Guava (High pectin content) most suitable for jelly preparation
- ★ The optimum pH for jelly is 3.2
- ✦ 60°B of jelly pH: 3.0
- ✦ 65°B of jelly pH: 3.2
- ✦ 70°B of jelly pH: 3.4

12.1. Classification of fruits based on their pectin and acid content:

Rich in pectin and acid	Jamun, Grape, Sour guava, lemon, sour orange, sour plum, sour crab apple
Rich in pectin and low acid	Unripe banana & fig, sour cherry, ripe guava, pear, peel of orange and grape fruit
Low in pectin and rich acid	Pineapple, sweet cherry, sour peach, sour apricot
Low in pectin and acid P	omegranate, Raspberry, ripe apricot, ripe peach

- ★ Recommended acid for jelly: Citric acid (2g/kg of fruit)
- ★ Determination of pectin content by alcohol test and jelmeter test
- ★ Judging end point of jelly by sheet or flake test, drop test and temperature test
- ★ End point of jelly: 65% TSS @ 105°C
- ★ Ripe fruits: Protopectin (water insoluble) → Protopectinase → Pectin (water soluble pectinic acid)
- ★ Overripe fruits: Pectin→ Pectic Methyl Esterase (PME) → Pectic acid (water insoluble)
- ★ Firm ripe fruits is most suitable for jelly making
- ★ Firmness of fruits due to calcium pectate
- ★ Premature gelation is due to excess of pectin
- ★ Formation of crystals or crystallization of jelly is due to excess sugar
- ★ The phenomenon of spontaneous exudation of fluid from a gel is called syneresis or weeping jelly
- ★ Syneresis is due to excess of acid, low sugar, insufficient pectin, premature gelation and fermentation
- ★ Cloudy or foggy jellies is due to use of immature fruits, over cooking and over cooling
- ★ Failure to set jelly is due to lack of pectin
- ★ Tough jelly is due to high pectin content
- ★ Removal of scum or foam by addition of edible oil

13. Marmalade:

- ★ **Marmalade**: Fruit jelly with addition of its peel eg. Sweet Orange
- ★ Browning is common problem in marmalade
- ★ Browning of marmalade is prevented by addition of KMS (0.09g/kg of marmalade)
- ★ Cooking temperature for marmalade is 103 to 105°C
- ★ Cooling temperature for marmalade is 82-88°C
- ★ End point of marmalade: 65% TSS @ 105°C
- ★ Sweet orange peel is most commonly used in preparation of marmalade
- ★ Jelly marmalade: Clarified extract of pectin is used
- ★ Jam marmalade: Whole pulp is used

14. Other products:

- ★ **Preserve**: Mature fruit/vegetable impregnated with heavy sugar syrup till it become tender and transparent
- ★ **Candy**: Mature fruit/vegetable impregnated with cane sugar syrup and subsequently drained free of syrup and dried *eg.* Ginger, Ber, Aonla
- ★ **Glazed fruit**: Covering of candied fruit/vegetable with a thin transparent coating of sugar for glossy appearance
- ★ **Crystallized fruits**: candied fruit when covered or coated with crystals of sugar
- ▪ Zero Energy Cool Chamber (ZECC) (Working under Principle of Evaporative Cooling) was developed by IARI.
- ▪ Production of vinegar by acetic acid fermentation
- ▪ Production of wine by alcoholic fermentation

Vinegar:

- ♣ The word vinegar is derived from French "vinaigre" means 'sour wine'
- ♣ Vinegar contains 5% acetic acid and has germicidal and antiseptic properties
- ♣ Vinegar is obtained through alcoholic and acetic acid fermentation
- ♣ Amount of acid in vinegar is expressed as 'grain strength'
- ♣ 1% acetic acid is termed as 10 grain strength
- ♣ Fermentation temperature for vinegar production: 21-27oC
- ▪ Production of pickles and Sauerkraut (Cabbage fermented product) by Lactic acid fermentation
- ▪ Asepsis: Absence of infection
- ▪ Pasturization is partial destruction of microbes by temperature below 100°C e.g. Fruit Juices
- ▪ Sterilization is complete destruction of microbes by temperature above 100°C e.g. Canning
- ▪ Fruit juice beverages are bottled with CO_2 @ 1-8g/Litre
- ▪ 14% alcohol act as preservative in wines
- ▪ **Cordial**: Sparkling clear, sweetened fruit juice from pulp- Suit for blending with wine
- ▪ Vinegar contains 5% Acetic Acid
- ▪ Ethylene inhibitors Silver thiosulphate ($AgNo_3$)
- ▪ Geotrophic bending problem in Gladiolus, Snapdragon
- ▪ Fruit Product Order-1955
- ▪ Father of Canning: Nicolas Appert

Unfermented beverages:

- ♣ Synthetic drinks contain only water (88%) and total carbohydrates (12%)
- ♣ Fruit juices, RTS and nectars are preserved by pasteurization
- ♣ Cordial: Sparkling clear, sweetened fruit juice from pulp
- ♣ Cordial is most suitable for blending with wines
- ♣ Squashes, crushes and cordials are preserved by only by chemicals
- ♣ Mango, orange and pineapple are used for making squash commercially
- ♣ Phalsa, Aonla, jamun, pomegranate, grape, lemon orange and ginger are used for the preparation of syrup
- ♣ Synthetic syrup contains 70-75% of sugar syrup
- ♣ Barley water is prepared from citrus fruits such as lime, lemon, grapefruit and orange
- ♣ Mostly widely used citrus fruits in barley water: Lime and Lemon

Fermented beverages:

- ★ Development of biochemical principles of fermentation was originated by Lavoisier (1789) in France
- ★ Wine is a beverage resulting from the fermentation of grape juice by yeasts
- ★ Wines prepared from Fruits: Perry-Pears, Berry- Strawberry, Blackberry, Elderberry, Nira-Palmyarah, Feni-Cashew, Cider-Apple, Champaigne-Grapes
- ★ Alcohol content of wine ranges from 7 to 20%
- ★ Optimum temperature for fermentation of grape wine is 22-28°C
- ★ Aging or maturation time for wing: 6-8months
- ★ Common yeast used in wine: *Saccharomyces cerevisiasae* var. *ellipsoideus* (20ml/kg of grapes)
- ★ Common yeast used in cider: *Saccharomyces carlbergensis*
- ★ Generally wines are pasteurized at 82-88°C for 1-2minutes
- ★ Wine made from pears is known as *perry*
- ★ *Feni* is a fermented wine made from cashew apple in Goa
- ★ Bael, jamun, phalsa and aonla are most widely used for preparation of cider
- ★ Common preservative used in cider: SO_2@ 100ppm or KMS (0.22g/kg)
- ★ Cider is mostly prepared from fermentation of special grades of apples
- ★ For cider preparation of apple should have 0.1-0.3%
- ★ *Port* is a fortified sweet red wine made originally in Portugal
- ★ *Champaigne* is a sparkling wine made chiefly in France
- ★ Suitable grape varieties for *Champaigne*: Chardonay and Pinot Noir
- ★ Nira is prepared from the juice of Palmyrah
- ★ Wines prepared from berries like strawberry, blackberry and elderberry are known as berry wines
- ★ Famous fortified wine in Hungary: *Tokay*
- ★ Syphoning off the fermented wine to separate it from the solid deposits is known as "raking"

Types of wine	% alcohol
Light wine	7-9%
Medium wine	9-16%
Strong wine	16-21%
Sparkling wine	CO_2
Still wine	No. CO_2
Fortified wine	Addition of alcohol in the form of brandy
Red wine	From coloured grapes (skin intact)
White wine	From white/green grapes (skin removed)
Dry wines	Very little/no sugar is detected
Sweet wines	High sugar content can be detected

Processing of tomato:

- ★ Tomato sauce and apple sauce are quite popular in India

FPO specification for tomato products

Tomato product	TSS	Other specifications
Tomato juice	5	Sugar 1%, salt 0.5%, citric acid 0.4%
Tomato soup	7	
Tomato puree		Tomato pulp without skin or seeds
Medium tomato puree	9	Benzoic acid @ 250ppm
Heavy tomato puree	12	Benzoic acid @ 250ppm
Tomato paste	25	Made from strained tomato juice or pulp, benzoic acid @ 250ppm
Concentrated tomato paste	33	
Tomato sauce/ketchup	25	Tomato solids 12%, , 1% acetic acid, benzoic acid @ 750ppm

- ★ Generally sugar content of tomato ketchup/sauces:10-26%
- ★ Salt content of tomato ketchup/sauces: 1.3-3.4%
- ★ Tomato sauce/ketchup contains 1.25-1.5% acetic acid

Important terms used in Post-harvest treatments for flowers

Pulsing/loading:

- ★ Preshipment treatment
- ★ Short term treatment
- ★ Pulsing given to the cut flowers before packing and transportation of flowers
- ★ Time for pulsing: 12-24 hrs, under a light intensity of 1000lux and 20 to 27°C
- ★ Pulsing treatment promote opening and increase the shelf life of flower
- ★ Sugar and other chemicals used for pulsing
- ★ Sucrose is most commonly used for pulsing (Osmo-regulation)

Bud opening solutions:

- ★ It is a procedure for harvesting flowers at storage earlier than normally coincided by cutting stage
- ★ Bud opening method is identical to pulsing, longer duration and low concentration of sugar
- ★ 8-HQC, Silver thiosulphate (STS), KCl, Al_2SO_4, 4% sucrose are used for bud opening

Vase solutions:

- ★ Holding solutions used in the vases to keep flowers for extending their vase life
- ★ Solutions: Sugar + Germicide + Growth regulator +Organic acid + Ethylene inhibitor
- ★ Commonly used: Sucrose + Citric acid + Quinoline salt (8-HQC, 8-HQS)

Holding solution:

- ★ Use of preservative in the form of tablets
- ★ Prepared by mixture of chemicals (Sugar, germicides, salt, growth regulator)

Conditioning/Hardening:

- ★ Flowers are kept standing loosely in a big container, so that air can circulate around the stems
- ★ Purpose: Restore the turgidity of cut flower from water stress during storage and transportation
- ★ Water + Germicides + Citric acid @ 500ppm pH: 4.5-5.0

Impregnation:

★ Ends of the cut flower stems are impregnated for short time with chemicals

★ Prevents blockage of xylem vessel in the stem by microbial growth and stem decay

★ Commonly used chemicals for impregnation: $AgNO_3$, $NiCl_2$, $CoCl_2$ @ 10-15minutes

Important fruit beverages specification:

Products	Fruit Juice (%)	TSS (%)	Acidity (%)	SO_2 (ppm)
1. Not Diluted Before Serving				
Unsweetened juices	100	Natural		
Sweetened juices	85	10		
RTS	10	10	0.3	-
Nectar	20	15		-
Cordial (Lime and Lemon)	25	30	1.5	350
Barley water	25	30	1.0	0.25% barley starch
Fruit juice concentrate		32%		
2. Diluted Before Serving				
Squash	25	40-50	1	350ppm SO_2 or 600 ppm sodium benzoate
Crush		55		-
Syrup		65	1.3-1.5	-

Maturity index of horticultural crops:

Horticultural Crops	Maturity Index
Mango	Tapka
Banana	Finger Filling
Jackfruit and Watermelon	Tapping
Muskmelon	Netting or Full slip stage
Onion and Garlic	Neck fall (50%)
Citrus	Juice content (50%)
Avocado	Oil content
Apple	T Stage
Pineapple	Flattening of Eyes
Carnation	Paint brush stage
Narcissus	Goose neck stage

10.9. HORTICULTURAL INSTITUTES

- Central Institute of Horticulture (CIH), Mediziphema, Nagaland
- Institute of Horticulture Technology (IHT), New Delhi
- Indian Institute of Horticultural Research (IIHR), Hessaraghatta, Bangalore, Karnataka
- National Bureau of Plant Genetic Resource Center (NBPGR), New Delhi
- National Horticultural Board (NHB) established in 1984, HQ in Gurgaon, Haryana
- National Horticultural Mission (NHM): 2005-06
- The Ministry of Agriculture has announced 2012 as the Year of Horticulture

Pomological Research Institutes:

- Central Institute of Subtropical Horticulture (CISTH), Lucknow, Uttar Pradesh
- Central Institute of Temperate Horticulture (CITH), Srinagar, Jammu and Kashmir
- Central Arid Zone Research Institute (CAZRI), Jodhpur, Rajasthan
- Central Institute of Arid Horticulture (CIAH), Bikaner, Rajasthan
- National Research Centre for Banana (NRCB), Trichy, Tamil Nadu
- National Research Centre for Citrus (NRCC), Nagpur, Maharashtra
- National Research Centre for Grapes (NRCG), Pune, Maharashtra
- National Research Centre for Pomegranate (NRCP), Solapur, Maharashtra
- National Research Centre for Litchi (NRCL), Muzaffarpur, Bihar
- National Research Centre for Makhana (NRCM), Darbhanga, Bihar

Olericulture research institutes:

- Indian Institute of Vegetable Research (IIVR), Varanasi, Uttar Pradesh
- Central Tuber Crops Research Institute (CTCRI), Sreekariyam Thiruvananthapuram, Kerala
- Central Potato Research Institute (CPRI), Shimla, Himachal Pradesh
- National Horticultural Research and Development Foundation (NHRDF), Nasik, Maharashtra
- Directorate on Onion and Garlic Research (DOGR), Nasik, Maharashtra
- Directorate of Mushroom Research (DMR), Solan, Himachal Pradesh

Floricultural research institutes:

- Directorate of Floricultural Research (DFR), Pune, Maharashtra
- National Research Centre for Orchids (NRCO), Pakyong, Gangtok, Sikkim
- National Botanical Research Institute (NBRI), Lucknow, Uttar Pradesh
- Institute of Himalayan Bio-resource Technology (IHBT), Palampur, Himachal Pradesh

Plantation crops research institutes

- Central Plantation Crops Research Intitue (CPCRI), Kasargod, Kerala
- Central Coffee Research Institute (CCRI), Chikmagalur, Karnataka
- Directorate of Cashew Research (DCR), Puttur, Karnataka
- Directorate of Oil Palm Research (DOPR), Pedavegi, Eluru, Andhra Pradesh
- United Planters' Association of Southern India (UPASI), Glenview, Conoor, Nilgiris District, Tamil Nadu
- UPASI Tea Research Foundation (UPASI TRF), Valparai, Coimbatore District, Tamil Nadu

- Tea Research Institute (TRI), Nirar Dam, Valparai, Tamil Nadu
- Central Arecanut and Cocoa Marketing and Processing Co-operative Limited (CAMPCO), Mangalore, Karnataka
- Directorate of Cashew and Coconut Development (DCCD), Cochin, Kerala
- Directorate of Arecanut and Spices Development (DASD), Calicut, Kerala

Boards:

- Coconut Development Board (CDB), Cochin, Kerala
- Tea Board of India, Kolkata, West Bengal
- Coffee Board of India, Chikmagalur, Karnataka
- Cashew Export Promotion Council of India (CEPC), Ernakulam, Kerala

Spices research institutes:

- Indian Institute of Spices Research (IISR) (Previously, NRC on Spices), Calicut, Kerala.
- National Research Centre for Seed Spices (NRCSS), Tabiji, Ajmer, Rajasthan
- Indian Cardamom Research Institute (ICRI), Myladumpara, Idukki, Kerala

Boards:

- Spices Board established in 1987, Cochin, Kerala (Ministry of commerce and industry)

Medicinal and aromatic plants research centres:

- Central Institute of Medicinal and Aromatic Plants (CIMAP), Lucknow, Uttar Pradesh
- Directorate of Medicinal and Aromatic Plants Research (DMAP), Anand, Gujarat (Previously known as National Research Center for Medicinal and Aromatic Plants (NRCMAP))
- National Medicinal Plants Board (NMPB), New Delhi

Post harvest research centers in India:

- Central Food Technological Research Institute (CFTRI), Mysore, Karnataka
- Central Post Harvest Engineering and Technology (CIPHET), Ludhiana, Punjab
- Fruit Preservation and Canning Institute (FPCI) is located at Lucknow, UP
- Regional Research Laboratory (R.R.L.) Jammu
- Bhabha Atomic Research Centre (BARC), Trombay, Bombay
- CSIR Laboratory, Palampur, DFRL, Mysore

Central Food Laboratories:

- Central Food Laboratory (CFL), Kolkata, West Bengal
- Food Research and Standardization Laboratory (FRSL), Ghaziabad, Uttar Pradesh
- Public Health Laboratory (PHL), Pune
- Central Food Technological Research Institute (CFTRI), Mysore, Karnataka
- Defence Food Research Laboratory (DFRL), Mysore, Karnataka
- Central Institute of Post-Harvest Engineering Technology (CIPHET), Ludhiana, Punjab
- National Institute of Food Technology Entrepreneurship and Management (NIFTEM), Kundli, Haryana
- Indian Institute of Crop Processing Technology (IICPT), Thanjavur, Tamil Nadu
- National Agricultural Cooperative Marketing Fedration of India Ltd. (NAFED), New Delhi

ANIMAL HUSBANDRY

11

M. Nagalingam

1. Cattle
2. Buffaloes
3. Sheep
4. Goat
5. Swine
6. Camel
7. Rabbit
8. Poultry

Animal husbandry

Animal husbandry is the branch of science concerned with the care and breeding of domestic animals such as cattle, hogs, goat, sheep and horses.

India supports nearly 20% of the world livestock and 16.8% human population with a land area of only 2.3%. India leads in cattle and buffalo population and has world's second largest goat and third largest sheep population. As per the 19th Livestock census (2012) cattle (191 millions), buffalo (109 m), Sheep (65 m), goats (135 m), pigs (10.29m), camels (0.4 m) and poultry (729.2 m) population was reported in India. The total livestock population has decreased by about 3.33% over the previous census. India's milk production increased from 43 million tonnes in 1983 to 155.5 million tonnes in 2015-16 (annual growth of 6.27%) and stands first in milk production in world.. The egg production is around 82.93 billion during 2015-16. The per capita availability is around 66 eggs per annum. The poultry meat production is estimated to be 3.26 million tonnes. Wool production (43.6 million Kgs) and meat production is 7.0 million tonnes in 2015-16.

Institutes

- ICAR-Indian Veterinary Research Institute (IVRI), Izatnagar
- ICAR-National Dairy Research Institute (NDRI), Karnal, Haryana
- ICAR-Central Institute for Research on Goats, Makhdoom
- ICAR-Central Institute for Research on Buffaloes, Hisar, Haryana
- ICAR-Central Avian Research Institute, Bareilly, Uttar Pradesh
- ICAR-Central Institute for Research on Cattle, Meerut, Uttar Pradesh
- ICAR-Central Sheep and Wool Research Institute, Avikanagar, Rajasthan
- ICAR-National Institute of Animal Nutrition and Physiology, Bengaluru, Karnataka
- ICAR-National Institute of High Security Animal Diseases, Bhopal, Madhya Pradesh
- ICAR-National Institute of Veterinary Epidemiology and Disease Informatics, Bengaluru, Karnataka
- ICAR-National Research Centre on Camel, Bikaner, Rajasthan
- ICAR-National Research Centre on Equines, Hisar, Haryana

- ICAR-National Research Centre on Meat, Hyderabad, Telangana
- ICAR-National Research Centre on Mithun, Jharnapani, Nagaland
- ICAR-National Research Centre on Pig, Guwahati, Assam
- ICAR- National Bureau of Animal Genetic Resources, Karnal, Haryana
- ICAR- National Research Centre on Yak, Dirang, Arunachal Pradesh
- ICAR- Directorate on Foot and Mouth Disease, Mukteshwar, Uttarkhand
- ICAR- Project Directorate on Poultry, Hyderabad, Telangana
- Central Frozen Semen Production and Training Institute - Hessarghatta (Bangalore)
- Central Poultry Development Organisations (CPDOs) located at four regions viz. Chandigarh, Bhubaneswar, Mumbai and Bengaluru
- National Veterinary Biological Products Quality Control Centre, Baghpat National Diary Development Board (NDDB) - Lali , Kheda - Gujarat)

Veterinary & Animal Science Universities

- Chhattisgarh Kamdhenu Vishwavidyalaya, Anjora, Durg, Chhattisgarh
- Guru Angad Dev Veterinary and Animal Sciences University, Ludhiana, Punjab
- Kamdhenu University, Gandhinagar, Gujarat.
- Karnataka Veterinary, Animal & Fisheries Sciences University, Bidar, Karnataka
- Kerala Veterinary and Animal Sciences University, Pookode, Wayanand, Kerala
- Lala Lajpat Rai University of Veterinary & Animal Sciences, Hisar, Haryana
- Maharashtra Animal & Fishery Sciences University, Nagpur, Maharashtra
- Madhya Pradesh Pashu-Chikitsa Vigyan Vishwavidyalaya, Jabalpur, Madhya Pradesh.
- Nanaji Deshmukh Pashu Chikitsa Vigyan Vishvavidyalaya, Jabalpur, Madhya Pradesh
- Prof. Jayashankar Telangana State Agricultural University, Hyderabad, Telangana
- Rajasthan University of Veterinary & Animal Sciences, Bikaner, Rajasthan.
- Sri Venkateswara Veterinary University, Tirupati, Andhra Pradesh
- Tamil Nadu Veterinary & Animal Sciences University, Chennai, Tamil Nadu
- UP Pandit Deen Dayal Upadhaya Pashu Chikitsa Vigyan Vishwa Vidhyalaya, Evam Go Anusandhan Sansthan,Mathura, Uttar Pradesh
- West Bengal University of Animal & Fishery Sciences, Kolkata, West Bengal

Department of Animal Husbandry, Dairying and Fisheries

- E-Pashuhaat is a portal connecting the breeders and farmers for sale and purchase of germplasm.
- The National Programme for Bovine Breeding and Dairy Development (NPBBD) has been initiated in February 2014 by merging four ongoing schemes of the department of Animal Husbandry, Dairying and Fisheries in the dairy sector, viz., National Project for Cattle and Buffalo Breeding (NPCBB), Intensive Dairy Development Programme (IDDP), Strengthening Infrastructure for Quality & Clean Milk Production (SIQ & CMP) and Assistance to Cooperatives (A-C).
- MAITRI-Multi-purpose AI Technician in Rural India
- Rashtriya Gokul Mission has been initiated as a part of National Programme for Bovine Breeding (NPBB) in December 2014 with the aim to conserve and develop indigenous breeds in a scientific and holistic manner.
- Livestock health and disease control components

Recent Livestock health and disease control components

(a) Assistance to States for Control of Animal Diseases (ASCAD)

(b) Professional Efficiency Development (PED)

(c) National Project on Rinderpest Surveillance & Monitoring (NPRSM)

(d) Foot and Mouth Disease Control Programme (FMD-CP)

(e) National Animal Disease Reporting System (NADRS)

(f) Peste des Petits Ruminants Control Programme (PPR-CP)

(g) Brucellosis Control Programme (Brucellosis - CP)

(h) Establishment and Strengthening of existing Veterinary Hospitals and Dispensaries (ESVHD).

(i) Classical Swine Fever Control Programme (CSF-CP).

✓ Robert Bakewell referred as Father of Animal breeding

✓ First animal to be domesticated is dog (Egypt)

✓ Last animal to be domesticated is horses (Europe and Asia)

✓ A breed is a group of animals related by descent and similar in most characters like general appearance, features, size, configuration etc.,

✓ National Milk Day on 26th of November

11.1. CATTLE

- There are seven Central Cattle Breeing breeding Farms is fuctioning under Department of Animal Husbandry, Diarying and Fisheries, Ministry of Agriculture, Government of India.

Name & location of the farm	Breed maintained
Suratgarh (Rajasthan)	Tharparkar
Dhamrod (Gujarat)	Surti Buffalo
Chiplima (Orissa)	Red Sindhi (RS) and RS x Jersey
Hessaraghatta (Karnataka)	Holstein Friesian
Sunabeda (Orissa)	Jersey
Alamadhi (Tamil Nadu)	Murrah Buffalo
Andesh Nagar (Uttar Pradesh)	Holstein Friesian x Tharparkar

- India possesses 27 acknowledged indigenous breeds of cattle and seven breeds of buffaloes.
- A Central Herd Registration Scheme for identification and location of superior germ plasm of cattle and buffaloes, propagation of superior germ stock, regulating the sale and purchase, help in formation of breeder's society and to meet requirements of superior bulls in different parts of the country is also being implemented.
- Around 70.6 million draught animals contribute 20% of energy input into crop farming.

I. Exotic cattle breeds

i. Jersey (from Islands of Jersey in English channel): Fawn with white markings, withstand tropical and humid climate better than Holstein Friesian (HF)

ii. Holstein Friesian (Holland): Black and white, large feeding capacity and udder. Higher milk yield. (Fat % in milk-3.5%) (On an average HF gives 25 litres of milk per day, whereas a cross bred cow gives 10 - 15 lts per day).

iii. Ayreshire (Scotland): most beautiful dairy breed and over active.

iv. Brown Swiss (Switzerland): Dual purpose breed for dairy and beef production, more heat tolerant than Jersey and docile.

II. Important Indian Cattle breeds

i. Milch breeds

- *Sahiwal* (from Montgomery District in Pakistan): Best dairy breed of Indian farm conditions with milk yield of 1000-5000 kgs
- *Sindhi* (also called Red Sindhi, Red Karachi from Karachi and Hyderabad of Sind): Heavy milkers, used for grading up local stock
- *Gir* (from Gir forests of South Kathiawar): good milkers
- *Deoni* (from N & W Portion of Hyderabad): resembles Gir but ears have no notch near the tips of ear

ii. Dual purpose breeds: Breeds that provide both milk and meet

- *Haryana* (from east Punjab): Cows are average milkers (1400 kg) and bullocks are good working
- *Ongole* (also called Nellore from Nellore tract of Andhra Pradesh)
- *Tharparkar* (also called Thari, Grey Sindhi, White Sindhi from Tharparkar district of Hyderabad)
- *Kankrej* (also called Wadhair, Bannai from North Gujarat and SE Region of Kutch): Heaviest of the Indian breeds (Energetic, vigorous, excitable), well suited for draught purposes

iii. Draught breeds: Common characteristic of draught breeds are deep chested with strong limb

- *Amrit Mahal* (from Karnataka): One of the best draught breeds of India
- *Hallikar* (from Karnataka): Excellent for road and field works.
- *Kangeyam* (from Tamilnadu along tracts of Kangeyam, Coimbatore and Pollachi): Excellent draught
- *Malvi* (from Rajasthan and Madhya Pradesh)
- *Siri* (from Darjeeling, Sikkim and Bhutan): Well suited for cart purpose

New Breeds (Cattle)

- *Taylor breed:* Crosses of Taurus bulls (Ayrshire bulls from UK) with local cows by Mr. Taylor
- *Jersind:* Red sindhi x Jersey
- *Karan swiss:* Brown swiss with Sahiwal & Red Sindhi
- *Karan Fries:* Tharparkar x HF
- *Sunandhini:* In Kerala, local non-descript with brown Swiss bulls resulted from Indo-Swiss project
- *Frieswal:* Friesian x Sahiwal cross breds
- *Vechur* (Kerala): Smallest cow

Other Breeds of Cattle

- *Kherigarh* - Kheri dist of Uttar Pradesh
- *Bachaur* - Bihar
- *Gaolao Nimari* - Maharashtra and Madhya Pradesh(closely resembles to Ongole)

- *Krishna valley* - Karnataka
- *Mewati, Rathi, Nagauri* - Rajasthan
- *Dangi* - Maharashtra
- *Mysore Breeds* - Amritmahal, Hallikar, Khillari
- *Alambadi* - Salem to Coimbatore of Tamil Nadu (Resembles to Hallikar)
- *Hilly breeds* - Ponwar, Siri (Himalayas)

Projects

- Indo Swiss Project- Initiated in Kerala and Punjab using Brown Swiss on local cows
- Indo Danish Project- Started in Karnataka in 1964- Red Dane used on Red Sindhi
- Indo German Project- Cross breeding with German mountain spotted bulls taken up in Himachal Pradesh

Determination of age in cattle

- Within 2 months all eight temporary incisors appears on lower jaw (no incisor on upper Jaws)
- 2 years (24-30 months) - central pair of temporary incisor replaced with permanent
- 3 years (36 months) - II pair permanent
- 4th year (48 months) - III pair permanent
- 5th year (54-60 months) - IV pair permanent

Terminologies

- Calf- Young one
- Bull - Male
- Bullock -Castrated male
- Cow- Female
- Heifer- Female not calved yet
- Herd- Group of animals
- Calving- Act of giving birth
- Springer - a cow or heifer close to calving
- Freemartin- a female born co-twin with a male
- Beef -Cattle meat
- Veal- Calf meat

Management of cattle

- Selection of a cow: The maximum yield by dairy cows is noticed during the first five lactations. So generally selection should be carried out during first or second lactation and more specifically a month after calving.
- Identification of farm animals: Ear tagging, tattooing, number tag, branding and ear notching.
- Calf is weaned on the day of birth or 2-3 days afterwards.
- Calf starter is mixture consisting of ground farm grains, protein feeds minerals, vitamins and antibiotics given after 2 weeks of age.
- Dehorning is done before 10 days old.

- Castration is unsexing of male
- Ringing is putting ring in nose of bull (Between 8 months to 1 yr old)
- Throwing or casting is to make the animal lying down
- Vaccination for FMD, HS, Anthrax, BQ, Brucellosis (S19 strain)
- Culling is removal of animal from the herd or farm
- Oestrous cycle is 21 days
- Pregnancy diagnosis should be made within 60-90 days after service
- Gestation period is 270 days
- The placenta is discharged within 8-12 hrs after calving
- Birth weight of calf in generally 20-25 kg
- Proper feeding of calf along with regular deworming will achieve a growth rate of 10-15 kg/month
- A common misconception about cattle (particularly bulls) is that they are enraged by the color red. This is incorrect, as cattle are red-green color-blind. In bull fight it is not the color of the cape that angers the bull, but rather the movement of the fabric that irritates the bull and incites it to charge.
- The average sleep time of a domestic cow is about 4 hours a day

S. No.	Particulars	Cows	Buffaloes
1.	Age at maturity	18-24 months	36-48 months
2.	Age at 1st calving	30-36 months	52-60 months
3.	Lactation period	300 days	270-280 days
4.	Dry period	90-120 days	120-150 days

Housing

- Each animal should be provided with 1-1.2 m width and 1.5 - 1.7 m length as standing space
- The roofing materials should preferably be asbestos sheet or galvanized iron sheets
- The roof should be 8' high at sides and 15' high at center.
- The height, depth and width of the manger (feed trough) should be 60, 50 and 40 cm respectively for each animal
- The width and depth of the gutter should be 30 cm and 7.5 cm respectively

Feeding

Cattle have one stomach with four compartments. They are the rumen, reticulum, omasum, and abomasum, with the rumen being the largest compartment. The reticulum, the smallest compartment, is known as the "honeycomb". Molar percentage of VFA (Volatile Fatty Acids)in rumen is acetic (65%), propionic (20%), butyric (12%) and others (3%).

- Body weight in pounds $= LG^2/300$ (L = length in inches ; G = girth in inches)
- Metabolic body weight $= W^{0.75}$
- Cattle will generally consume 2 to2.5 kg of dry matter per 100 kg body weight
- Buffaloes, cross bred animal and heavy yielders consume 2.5 to 3 kg dry matter per 100kg body weight
- Total dry matter required is 2/3 roughages (2/3 dry roughage and 1/3 green roughage) and 1/3 concentrates
- Total digestible nutrients (TDN) requirement for maintenance = 34g TDN/Kg $W^{0.75}$

- Digestible crude protein (DCP) requirement for maintenance = 2.84gDCP/Kg $W^{0.75}$
- TDN requirement for lactation = 330g TDN/kg fat corrected milk
- DCP requirement for lactation = (1.9+0.4F)/65 x100 DCP/kg
- Cost of feeding of cattle is 60% of the total cost of maintenance of cattle and buffaloes
- Nutritive ratio for the work cattle is 1:9 to 1:11 (digestible protein: total digestible nutrients)
- Roughages are divided into 2 groups: Succulent (Contain moisture from 60-90 percent) and dry (10-15 %)
- Succulent are again classified into pasture, cultivated fodder crops, tree leaves, silage and root crops.
- Dry roughages are hay and straw
- A concentrate supplies primary nutrients at higher level but less than 18% crude fibre with low moisture.
- Energy rich concentrates - CP (Crude protein) content is less than 18%
- Protein rich concentrates - CP value exceeds 18%
- Molasses brix is used to refer the amount of sugar content of molasses
- Concentrates are an important and integral part of a ration for a dairy cow as they supply energy (TDN) and proteins (DCP) in right proportion. They should have 15% DCP and 70% TDN. Concentrates should not constitute more than 60% of the ration. It may lead to milk fat depression
- Pregnant animals above 6 months of gestation should be given an extra 1-2 kg of concentrate mixture
- For every 2 kg of milk produced, buffaloes require 1 kg of concentrate feed
- For every 3 kg of milk produced, cows require 1 kg of concentrate feed
- Cows in first and second lactation will still be growing so they are provided 20% and 10% more ration over the maintenance allowance respectively
- National Research Council (NRC) of United States recommends 17% crude fibre (CF) in the total ration DM to prevent milk fat depression
- Provide an extra allowance of 1-1.5 kg concentrate mixture during first 2 months after calving
- Plenty of fresh water should be provided as its requirement is 4 times that of DM consumption
- If a dairy cow consumes 2% DM when dry, it will consume 4-5% DM while yielding 20-30 of milk / day
- If the milk yield is 15-30 kg / day then the DCP, TDN requirements are more by 10%
- During the last 2 months of gestation the animal should be dried off. During this period it should be fed 1-11/2 kg concentrates in addition to its maintenance ration

Diseases

i. Viral

- *Rinderpest was* most destructive of the virus diseases of cloven-footed animals, such as cattle, buffaloes, sheep, goats, pigs and wild ruminants. **Eradicated.**
- *Foot-and-mouth disease* is a highly communicable viral disease affecting cloven-footed animals characterized by fever, formation of vesicles and blisters in the mouth, udder, teats and on the skin between the toes and above the hoofs resulting in low milk yield and infertility
- Others: *Bovine Viral Diarrhoea (*BVD), *Infectious bovine rhinotracheitis (*IBR), cow pox, epemeralfever

ii. Bacterial

- *Anthrax* is an acute, infectious febrile disease of virtually all animals and man. It is caused by *Bacillus anthracis*
- *Black-quarter (Black-leg)* is caused by *Clostridium chauvoei* affecting mostly start and sturdy animals Crepitating sound on pressing the affected part of muscle. Death may occur in severe cases in 1 or 2 days
- *Haemorrhagic septicaemia* is acute septicaemic disease of cattle and a buffalo caused by *Pasteurella multocida* occurs generally in low-lying areas and also stress may predispose.

- *Bovine mastitis* is characterized by the inflammation of the udder, resulting in changes in the udder tissue and its secretion
- Other diseases include brucellosis, leptospirosis, TB, Johnes disease etc.,

iii. Parasitic

- Internal and external parasites,
- Protozoal- Babesiosis, Theilireosis and Trypanosomiasis- Mainly affecting crossbreds

iv. Others

- *Metabolic disease:* Milk fever also known as parturient hypocalcaemia and parturient paresis due to Decrease in the levels of ionized calcium in tissue fluids because of high milk production.
- *Impaction*:Due to over feeding of fodder without water intake or due to intake of grain rich diet will lead to impaction of rumen.
- Bloat, acidosis
- Yoke gall is condition characterized by local swelling upon neck caused due to constant irritation of yoke or plough.

Milk

Average composition of cow milk: Lactose 4.8%, Fat 3.7 % , Protein 3.4%, non-protein nitrogen 0.19% and ash 0.7%. Milk protein constitute 80% casein and 20% whey proteins (á lactalbumin and â lactoglobulin). Fat contains 97-98% Triacyl glycerol.

Lactose is the major solid constituent of milk. Lactose makes a major contribution to colligative properties of milk; osmotic pressure, freezing point depression and boiling point elevation. Milk is important source of calcium but poor source of Iron and vitamin C. Plasmin is major proteinase and lipoprotein lipase is major lipolytic enzyme in milk

- Percapita availability of milk 322g/day (2014-15) in India, while recommended is 230g/day
- Maillard reaction occurs by non enzymatic reaction between lactose and proteins on heating
- Pasteurization-HTST (High temperature short time) 72 degree 15 sec
- Pasteurization- LTLT (Low temperature long time) 63 degree 30 min
- After pasteurization phosphatase test to be negative
- Pasteurization destroys TB organism (*Mycobacterium tuberculosis*)
- In homogenized milk, the fat globules present in the milk are split in to less than 2 microns in size.
- *Kefir* a self carbonated milk beverage containing 1 per cent lactic acid and 1 per cent alcohol
- Standardized Milk is milk whose fat and / or solids not fat content have been adjusted to certain pre determined level. As per Prevention of Food Adulteration (PFA) Rules (1976), the standardized milk for liquid consumption should contain a minimum of 4.5 per cent fat and 8.5 per cent solids not fat.
- Recombined Milk is product obtained when butter oil (otherwise known as dry / anhydrous milk fat), skim milk powder and water are combined in the correct proportions to yield fluid milk. As per PFA Rules (1976), the recombined milk should contain a minimum of 3.0 per cent fat and 8.5 per cent solids not fat.
- Reconstituted /Rehydrated Milk is milk prepared by dispersing whole milk powder (also called dried whole milk) in water approximately in the proportion of 1 part powder to 7-8 parts water.
- Toned milk should contain a minimum of 3.0 per cent fat and 8.5 per cent solids not fat.
- Double toned milk should contain a minimum of 1.5 per cent fat and 9.0 per cent solids not fat.
- Skimmed milk is not more than 0.5% fat; not less than 8.7% Solid Not Fat (SNF).

11.2. BUFFALOES

There are two main categories of buffaloes namely Swamp and river buffalo

- Swamp buffalo (chromosomes 2n=48) permanent inhabitant of marsh land
- River buffalo (chromosomes 2n=50) found in India. Primarily milch animals.
- Buffaloes are nocturnal in habit
- Buffaloes Wallowing- rolling or floundering in mud or in water for regulation of body temperature and control of parasites
- The concentration of buffaloes was highest in northern region (46%) and lowest in eastern region (11%)

Breeds of Buffaloes

1. *Murrah* : Punjab and Delhi - Jet Black colour
2. *Jaffarabadi* : Gir forests (Gujarat) - black - Bulls hauls heavy loads, Heavy horns
3. *Nili R*avi : (looks like Murrah) - Montogmery (Punjab) - Black, Males used for draught purposes
4. *Bhadawari* : Uttar Pradesh and Madhya Pradesh, Copper colour
5. *Surti* : Gujarat - Horns are sickle shaped
6. *Mehsana* : Gujarat, JISM - Gujarat
7. *Nagpuri* : Maharahtra and Andhra Pradesh

   Dual purpose breed: Jaffarabadi

   Other buffalo breeds: Toda - Nilgiris (Tamil Nadu); Parlakmedi – Orissa; Pandharkiri and Maratharada – Maharashtra.

- Buffalo meat **c**alled as *cara beef,* it contains white fat because of absence of carotene
- Dressing % of old buffaloes- 43-47% (The percentage of the live animal weight that becomes the carcass weight at slaughter)
- Dressing % of 16-20 months buffaloes- 57- 60%
- Buffaloes come to heat regularly but well exhibited in October and lowest in April. Hence the name seasonal breeders or *silent heat*
- Body weight in seers (1 seer = 0.93kg) = GL/Y(L=length in inches; G=girth in inches; Y=constant)
- Calf mortality under one month on average is 10%
- Calves at birth should be fed colostrum within 2 hrs and feeding should continue @1 to 1.5 kg/day for 5 days
- Milk requirement for calf is half Kg less than one tenth of body weight
- Oestrous cycle is 21 days
- Gestation period is 316 ± 5 days (6 weeks less than a year)
- Calving interval is 420 days
- Coarse fodder is readily eaten by Buffaloes than cattle
- Buffaloes are sensitive to diseases compared to cattle
- Most of the diseases of cattle are applicable to buffaloes

11.3. SHEEP

Sheep along with goats are first ruminants to be domesticated

Important Indian Breeds of sheep

1. Temperate Himalayan
   - Gaddi - Jammu
   - Bhakarual - Himalayas
   - Gurez - Himalayas (Biggest among Kashmir breeds)
2. Northwestern Region
   - Hissar Dale (Highest wool production) - Hissar – Resistant to adverse climatic condition
   - Chokla - Rajasthan - finest wool among Rajasthan breeds)
   - Marwari - Rajasthan
3. Southern Region
   - Mandya - Mysore (Dual purpose)
   - Nellore (Biggest) - Nellore (famous for mutton)
   - Deccani - Maharashtra and Andhra Pradesh (making rough blankets)
4. Eastern Region
   - Shahabadi - Bihar and West Bengal

Important sheep breeds of Southern regions

- Mecheri - Salem, Erode and Coimbatore district of Tamil Nadu. Preferred for mutton (coarse woolly breed)
- Ramnad - Mutton type breed
- Ramnad and Tirunelveli
- Trichy Blacks - mutton type breed

Exotic breeds of sheep

- Merino - Most popular fine wood breed, Native of Spain
- Rambouillet - France
- South Down - NZ, Australia and USA
- South Down - One of the oldest of the English breeds, Second best after merino
- Lincoln - England, Largest of the British breeds, successfully used in cross breeding
- Leicoster - England - One of the purest breeds of British, excellent type of mutton
- Corriedale - Ancestry from Lincoln and Merino and some genetic contribution from Leicoster

There are 44 descript breeds of sheep

- Fine wool breeds - Merino, Rambouillet
- Medium wool breeds - Dorset horn, Suffolk - England
- Long wool breeds – Lincoln, Leicoster, Karkul (USA)
- Newly evolved breeds- Hissardale, Kashmir merino, Nilgiris, Avivastra, Avikalin and Bharat Merino
- Finewool exotic breeds- Merino, Ram bouillet, Polworth
- Mutton breeds- Suffolk, Dorset, Southdown

- Dual purpose- Corriedale
- Pelt breeds- Karakul

- In India, most of the sheep breeds produce fleece with higher average fibre diameter and medullation percentage
- Among native wools, chokla has overall best performance followed by Magra, Jaisalmeri, Malpura, Nali and Sonads
- Avikalin - Rambouillet x Malpura

Technical norms

- One Ram (male) sufficient for breeding 20 Ewes (female) 1:20
- First lambing is at age of 2 years
- Gestation period is 5 months (147 days)
- Lambing rate of sheep is 60-70% and goat is 100%
- Ewes give birth for about 8 times in (10-11 years) life span
- Sheep requires 10 hours of grazing
- Concentrated feed - 200 g / Eve / day for 50 days (20 days before lambing and 30 days after lambing)
- Water requirement is 2 litres/ day in winter and 3.5-4 litres/ day in summer
- Floor space : 10-15 sq.feet in covered area and 20sq.feet per animal in open area
- Tannic acid in tree leaves is a limiting factor for proper utilization of protein and carbohydrates in ruminants
- Marketable age (6 months) weight of sheep and goat is 18-20 kg

Wool production

- Exotic breeds: 5-6 kgs (wool) - 60-80 kgs (body wt)
- N.I breeds - 1-2 kgs(wool) - 40-50 kgs (body wt)
- S.I breeds - 0.6-1 kgs(wool) - 30-40 kgs (body wt)
- Fibre fineness is important quality parameter. Wool classified as
  - i. Fine wool (< 25 μ), used for suiting, knit wear
  - ii. Medium quality (< 40 μ and > 25 μ) used for carpets and blankets
  - iii. Coarse wool (> 40 μ) used for coarse carpet and blanket
- Skirting: After shearing, removing objectionable parts such as tags, leg pieces, neck pieces from body of fleece.
- Indian wools are ideal for carpet.

- Minimum floor space of 213 cm^2 /sheep must be ensured during transportation
- Sheep should be rested atleast 16 hours prior to slaughter with continuous access to water

Terminologies

- *Black wool*: Any wool that is not white, but not necessarily black
- *Bottle lamb* or *cade lamb*: An orphan lamb reared on a bottle. Also *poddy lamb/pet lamb*
- *Clipping*: Cutting off the wool
- *Shearing*: Cutting off the fleece, normally done in two pieces by skilled shearers
- *Crimp*: The natural wave formation seen in wool. Usually the closer the crimps, the finer the wool
- *Dipping*: Immersing sheep in a plunge or shower dip to kill external parasites
- *Drench*: An oral veterinary medicine administered by a drenching gun (usually an anthelmintic)
- *Fleece*: The wool covering of a sheep.

- *Flushing*: Providing especially nutritious feed in the few weeks before mating to improve fertility, or in the period before birth to increase lamb birth-weight.
- *Gimmer*: A young female sheep, usually before her first lamb
- *Stag*: A ram castrated after about 6 months of age
- *Tup*: an alternative term for *ram*
- *Tupping*: mating in sheep, or the mating season
- *Wether*: A castrated male sheep
- *Sheep folding*: Practice of utilizing sheep manure for fertilizing the cultivated field
- *Dagging* or *crutching*: Cutting away of dirty, wet wool from around the tail and anus (crutch) of the sheep
- *Mutton*: Sheep meat
- *Culling*: Removal of unwanted animal from the farm

Diseases of sheep

Viral diseases

i. *Sheep pox* is highly contagious disease causing a mortality of 20 to 50 per cent in animals below the age of 6 months, and causes damage to the wool and skin in adults.

ii. *Blue tongue* is a non-contagious, non-zoonotic, insect-borne (culicoides), viral disease of ruminants, caused by the Bluetongue virus (BTV).-coronitis found

iii. *Peste de petits ruminants* (PPR) is involve respiratory and digestive system- bran like lesions in mouth

Bacterial

- Anthrax is Caused by Bacillus anthracic-sudden death
- Enterotoxaemia is called as Pulpy kidney disease. Caused by *Clostridium perfringens* Type D
- *Hemorrhagic septicemia (HS)*
- Brucellosis - zoonotic

Parasitic

- Fasciola, Immature Amphistomes, Strongyloides, Lungworms etc.

Vaccination is generally practised against the diseases such as Sheep pox, Enterotoxaemia, Anthrax, FMD

11.4. GOATS

Goat was first domesticated first among all ruminants followed by sheep. Provide a dependable source of income to 40 percent of rural population below poverty line.

- Goat milk contains lower fat as compared to buffalo milk, with smaller fat globules
- Ghee from goat milk fat is considered as efficient decongestant for respiratory disease and pneumonia
- Goat hairs have higher tensile strength preferred for rope making
- Mohair from Angora goats are suitable for fancy bags.
- Chegu and Changthangi goats of northern Himalayan region produce the finest under coat (Pashmina)
- The average Pashmina production from the Indian breeds is low (about 100g in chagu and 250g in changthangi per yr) but fibre is superior in quality (8-14 µm in diameter)
- Glaced kid feather of Black Bengal is quite famous in foreign countries.

Breeds of goat

1. Meat and Skin

i. Black Bengal - West Bengal; Jet Black - skin is superior quality; excellent quality meat

ii. Ganjam - Orissa

iii. Kannaiadu - Tamilnadu

2. Meat, Milk and skin

i. Barbari

ii. Malabari (Tellichery) - Kerala

iii. Osmanabadi - Maharastra

iv. Sirohi and Jhakra - Rajasthan

v. Surti, Mehsana and Zalawadi - Gujarat

3. Milk, meat and skin

i. Beetel- Punjab - Red / tan - Resembles Jamunapari, but small twins are common

ii. Jamunapari- Uttar Pradesh, average milk yield/day is 1.03Kg. White with tan markings, biggest and majestic breed of goat in India

4. Meat, hair and skin

i. Bakharwal- Kashmir

ii. Gaddi, Kangan- Himachal Pradesh

iii. Marwri - Rajasthan

iv. Sangamneri - Maharashtra

5. Meat, Pashmina and skin

i. Chanthangi- Jammu & Kashmir

ii. Chegu- Himachal Pradesh

Breeds of Exotic Goat reared in India

- *Alpine* - France - Milch type, tropical
- Anglo Nubian - (England) is called "Jersey cow" of the goat world-Tropical
- Saanen - Switzerland - milk queen of the goat world,sensitive to direct sunlight, presence of beards in both sexes
- Doggenburg - Switzerland - tropical climate
- Angola - Turkey - Chocolate colour

➢ Mohair obtained from Angora and Sangamneri

➢ Highly mobile upper lips of goats enable them to graze as close to the ground as the sheep whilst being more selective than cattle.

Technical norms

- goat meat is called *chevon*
- Breeding ratio - 1:25 (Bucks (M): Doe's (F)

- Gestation period is 147-149 days (5 months)
- Kidding percentage is 100% in 1st year and 120% in 2nd year onwards
- Lactation period is 210 days
- Milk yield of Indian breeds is 1-2 litres / day and Exotic breeds is 3 litres / day

Age in Goat and Sheep (Permanent Incisors)

- 1st pair - 1-1½ years (1½ yrs) - 12-18 months
- 2nd pair - 1½ - 2 years (2 yrs) - 19-25 months
- 3rd pair - 2-2½ years (2½ yrs) - 26-30 months
- 4th pair - 2½ - 3 years (3 yrs) - 30-36 months

Economic life of sheep and goat is 6-7 years

Goat mean gestation period is 147 days

Floor space requirement

- ➢ Adult goat 1.25 to 1.5 m^2
- ➢ Lactating and pregnant does 2 m^2
- ➢ Kids 7-90 days 0.5-0.6 m^2
- ➢ 3-6 months 0.7-0.9 m^2
- ➢ 9-12 months 1 m^2

Diseases of goat

Viral disease: Peste de petits ruminants (PPR), Goat pox, Food and Mouth Disease (FMD)

Bacterial disease: Enterotoxaemia, *Hemorrhagic septicemia (HS)*, Contagious pleura pnemonia

11.5. SWINE (PIG)

Breeds of swine

- large white yorkshire - popular English bacon breed (England)
- Middle white yokshire - England
- Berkshire - "Britain's oldest pig breed"
- Others important breeds
  - land race – Denmark
  - Tomworth – England
  - Duroc – USA
  - Chester white

Norms for piggery

- Breeding ratio; 1:10 (Boar : Sows)
- Gilts become sexually mature in 5-6 months age
- Sexualcycle is 21 days
- Gestation period is 114 days (3 months, 3 weeks and 3 days)
- 10 piglets / farrowing (Act of birth)

- Marketing weight - - 80 kgs (9 months age)
- Pork - meat of pig
- Pig, hog or swine - the species as a whole, or any member of it. The singular of "swine" is the same as the plural.
- Shoat, piglet - unweaned young pig, or any immature pig.
- Sucker- a pig between birth and weaning.
- Runt- an unusually small and weak piglet, often one in a litter.
- Boar or hog- male pig of breeding age.
- Barrow- male pig castrated before puberty.
- Stag- male pig castrated later in life, (that is, an older boar after castration).
- Gilt- young female not yet mated, or not yet farrowed, or after only one litter
- Sow- breeding female, or female after first or second litter.
- Suckling pig- a piglet slaughtered for its tender meat.
- Feeder pig- a weaned gilt or barrow weighing between 18 kg (40 lb) and 37 kg (82 lb) at 6 to 8 weeks of age that is sold to be finished for slaughter.
- Porker- market pig between 30 kg (66 lb) and about 54 kg (120 lb) dressed weight.
- Baconer- a market pig between 65 kg (140 lb) and 80 kg (180 lb) dressed weight. The maximum weight can vary between processors.
- Grower- a pig between weaning and sale or transfer to the breeding herd, sold for slaughter or killed for rations
- Finisher- a grower pig over 70 kg (150 lb) liveweight.
- Butcher hog- a pig of approximately 100 kg (220 lb), ready for the market.
- Backfatter- cull breeding pig sold for meat; usually refers specifically to a cull sow, but is sometimes used in reference to boars.
- Herd- a group of pigs, or all the pigs on a farm or in a region.
- Sounder- a small group of pigs (or wild boar) foraging in woodland
- Trotters, the feet of pigs (they have four hoofed toes, walking mainly on the larger central two).
- In pig- pregnant.
- Farrowing- giving birth.
- Hogging- a sow when on heat (during estrus).
- Sty- a small pig-house, usually with an outdoor run or a pig confinement.
- The dental formula of adult pigs is $\frac{3.1.4.3}{3.1.4.3}$, giving a total of 44 teeth
- The teeth of the young pig (needle teeth) are clipped as soon as possible after birth. The piglet is born with 8 teeth.
- Young pigs should be castrated at 2 to 3 weeks of age.
- Identification by ear tagging or notching
- Most efficient feed converting animal species
- Piglet anaemia (Thumps)- prevented by Pasting Ferrous sulphate paste in udder of sow

Diseases

- ➢ Viral- vesicular stomatitis, psedo rabies (Aujesky's disease), swine pox, , Classical swine fever(Hog cholera), vesicular exanthema, Transmissable gastro enteritis, Japanese encephalitis
- ➢ Bacterial- Brucellosis, leptospirosis, colibacillosis, salmonellosis.
- ➢ Parasitic- trichinosis, *Taenia solium*, cysticercosis

11.6. CAMEL

Camelus dromedaries (Indian camel) is single humped or dromedary in dry hot arid lands of Africa and Asia and *Camelus bactrianus* is double humped or Bactrian in cold arid lands of central palaerctic, avialble in Ladhak area of Jammu and Khasmir. Camels have 37 pairs of chromosomes.

- ✓ India stand third in Camel population in the world
- ✓ Adult camel consumes dry matter equivalent to 1.8 to 2 % of body weight
- ✓ Camels can withstand upto 20% of body loss due to continuous dehydration
- ✓ They can consume 100 to 125 litres in one day
- ✓ Distinct Camel breeds are Bikaneri, Jaisalmeri, Kachchchi and Marwari. Other breeds include Mewari, Sindhi, Shekawati, riverine and double humped camels
- ✓ Camels are seasonal breeders and are induced ovulators and don't exhibit estrous cycle
- ✓ Pregnant females lift their tail and curl it up in a typical style when approached by a male
- ✓ Gestation period is 380-390 days
- ✓ Males come in rut during winter period
- ✓ A male in rut extrudes soft palate from its mouth and produces typical sound
- ✓ The ratio of breeding stud to females is 1:30-35
- ✓ Secretion from poll gland in the male camel is special feature during rutting season
- ✓ Viral diseases include Camel pox, Rabies and Rinderpest
- ✓ Bacterial diseases include Anthrax, Brucellosis, TB, Contagious skin necrosis, Salmonellosis and others
- ✓ In parasitic diseases, Mange is most common. Trypanosomiasis cause wide spread mortalitity and morbidity.
- ✓ Daily milk yield in camels ranges from 3.5 to 10 kg and peak yield is 20 kg/day

11.7. RABBIT

China has emerged as the lead country and has emerged as a major producer of rabbit meat. Can be bred any time of the year with a peculiarity that doe can be bred even at the day of parturition.

Breeds of rabits

i. Wool breed- Angora

ii. Meat/ fur skin- Newzealand white (from UK) , White California, Soviet chinchilla, grey giant and white giant from Russia

iii. Fancy breeds- Polish, Palwino, Hawana, Beveran, Newzealand red, English spot white

- Gestation period is 30 days
- Backyard meat rabbit is nick named as " Biological refrigeration" at farmers' house
- Rabbit reared for meat, fur and wool purpose.
- A doe selected for breeding should have eight functional teats.

- Rabbits of smaller breeds reach breeding age at 5-6 months; medium breeds 6-7 months and large breeds at 8-10 months.
- Female rabbit is always taken to male rabbit cage for mating purposes.
- Process of giving birth is known as kindling
- New born open eyes between 10-14 days
- Weaning age is 6-8 months
- The peculiarity in the rabbit digestion is the process called coprophagy or pseudo rumination.
- Common ailments of rabbits are snuffles (rhinitis), weepy eyes (conjunctivitis), bronchopneumonia, wry neck (torticollis), coccidiosis, mucoid enteritis, dysentery, mange, ear canker, sore hocks, paraplegia, heat prostration and cold shock.
- Rabbit manure can be applied as raw manure
- Male female ratio is 1:4
- Space requirement is 4 sq.feet for breeding rabbit and 3 sq.feet for grower
- Feed requirement is 120g/adult/day, 80-100g/grower/day and 50g/weaner/day

Diseases

i. Myxomatosis- by Virus- Inflammation and oedema of eyelids, ears, anal and genital orifices- The disease causes 100 % mortality Pasteurellosis- caused by bacteria Pasteurella multocida

ii. Coccidiosis- Protozoa (Eimeria species)

iii. Ear canker (ear mange)- Psoroptes cuniculi – treated by Benzyl benzoate

iv. Ring worm- Fungus (Trichophyton microsporon)- Treated by Griseofulvin

v. Body mange- Notedres cati

11.8. POULTRY

Poultry includes chicken, ducks for both egg and meat. Turkeys, Geese, Guinea fowls and pea fowls are only for meat purpose.

Chicken- Common poultry breeds

- American class (Dual purpose) - Rhode Island Red (RIR), New Hampshire (NHS), Plymouth Rock
- Others Wyandotic -Yellow skin
- Asiatic class - Brahma, Cochin, Langshan - Feathered - Shanks
- Mediterranean class - Leghorn-Egg purpose, Minorca - White earlobes -Yellow/white skin
- English class (Dual purpose) - Australop, Orpington, Sussex, Cornish - White skin
- Dual purpose - American class, English class
- Egg purpose - Mediterranean class

Strains of commercial poultry breeds

1. **Broilers:** Samrat, Petersen, Indian Kiser, Marshall, Lehmann (B), Pearl Vikram, Unibro and Ross.
2. **Layers:** Babcock, Lehmann (L), Star cross, Pearl sonali, Pearl Tara – 287, H & N, Colonial, Unisex white and Kalinga Hybrid Brown.

Indigenous breeds (Poultry)

- Aseel - A.P, U.P, Rajasthan

- Busra - Gujarat and M.H
- Chittagong (Malay) - E. India
- Karaknath (Kalamasi) - M.P

Layer farming

1. Deeplitter system (1 labourer for 1000 layers)
2. Cage system - (1 labourer for 2500 layers

- A **pullet** is a sexually mature when she lays her first egg
- Production per hen housed is 275 eggs / hen

Brooder management (0- 8 weeks)

- Chick mash contains Crude protein (CP)-20%, ME 2600 k cal/kg
- Photo period for chicks is 23 hours
- Temperature requirement for 1st week to 5th week is 90° F to 70° F with weekly decrement of 5° F
- Brooder require 0.5 square feet on floor
- Beak trimming (to avoid feed wastage and cannibalism), toe clipping (for breeder males) and dubbing (Dubbing is a procedure used by poultry farmers to remove the comb from the head of the bird in an attempt to limit later damage by injury, freezing or cannibalism) are (done when they are brooders)
- Problems include stress, coccidiosis, unabsorbed yolk, hotweather and cold weather

Grower (8-20 weeks old)

- Grower mash contains Crude protein (CP) is16%, ME (Metabolizable Energy)2500 k cal/kg
- No artificial light only natural light
- Require 1 square feet

Layer (21-72 weeks)

- Layer mash contais Crude protein (CP) is18%, ME 2600 Kcal/Kg
- Light should be provided from 20 weeks an increment of 30 minutes light every week till it reaches 16 hours, after 6 months 17 hours
- Require 2 square feet

Broiler farming: "All in All out" system

- 1 labourer for 800 birds
- Weight at 8 weeks is to be 1.35-1.50 kgs / bird (marketable) (Nowadays 6wks-marketable age)
- Feed consumed upto 8 weeks is 3.75 kgs/bird
- Broiler starter mash (0-4 wks) contains CP 23%, ME 2800 k cal/kg
- Broiler finisher mash (5-8 weeks) is CP 20%, ME 2900 k cal/kg

Hatcheries

- Male: female ratio required is 1:10 to 15 for egg type and meat type
- Hatchery should be situated at least 1000 metres from the houses wherein birds are kept
- Hatchery rooms should have egg receiving room, egg storage room, chick holding room, wash room, storage room, fumigation chamber, egg holding room, incubator room, sexing room
- Candling should be done on the 7th day to remove infertile eggs and also on 18th day
- Large end of the egg should be kept up where the egg shell is located for incubation
- Optimum temp for holding the fertile eggs before incubation varies from 50-70° F (10° C to 21.1° C)

- For Better results 60° F (15.6° C) and ralative hunidity of 70-80% (less humidity promotes loss of water from Eggs) is ideal
- Incubator temperature; 99.5° F - 100.5° F (37.2° C - 37.8° C) for Forced draft -type incubator. 1° F higher for still-air incubators
- Low temp slows down the development of embryo and higher than optimum temp hastens embryonic development.
- Sexing of chicks by identification of rudimentary testes in the cloaca of male chicks on day one old chick.
- Egg contains albumin (56-61%), yolk (27-32%) and egg shell (8-11%)
- NAFED (National co-operative marketing federation)

Diseases

i. Viral diseases: Ranikhet disease (New castle disease), fowl pox, infectious bronchitis, infectious laryngo tracheitis, infectious bursitis, avian encephalomyelitis, marek's disease,leukosis, chronic respiratory diseases, duck viral enteritis, hepatitis etc.,

ii. Bacterial diseases: Pullorum, Typhoid, Paratyphoid, fowl cholera, Arizona infection in turkey

iii. Protozoal diseases: Coccidiosis in chicken, blackhead in turkey and leucocytozoonosis

iv. Fungal diseases: Aspergillosis, moniliasis

Nutritional Deficiency diseases/symptoms

(if diseases made bold make it uniform)

Deficiency of

- Calcium, phosphorus and vitamin D in young growing animals- Rickets
- Calcium, phosphorus and vitamin D in adult animals-osteomalacia
- Calcium commonly occuring in dairy cows shortly after calving- **Milk fever (parturient paresis)**
- Phosphorus -pica or depraved appetite
- Magnesium -hypomagnesaemic tetany ,magnesium tetany, lactation tetany and grass staggers.
- Iron **-Anaemia** (Anemia in piglets –**thumps)**
- **Copper in adult cattle- Falling disease**
- **Copper - Peat Scours**
- **Zinc - Parakeratosis in pigs**
- Manganese - Perosis or slipped tendon in young chicks
- Cobalt - Wasting disease or coast disease or Pining or Enzootic marasmus
- Selenium, vitamin E - Nutritional myopathy / white muscle disease / stiff lamb disease / mulberry heart disease
- Vitamin E - **Nutritional encephalomalacia (crazy chick disease) and exudative diathesis**
- Iodine – Goitre
- Vitamin A - **Bitot's spots, xerophthalmia (dry eye), congenital blindness, nutritional roup**
- Vitamin K - Sweet clover disease
- Vitamin C – Scurvy
- Vitamin B1 - star gazing in chicks, Chastek paralysis in foxes
- Vitamin B2 - curled toe paralysis / clubbed down condition in chicks

- Niacin (nicotinamide) –Pellagra in human
- Pantothenic acid - goose-stepping' gait in chicks
- Folic acid-anemia
- Biotin - fatty liver and kidney syndrome (FLKS) in chicks
- Choline - perosis or slipped tendon in chicks

Excess

- Excess of dietary phosphorus in relation to calcium (optimum calcium phosphorus ratio is between 1:1 and 2:1) - **Nutritional secondary hyperparathyroidism (NSH)**. **(miller's disease or bran disease or big head disease in horses)**
- Selenium toxicity - Alkali disease and Blind staggers
- Molybdenum toxicity- '**teart**' or **'peat scours'**
- Fluorine toxicity- fluorosis.
- Vitamin A toxicity- hypervitaminosis A

Meat Science

- Heated beef or sour slides- white discolouration caused by inability of freshly killed carcass to dissipate heat rapidly
- Green struck- Similar condition in hares and rabbits
- Sulphiding/psedomelanosis- occurs in carcasses especially in lamb carcassesnand in pigs when evisceration is delayed after killing
- Bone taint- occurs when the carcasses are not cooled rapidly
- Taint in hams (smoked product of hip of pig)- souring due to contamination of proteolytic organisms like Clostridium
- Phosphorescence- Caused by *Psedomonas phosphorescens*
- Torrymeter- electronic freshness meter for meat

Zoonosis

The term zoonosis (zoonoses, plural) was coined by Rudolf Virchow to describe the disease that was transmitted from animal to man and vice versa.

1.	*Bacterial zoonoses*	e.g., anthrax, brucellosis, plague, salmonellosis,leptospirosis etc.
2.	*Viral zoonoses*	e.g., rabies, influenza, Japanese encephalitis, yellow fever, etc.
3.	*Rikettsial zoonoses*	e.g., Q-fever, typhus fever, etc.
4.	*Fungal zoonoses*	e.g., cryptococcosis, candidiasis, ringworms, etc.
5.	*Protozoan zoonoses*	e.g., toxoplasmosis, leishmaniasis, trypanomiasis, etc.
6.	*Helminthic zoonoses*	e.g., echinococcosis (hydatid disease), taeniasis, guineaworm infection, etc.
7.	*Ectoparasitic zoonoses*	e.g., scabies, myiasis, etc.
8.	Zoonoses caused by unconventional agents(Prions)	e.g., transmissible spongiform encephalopathies, bovine spongiform encephalopathy (TSE, Mad cow disease), etc.

BASIC FACTS

P. KANNAN and M. THIRUNAVUKKARASU

12.1. BASIC AND CURRENT FACTS OF AGRICULTURE

Basic information related to agriculture and allied sciences in India

Average Annual Rainfall	:	1190 mm
Total Geographical Area	:	328.7 mha
The Generated Rainfall Volume	:	400 mha-m
Maximum Area under Irrigation	:	Ganga basin
Maximum Avg. Annual Runoff occurs in	:	Brahmaputra basin
Gross Irrigated Area	:	95.8 mha
Net Irrigated Area	:	68.1 mha
Gross Cropped Area/total cropped area	:	200.8 mha
Gross Cropped Area /Total Cropped Area	:	61.1 % of Geographical area
Net area sown	:	141.4 mha
Area sown more than once	:	59.4 mha
Cropping Intensity	:	142 m ha
I.Fallow Lands	:	24.8 m ha
Ia.Fallow Lands other than Current Fallows	:	10.7 m ha
Ib.Current Fallows	:	14.1 m ha
II.Other Uncultivated land excluding Fallow Land	:	25.8 mha
IIa. Permanent Pasture & other Grazing Land	:	10.3m ha
IIb. Land under Miscellaneous Tree Crops & Groves not included in Net Area Sown	:	3.2 m ha
IIc.Culturable Waste Land	:	12.4 mha
III. Not Available for Cultivation	:	43.9 m ha
IIIa. Area Under Non agricultural Uses	:	26.9 m ha
IIIb. Barren & Un-culturable land	:	16.9 m ha
Total Degraded Land	:	187.7 mha
Severely Degraded Land	:	65.0 mha

Total Area Under Forest	:	71.8 mha
Optimum Forest Area Required	:	33 % of total geographical area
Total Area Under Forage Crops	:	11 mha
Total Area Under Fodder Crops	:	4.9 % of cropped area
Total Area Under Horticultural Crops	:	23.8 m ha (11.8% of total cropped area in 2005)
Area Under Pasture cum Grazing Lands	:	0.97% of total Geographical area
Maximum Production Under Fruit Crops	:	Banana
Maximum Area Under Fruit Crops	:	Mango
Maximum production and area under vegetable crops	:	Potato
Maximum production and area under plantation crops	:	Coconut
Maximum production under spices		Garlic
Maximum Area Under spices		Dried chillies followed by Cumin
Total Milk Production in India (2015-16)	:	155.5 mt
Milk Availability in India	:	337 g / person/ day
Average Availability of Milk in the World	:	300 g / person / day
Maximum Requirement of Milk	:	240 g / person / day
Availability of Animal Protein in Indian diets	:	10 g / person / day
Area Under Major / Medium Irrigation	:	55.0 mha
Area Under Micro Irrigation	:	8.6 mha (8626784 ha)
Total area under drip irrigation system	:	3919780 ha
Total area under sprinkler irrigation system		4707004 ha
Maximum area under drip irrigation is in	:	Andhra Pradesh
Maximum area under sprinkler irrigation is in		Rajasthan
Irrigated area in India	:	47.7% of the total cultivated area
Maximum area under irrigation is in	:	Punjab (98.5 % of cultivated area)
Major source of irrigation in India is	:	Canal (45 %), Tube Wells (34 %)
Per capita availability of arable land	:	0.15 ha
Average size of land operational holdings	:	1.15 ha
Maximum operational holdings size	:	Nagaland (6.82 ha)
Minimum operational holding size	:	Kerala (0.33 ha)
Number of farm operational holdings	:	106.6 million
Annual soil loss due to erosion	:	5344 million tones
Per capita water availability	:	2200 cubic meter / person
Percentage of fresh water on the earth	:	2.8%
Agricultural & livestock census conducted	:	Every 5 years

Highest population below poverty line	:	Chhattisgarh (39.9%)
Maximum no. of persons below poverty line	:	Uttar Pradesh
Percentage of agril. Exports to total exports	:	12.5 %
Percentage of agril. imports to total imports	:	5.6 %
Share of agrl. and allied sectors	:	17.32 % (2016-17)
Area covered under soil conservation	:	5.46 lakh hectares (2013-14)

Percentage share of area under major crops (2015-16)

Major Crops	Area (%)	Remarks
Rice	22.9	Total food grains -64.7%
Wheat	15.96	
Coarse cereals	12.5	
Pulses	13.3	
Oilseeds	13.8	
Sugarcane	2.6	
Cotton	6.3	----------
others	12.6	

Percentage distribution of Gross cropped area (2013-14)

Crops	Area (%)	Others category
Rice	22.1	Total Cereals & Millets - 50.5%
Sorghum/Jowar	2.9	
Cumbu/Bajra	4	
Maize	4.5	
Ragi	0.6	
Wheat	15.6	
Barley	0.3	
Other cereals& millets	0.4	
Bengalgram	4.7	Total pulses -11.8%
Redgram	1.8	
Other pulses	5.4	
Sugarcane	2.8	
Total Condi-ments & Spices	1.6	

Crops	Area (%)	Others category
Total Fruits & Vege-tables	4.9	
Groundnut	2.7	Total oilseeds - 15%
Castor -seed	0.5	
Sesamum	0.8	
Rapeseed & Mustard	3	
Linseed	0.1	
Coconut	0.9	
other oilseeds	6.9	
Cotton	5.9	Total Fibres - 6.4%
Jute	0.4	
Others fibre	0.1	
Fodder Crops	4.9	-
Total food crops	71.7	-
Total Non food crops	28.3	-

Size of land holding (2010-11)

Category of holdings	Average size of holdings (Ha)
Marginal (Less than 1 hectare)	0.39
Small (1.0 to 2.0 hectares)	1.42
Semi- Medium (2.0 to 4.0 hectares)	2.71
Medium (4.0 to 10.0 hectares)	5.76
Large (10.0 hectares and above)	17.4
All holdings	1.15

Agricultural Research and Education

- Agricultural Research Institute started at PUSA, Bihar (renamed later as Imperial Agricultural Research Institute) in 1905 and shifted to New Delhi in 1936
- Indian Council of Agricultural Research (ICAR) was established in 1929
- Krishi Vigyan Kendra in India is 660 KVKs and in Tamil Nadu is 30 KVKs
- Number of KVK zones in India : 8
- Total number of Agricultural University- 71

ICAR Institutes

- Central Research Institute – 64
- National Bureau -6

- National Research Centre – 15
- Directorate and Project directorate -13
- Soil climatic zone – 40 (ICAR)
- Agroclimatic zone -15
- Agroecological region-20
- State Agricultural University – 39 and 1 Central Agricultural University at Imphal
- Deemed Universities – 4
- Agricultural Universities- 71

New agriculture policy has been described as **"Rainbow Revolution"** which includes following revolutions

Green	Food grain production
White	Milk
Yellow	Oil seeds
Blue	Fisheries
Red	Meat/tomato
Golden	Fruits-apple
Grey	Fertilizers
Black/Brown	NON-conventional energy sources
Silver	Eggs
Round	Potato
Pink	Shrimp/Onion

12.2. AGRICULTURAL RESEARCH ORGANIZATIONS

ICAR Institutions: 64

1 ICAR-Central Island Agricultural Research Institute , Port Blair

2 ICAR-Central Arid Zone Research Institute, Jodhpur

3 ICAR-Central Avian Research Institute, Izatnagar

4 ICAR-Central Inland Fisheries Research Institute, Barrackpore

5 ICAR-Central Institute Brackishwater Aquaculture, Chennai

6 ICAR-Central Institute for Research on Buffaloes, Hissar

7 ICAR-Central Institute for Research on Goats, Makhdoom

8 ICAR-Central Institute of Agricultural Engineering, Bhopal

9 ICAR-Central Institute for Arid Horticulture, Bikaner

10 ICAR-Central Institute of Cotton Research, Nagpur

11 ICAR-Central Institute of Fisheries Technology, Cochin
12 ICAR-Central Institute of Freshwater Aquaculture, Bhubneshwar
13 ICAR-Central Institute of Research on Cotton Technology, Mumbai
14 ICAR-Central Institute of Sub Tropical Horticulture, Lucknow
15 ICAR-Central Institute of Temperate Horticulture, Srinagar
16 ICAR-Central Institute on Post harvest Engineering and Technology, Ludhiana
17 ICAR-Central Marine Fisheries Research Institute, Kochi
18 ICAR-Central Plantation Crops Research Institute, Kasargod
19 ICAR-Central Potato Research Institute, Shimla
20 ICAR-Central Research Institute for Jute and Allied Fibres, Barrackpore
21 ICAR-Central Research Institute of Dryland Agriculture, Hyderabad
22 ICAR-National Rice Research Institute, Cuttack
23 ICAR-Central Sheep and Wool Research Institute, Avikanagar, Rajasthan
24 ICAR- Indian Institute of Soil and Water Conservation, Dehradun
25 ICAR-Central Soil Salinity Research Institute, Karnal
26 ICAR-Central Tobacco Research Institute, Rajahmundry
27 ICAR-Central Tuber Crops Research Institute, Trivandrum
28 ICAR-ICAR Research Complex for Eastern Region, Patna
29 ICAR-ICAR Research Complex for NEH Region, Barapani
30 ICAR-Central Coastal Agricultural Research Institute, Ela, Old Goa, Goa
31 ICAR-Indian Agricultural Statistics Research Institute, New Delhi
32 ICAR-Indian Grassland and Fodder Research Institute, Jhansi
33 ICAR-Indian Institute of Agricultural Biotechnology, Ranchi
34 ICAR-Indian Institute of Horticultural Research, Bengaluru
35 ICAR-Indian Institute of Natural Resins and Gums, Ranchi
36 ICAR-Indian Institute of Pulses Research, Kanpur
37 ICAR-Indian Institute of Soil Sciences, Bhopal
38 ICAR-Indian Institute of Spices Research, Calicut
39 ICAR-Indian Institute of Sugarcane Research, Lucknow
40 ICAR-Indian Institute of Vegetable Research, Varanasi
41 ICAR-National Academy of Agricultural Research & Management, Hyderabad
42 ICAR-National Institute of Biotic Stresses Management, Raipur
43 ICAR-National Institute of Abiotic Stress Management, Malegaon, Maharashtra
44 ICAR-National Institute of Animal Nutrition and Physiology, Bengaluru
45 ICAR-National Institute of Research on Jute & Allied Fibre Technology, Kolkata
46 ICAR-National Institute of Veterinary Epidemiology and Disease Informatics, Hebbal, Bengaluru

47 ICAR-Sugarcane Breeding Institute, Coimbatore

48 ICAR-Vivekananda Parvatiya Krishi Anusandhan Sansthan, Almora

49 ICAR-Central Institute for Research on Cattle, Meerut, Uttar Pradesh

50 ICAR-National Institute of High Security Animal Diseases, Bhopal

51 ICAR-Indian Institute of Maize Research,New Delhi

52 ICAR- Central Agroforestry Research Institute , Jhansi

53 ICAR-National Institute of Agricultural Economics and Policy Research, New Delhi

54 ICAR- Indian Institute of Wheat and Barley Research, Karnal

55 ICAR- Indian Institute of Farming Systems Research, Modipuram

56 ICAR- Indian Institute of Millets Research, Hyderabad

57 ICAR- Indian Institute of Oilseeds Research, Hyderabad

58 ICAR- Indian Institute of Oil Palm Research, Pedavegi, West Godawari

59 ICAR- Indian Institute of Water Management, Bhubaneshwar

60 ICAR-Indian Institute of Rice Research, Hyderabad

61 ICAR- Central Institute for Women in Agriculture, Bhubaneshwar

62 ICAR-Central Citrus Research Institute, Nagpur

63 ICAR-Indian Institute of Seed Research, Mau

64 ICAR-Indian Agricultural Research Institute, Hazaribag , Jharkhand

National Research Centres: 15

1. National Centre for Integrated Pest Management, New Delhi
2. National Research Centre for Banana, Trichi
3. National Research Centre for Grapes, Pune
4. National Research Centre for Litchi, Muzaffarpur
5. National Research Centre for Pomegranate, Solapur
6. National Research Centre on Camel, Bikaner
7. National Research Centre on Equines, Hisar
8. National Research Centre on Meat, Hyderabad
9. National Research Centre on Mithun, Medziphema, Nagaland
10. National Research Centre on Orchids, Pakyong, Sikkim
11. National Research Centre on Pig, Guwahati
12. National Research Centre on Plant Biotechnology, New Delhi
13. National Research Centre on Seed Spices, Ajmer
14. National Research Centre on Yak, West Kemang
15. National Research Centre on Integrated Farming (ICAR-NRCIF), Motihari

Directorates/Project Directorates : 13

1 ICAR-Directorate of Groundnut Research, Junagarh
2 ICAR-Directorate of Soybean Research, Indore
3 ICAR-Directorate of Rapeseed & Mustard Research, Bharatpur
4 ICAR-Directorate of Mushroom Research, Solan
5 ICAR-Directorate on Onion and Garlic Research, Pune
6 ICAR-Directorate of Cashew Research, Puttur
7 ICAR-Directorate of Medicinal and Aromatic Plants Research, Anand
8 ICAR-Directorate of Floricultural Research, Pune, Maharashtra
9 ICAR-Directorate of Weed Research, Jabalpur
10 ICAR-Project Directorate on Foot & Mouth Disease, Mukteshwar
11 ICAR-Directorate of Poultry Research, Hyderabad
12 ICAR-Directorate of Knowledge Management in Agriculture (DKMA), New Delhi
13 ICAR-Directorate of Cold Water Fisheries Research, Bhimtal, Nainital

Deemed Universities: 4

1. ICAR-Indian Agricultural Research Institute, New Delhi
2. ICAR-National Dairy Research Institute, Karnal
3. ICAR-Indian Veterinary Research Institute, Izatnagar
4. ICAR-Central Institute on Fisheries Education, Mumbai

All India Coordinated Research Projects - 60

All India Network Projects -19

International institutes

1. Africa Rice Center (WARDA),Cotonou,Benin
2. Bioversity International,Rome, Italy
3. CIAT - Centro Internacional de Agricultura Tropical, Cali,Colombia
4. CIFOR - Center for International Forestry Research, Nairobi,Kenya
5. CIMMYT - Centro Internacional de Mejoramiento de Maiz y Trigo, Mexico
6. CIP - Centro Internacional de la Papa, Lima, Peru
7. ICARDA - International Center for Agricultural Research in the Dry Areas, Aleppo, Syria
8. ILRI - International Livestock Research Institute, Nairobi, Kenya
9. IRRI - International Rice Research Institute, Los Benos, Phillipines
10. IFPRI - International Food Policy Research Institute, Washington, USA
11. IITA - International Institute of Tropical Agriculture, Ibadan, Nigeria
12. IWMI - International Water Management Institute,Colombo, Srilanka
13. World Agro forestry Centre (ICRAF), Nairobi, Kenya
14. World Fish Center, Penang, Malaysia

Fathers of different concepts of agricultural sciences

Fathers	Related science
Father of Sociology	Auguste Comette
Father of Extension	Paul Leagans
Father of Botany	Theopharastus and Linnaeus
Father of Genetics	Gregeor John Mendel
Father of Pure Culture Technique	Oscar Brefeld
Father of Tissue Culture	Gotlieb Haberiandt
Father of Agricultural Chemistry	Justus von Liebig
Father of Indian Rust	K. C. Mehta
Father of Medical Bacteriology	Robert Koch
Father of Indian Plant Pathology	E. J. Butler
Father of Modern Plant Pathology	Anto de Bary
Father of Mycology	Pler A. Micheli
Father of Nematology	N. A. Cobb
Father of Bacteriology	Anton van Leewenhoeck
Father of Forest Pathology	Robert Haring
Father of Ornamental Gardening	M. S. Randhwa
Father of Statistics	R. A. Fisher
Father of Modern Tillage	Jethro Tull
Father of Microbiology	Louis Pasteur
Father of Plant Physiology	Stephen Hales
Father of Economics	Adam Smith
Father of modern economics	J.M. Keynes
Father of Soil Science	Dokucheiav
Father of Agronomy	Peter De Cresenzi
Father of Nanotechnology	Richards Feyman

Availability of Biodegradable waste in India for composting

- Crop residues (350 million tonnes),
- Vegetable and fruit wastes (5 million tonnes),
- Press mud (5 million tonnes),
- Poultry litter (1 million tonne) and
- City waste (57 million tonness)
- 5-6 million tonnes of low-grade rock phosphate

12.3. CURRENT STATISTICS

- Overall Gross Domestic Product (GDP) growth rate was slow down to 6.3% during third quarter of 2017-2018 in India
- Agriculture provide livelihood to 54.6% of population
- Employment opportunity generated from agriculture is 54.6% as per 2011 census
- Livestock sector contribution to agricultural GDP = 25.6%
- Livestocks provide employment to about 8.8 % of the population in India
- Total geographical area of India is 328.7 mha, which is 2.3% of world area (7th place)
- India stands on 16.8% of world population, 4.2% of water resource in world
- India had 21.9% of population found below poverty line (As per 2012 census)
- Tamil Nadu account 11.3% of population found below poverty line (As per 2012 census)
- In India, Un employment rate is 5.6%
- In India, Uttar Pradesh stands first in total food grain production of 44 Million tones and sharing of 19 % total area with yield of 2278 kg ha-1.
- Tamil Nadu had total food grain production of 11.9 Million tones and sharing 3.86% of total area with yield of 3090 kg ha-1.
- Punjab stand first in productivity of total food grains of 4273 kg ha-1 with sharing of to 6.6 % of total area
- All India Annual Average Daily Wage Rate in Rural India for various Agricultural Operations (For male - Rs.244/- and female -Rs. 191.9/-)
- Number of soil health card issued in India (8.11.2016)-33707428
- Number of soil health card issued in Tamil Nadu (8.11.2016)-3507666
- In India, Consumption of Electricity for Agricultural Purposes- 168913 GWh
- In India, Percentage electricity share of agricultural consumption to total consumption -20.7%
- In India, total crop loan and term loan were disbursed during 2015-16 - 877527 crores

Crop-wise area insured under all Insurance Schemes

Crops	Area insured (Lakh hectares)	% of Area Insured
India (area under all crops)	524.5	26.1
Tamil Nadu (area under all crops)	12.5	21.1
Paddy	114.9	25.8
wheat	87.8	27.9
cotton	14.5	12.2
Sugarcane	2.29	4.1
pulses	71.2	29.9
oilseeds	132.5	44
vegetables	20.5	36.6
fruits	3.9	9.5

Crop-wise irrigated area in India

Crops	Irrigated area (%)
Total cereals	59.5
Total food grains	51.9
Total pulses	19.7
Total oilseeds	27.3
Cotton	32.5
Sugarcane	95.3
Rice	59.6
Wheat	93.6
Maize	27.2
Groundnut	25.8
Bengalgram	35.2

- In India, Total area under irrigated is 47.7 % for all crops
- Coverage of irrigated area in Tamil Nadu is 56.1%
- Punjab state hold 98.5% of irrigated area in total area of Punjab
- Forest occupies 21.8% of total area (71.8 Mha)

Sector wise	Contribution	Growth rate
Agriculture	17.3%	4.9%
Industry	29.0%	5.6%
Service	53.6%	7.7%

- Net sown area of India is 141.4 M ha (43%)
- Irrigated area is 47.7% and rainfed area is 52.7%
- Horticulture = 23.8 M ha, 11.8% of Gross cropped area
- Horticulture share in Agriculture GDP = 28%

Area, production and productivity of major crops in India during 2015-2016

Crop	Area (million ha)	Yield (kg/ha)	Production (million tonnes)
Total Food grain production	122.6	2056	252.52
Rice	43.1	2404	104.3
Sorghum	6.7	781	6.1
Cumbu	8.5	1155	9.0
Wheat	29.6	3093	93.5

Crop	Area (million ha)	Yield (kg/ha)	Production (million tonnes)
Maize	8.7	2510	21.8
Coarse cereals	26.5	1595	
Bengalgram	8.8	859	7.17
Redgram	3.9	656	2.46
Total Pulses	24.5	652	16.47
Total Oilseeds	30.2	968	25.3
Groundnut	5.4	1486	6.77
Sunflower	0.93	697	0.33
Rapeseed & Mustard	6.3	1184	6.82
Sugarcane	4.82	71.095 t ha-1	352.16
Cotton	11.5	432	30.15 million bales 1bale =170kg
Jute and Mesta	0.88		10.46 million bales 1bale =180kg

Minimum Support Price (2017-2018) for 24 crops

S.No.	Crop	Rs/Quintal
1	Paddy -Common	1550
	Paddy - Grade A	1590
2	Wheat	1735
3	Sorghum/Jowar-hybrid	1700
	Sorghum/Jowar-maldandi	1725
4	Bajra/cumbu	1425
5	Maize	1425
6	Ragi	1900
7	Redgram/ Arhar (Tur)	5450 +Bonus Rs.200/q above MSP
8	Green gram/Moong	5575 +Bonus Rs.200/q above MSP
9	Black gram/Urad	5400 +Bonus Rs.200/q above MSP
10	Cotton-Medium stable	4020
	Cotton-long stable	4320
11	Groundnut in shell	4450 +Bonus Rs.200/q above MSP
12	Sunflower seed	4100 +Bonus Rs.100/q above MSP
13	Soyabean	3050 +Bonus Rs.200/q above MSP

14	Sesame	5300 +Bonus Rs.100/q above MSP
15	Niger seed	4050 +Bonus Rs.100/q above MSP
16	Bengal Gram	4400 +Bonus Rs.150/q above MSP
17	Barley	1410
18	Mustard/Rapeseed	4000+ Bonus Rs.100/q above MSP
19	Lentil	4250 +Bonus Rs.100/q above MSP
20	Safflower	4100+ Bonus Rs.100/q above MSP
21	Jute	3500
22	Sugarcane	255
23	Milling Copra	6500
	Ball Copra	6785
	De-husked coconut	1760

Statutory Minimum Price

Sugarcane = 81.2 Rupees / Quintal

- MSP for cereals (24 numbers) is fixed by Commission for Agriculture Costs and Prices (CACP)
- MSP for pulses and oilseeds is fixed by National Agricultural Cooperative Marketing Federation of India (NAFED)
- Interest rate for agriculture loan is 4% upto 3 lakhs for 6 months
- Weather Based Crop Insurance Scheme (WBCIS) was started during 2007-2008
- Accelerated Irrigation Benefit Programme was started in 1996-1997
- National Agriculture Insurance Scheme implemented in 1999-2000
- Pradhan Mantri Fasal Bima Yojana (PMFBY) and Weather Based Crop Insurance Scheme (WBCIS) started in - 2016
- Centrally sponsored National Food Security Mission (NSFM) implemented in 2007-2008

Fertilizer statistics

Fertilizer (Lakh tones)	N	P	K	All (NPK)
Production	134.2	43.9	Nil	178.1
Import	50.7	28.9	20.5	100.1
Consumption	173.7	69.8	24.02	267.5

- Per hectare fertilizer consumption (2015-16) is 130.7 kg/ha
- Fertilizer Consumption ratio 6:2. 4:1 But standard ratio 4:2:1
- Consumption of pesticides – 50.4 Thousand tones (2015-16)

Maximum Fertilizer Consumption

Puduchery	-	1086 kg/ha
Punjab	-	194 kg/ha
Andhra Pradesh	-	158 kg/ha
Lowest Nagaland	-	1.46 kg/ha

Different crop production in India during 2015-2016 (Million tonnes ie.MT)

Total food grains (MT)			Rice (MT)			Wheat (MT)		
Uttar Pradesh	-	17.4	West Bengal	-	15.1	Uttar Pradesh	-	28.7
Madhya Pradesh	-	11.9	Uttar Pradesh	-	11.9	Madhya Pradesh	-	18.9
Punjab	-	11.3	Punjab	-	11.3	Punjab	-	17.2

Maize (MT)			Total Pulses (MT)			Total Oilseeds (MT)		
Karnataka	-	14.9	Madhya Pradesh	-	31.1	Madhya Pradesh	-	24.7
Madhya Pradesh	-	11.8	Rajasthan	-	11.9	Rajasthan	-	22.6
Bihar	-	10.9	Maharashtra	-	8.6	Gujarat	-	16.2

Groundnut (MT)			Sugarcane (MT)			Cotton (MT)		
Gujarat	-	38.4	Uttar Pradesh	-	41.3	Gujarat	-	32.2
Rajasthan	-	15.6	Maharashtra	-	20.5	Maharashtra	-	21.6
Tamil Nadu	-	13.0	Karanataka	-	10.9	Telangana	-	12.8

Percent share of different crop production in India during 2015-2016

Total food grains			Rice			Wheat		
Uttar Pradesh	-	17.5	West Bengal	-	15.1	Uttar Pradesh	-	28.7
Madhya Pradesh	-	11.9	Uttar Pradesh	-	11.9	Madhya Pradesh	-	18.9
Punjab	-	11.3	Punjab	-	11.3	Punjab	-	17.2

Maize			Total Pulses			Total Oilseeds		
Karnataka	-	14.9	Madhya Pradesh	-	31.1	Madhya Pradesh	-	24.7
Madhya Pradesh	-	11.8	Rajasthan	-	11.9	Rajasthan	-	22.6
Bihar	-	10.9	Maharashtra	-	8.6	Gujarat	-	16.2

Groundnut			Sugarcane			Cotton		
Gujarat	-	34.8	Uttar Pradesh	-	41.3	Gujarat	-	32.2
Rajasthan	-	15.6	Maharashtra	-	20.5	Maharashtra	-	21.6
Tamil Nadu	-	13	Karanataka	-	10.9	Telangana	-	12.8

Three Largest Producing States of Important Crops during 2015-16

Crops	Leading states
Rice	West Bengal >Uttar Pradesh >Punjab
Wheat	Uttar Pradesh> Madhya Pradesh >Punjab
Maize	Karnataka > Madhya Pradesh > Bihar
Total coarse cereals	Rajasthan >Karnataka > Madhya Pradesh
Total Pulses	Madhya Pradesh> Rajasthan> Maharashtra
Total food grains	Uttar Pradesh>Madhya Pradesh>Punjab
Groundnut	Gujarat > Rajasthan> Tamil Nadu
Rapeseed & Mustard	Rajasthan> Haryana >Madhya Pradesh
Soybean	Madhya Pradesh > Maharashtra >Rajasthan
Sunflower	Karnataka >Haryana >Andhra Pradesh
Total oilseed	Madhya Pradesh > Rajasthan >Gujarat
Sugarcane	Uttar Pradesh >Maharashtra >Karnataka
Cotton	Gujarat > Maharashtra >Telangana
Jute and Mesta	West Bengal >Bihar >Assam

Horticulture

- Total horticulture production - 286.1 MT with an area of 24.4 m ha (2015-16)
- Percentage share of horticultural output to total agriculture output- 30.9%
- Fruits (IInd rank) - 90.1 MT - 6.30 M.ha (Area)
- Vegetables (IInd rank) - 169.0 MT - 10.1 M. ha
- Fruits share to world -12.8%
- Vegetable share to world - 10.8%
- India is the largest producer of mango, banana, papaya, cashewnut, areacanut, potato and okra.
- India contributes 50% of mango and 11% of banana production in the world
- Maximum area in fruit production is in Maharashtra

Area, production and productivity of Horticultural crops

Crop category	Area (Million ha)	Production (million tonnes)	Productivity (metric tonnes ha-1)
Fruits	6.3	90.2	14.3
Vegetables	10.1	169.1	17.4
Plantation crops	3.7	16.7	4.2
Spices	3.5	6.9	1.95
flowers loose +cut flowers	0.28	2.2	3.94

Crop category	Area (Million ha)	Production (million tonnes)	Productivity (metric tonnes ha-1)
Aromatic	0.63	1.02	
Honey	-	0.089	-
Total Horticulture	23.8	283.4	11.94

Leading Horticultural crops in India

Crop category	Area	Production
Fruits	Mango>Citrus>Banana	Banana (31.8%)>Mango (20.6%)>Citrus (12.6%)
Vegetables	Potato>Onion>Tomato	Potato (26.3%)>Onion(12.5%)>Tomato (11%)
Plantations	Coconut>Cashewnut>Arecanut	Coconut(91.5%)>Arecanut(4.5)> Cashewnut (4.4%)
Spices	Dried chillies>Cumin>Coriander	Garlic (22.2%)>dried chillies (22.1%)> Ginger (17.5%)

Percentage share of production of various horticultural crops (2015-16) in India

Crops	Percentage share of Horticultural crops
Fruits	32.3
Vegetables	58.8
Flowers & Aromatics	1.2
Plantations	5.5
Spices	2.2

Animal Husbandry

- It contribute 29% of agricultural output
- Livestock sector is, contributing 25.6% to national GDP
- First in the total buffalo population in the world
- India is the largest livestock production in the world
- Second in the population of cattle and goats
- 16% cattle population to world population
- 57% Buffalo population of world population
- Second largest poultry market in the world - production of 82.9 billion eggs and 3.26 million tonnes poultry meat
- Cattle -190.9 million numbers (India stand first in the cattle population in the world)
- Buffaloes – 108.7 million numbers
- Sheep - 65.1 million numbers (Third in the population of sheep)

- ❖ Goats -135.2 million numbers
- ❖ Pigs – 10.3 million numbers
- ❖ Poultry - 729.2 million numbers
- ❖ Total livestocks population- 512.1 million numbers
- ❖ India is the largest producer of milk in the world
- ❖ Crossbred cows - average daily milk yield of 6.4 kg/animal
- ❖ Non-descript cows - average daily milk yield of 1.97 kg/animal
- ❖ Buffaloes - average daily milk yield of 4.3 kg/animal
- ❖ Per capita availability of milk is 337 g per day
- ❖ Per capita availability of meat about 5 kg per annum
- ❖ Per capita availability of eggs about 66 eggs per annum

Production status in 2015-16

- ❖ Milk - 155.5 million tonnes
- ❖ Egg - 82.9 billion numbers
- ❖ Wool - 43.6 million kgs
- ❖ Meat - 7.0 million tonnes
- ❖ Fish - 10.8 million tonnes (Inland -7.2 MT and Marine – 3.6 MT)

12th Plan Target		
Agriculture	-	4%
Livestock	-	6%
Fisheries	-	6%
Horticulture	-	5%
Overall growth	-	9%

Rank Forest area		
1st	-	Madhya Pradesh
2nd	-	Andhra Pradesh
3rd	-	Chattisgarh

Per-Capita availability		
Cereals	-	153.8 kgs / year
Pulses	-	16 kgs / year
Milk	-	337 gm / day
Food grains	-	169.8 kg/year
Rice	-	67.9 kg /year

Per-Capita availability		
Wheat	-	61.3 kg /year
Bengal gram	-	4.5 kg /year
Tea	-	752 gram
Coffee	-	100 gram
Edible oils	-	18.3 kgs
Sugar	-	20.3 kgs

Buffer stock position of food grains in India

Buffer stock	Million tonnes						
	Minimum norms				Actual stock in 2017		
	1st April	1st July	1st October	1st January	1st April	1st July	1st October
Food grain stock	21.04	41.12	30.77	21.41	31.14	53.31	42.17
Wheat	7.46	27.58	20.52	13.80	8.05	32.27	25.86
Rice	13.58	13.54	10.25	7.61	23.08	21.04	16.30

Agriculture credit flow 2015		
Co-operative	-	21%
RRB's	-	16%
Commercial Bank	-	63%

Irrigation Scheme (CCA)		
Major	-	>10,000 ha
Medium	-	2000-10,000 ha
Minor	-	Upto 2000 ha

Irrigation Scheme (CCA)		
Marginal farmers	-	<1 ha
Small farmers	-	1-2 ha
Medium farmers	-	2-5 ha
Large farmers	-	> 5 ha

Emission of green house gases

- World agriculture contribute 4% of total CO_2 emission
- Soil contribute 65% total N_2O emissions
- Global Warming Potential Index (GWPI) relative to CO_2, methane (CH_4) is 21 and nitrous oxide (N_2O) is 310

- India emit 585 Tera grams (Tg) of CO_2 (2.1%)
- CH_4 from Indian Agriculture is 17.8 Tg (3.2%)
- N_2O from India Agriculture is 2.7 Tg (0.24%)

New schemes in operation

- National Food Security Mission (NFSM) launched in 2007 with the aims of additional production of rice, wheat and pulses respectively in India
- Rashtriya Krishi Vikas Yojana (RKVY) was launched in 2007-08 with the objective of enhance public investment to achieve 4 per cent growth rate in agriculture and allied sectors
- Integrated Scheme of Oilseeds, Pulses, Oil Palm and Maize (ISOPOM) was started in 2004
- Kisan Call Centre (KCC) was started in 2004 and toll free number is 1551
- National Mission for Sustainable Agriculture (NMSA) was launched 2010 aimed to enhance food security and protection of resources such as land, water, biodiversity and genetic resources by developing strategies to make Indian agriculture more resilient to climate change.
- National Horticulture Mission (NHM)
- National Bamboo Mission (NBM) was launched in 2007
- Rainfed Area Development programme (RADP) was implemented in 2008

India's Position in World Agriculture & Horticulture in 2014

Items	India	% share area in world	India 's Rank	World first
Total area (M ha)	328.7	2.4	7	Russian federation, Canada
Land area (M ha)	297	2.3	7	Russian federation, Canada
Arable Land (M ha)	156	11	2	USA
Population (Total) Million	129.5	17.8	2	China
Population (rural) Million	87.5	25.5	1	India
Total Cereals	295	10.5	3	China
Rice	157	21.2	2	China
Wheat	96	13.1	2	China
Total pulses	20	25.8	1	India
Oilseeds				
Groundnut (Shell)	7	14.9	2	China
Rapeseed	8	10.7	3	Canada
Sugarcane	352	18.7	2	Brazil
Tea	1.21	21.7	2	China

Items	India	% share area in world	India 's Rank	World first
Coffee (green)	0.3	3.5	6	Brazil
Jute & Jute like				
Fibres	2.07	56.8	1	India
Cotton (Lint)	6.2	23.7	2	China
Tobacco				
Unmanufactured	0.72	10	3	Brazil
Fruits & Vegetables Production (Million Tonnes)				
Vegetables &				
Melons	127	10.8	2	China
Okra	6.4	73.6	1	India
Tomato	18	11.1	2	China
Cabbage & other Brassicas	8.5	11.9	2	China
Cauliflower & Broccoli	8	36.4	2	China
Brinjal	13	27.1	2	China
Banana	28	26.2	1	India
Mango, Mangosteen, Guava	18	41.9	1	India
Lemon &Lime	2.5	16.4	1	India
Papaya	5.5	44.4	1	India
Fruits excluding				
Melons	88	12.8	2	China
Potatoes	46	12.1	2	China
Onion (Dry)	19	21.9	2	China
Livestock (Million Heads)				
Cattle	187	12.7	2	Brazil
Buffaloes	110	56.6	1	India
Sheep	63	5.3	3	China
Goats	133	13.2	2	China
Chickens	725	3.4	7	China
Camels	0.38	1.4	12	Somalia
Animal Products (Million Tonnes)				
Milk total	155.5	18.3	1	India
Eggs	3.96	5.3	3	China
Meat total	7.0	2.1	6	China

12.4. TAMIL NADU STATISTICS AT A GLANCE

Geographical Position

North Latitude Between 8° 5' and 13° 35'
East Longitude Between 76° 15' and 80° 20'

Climatic data of Tamil Nadu

Temperature(Degree Celsius)	
Plains	
Maximum	33.18
Minimum	23.59
Hill Stations	
Maximum	20.73
Minimum	12.15
Rainfall (mm)	
Normal	911.6
Actual	859.7
Season	**Seasonal Contribution (%)**
South-West Monsoon	23.7
North-East Monsoon	63.5
Winter	1.2

- Current farm power availability hovers around 1.7 kw/ha which is much lower than that of Korea (7+ kw/ha), Japan (14+kw/ha) and the USA (6+kw/ha).
- Tamil Nadu has cold storage capacity of 316583 MT (2016)
- India has cold storage capacity of 34050359 MT (2016)

AGRICULTURE

Total Cultivated Area (Ha) Net Area Sown (Ha) Area Sown more than once (Ha)		6032718 5243839 788879
Area and Production of Principal Crops	**Area (Ha)(in '000)**	**Production ('000 Tonnes)**
Paddy (Rice)	2050	5209
Millets and Other Cereals	741	729
Pulses	525	177
Sugarcane (Cane)	335	35113
Groundnut	619	1097
Gingelly	65	31Cotton (bales of 170
Kg. lint)	109	168

Agricultural Land Holdings (as per 2000-01 Agricultural Census)	
Holdings (in Nos)	7858887
Area (Ha)	6971516
Average size of Holdings (Ha)	0.89

IRRIGATION

Net Area Irrigated (in Ha.)		2919545
Government Canals	799601	
Private Canals	560	
Tanks	575352	
Tube Wells	381441	
Open Wells	1155367	
Other Sources	7224	
Gross Area Irrigated		3396700

In Tamil Nadu Area, production and productivity of Horticultural crops (2015-16)

Crop category	Area (Million ha)	Production (million tonnes)
Fruits	0.31	6.1
Vegetables	0.26	6.3
Plantation crops	0.62	4.2
Spices	0.11	0.19
flowers loose	0.055	0.34
Aromatic & medicinal	0.014	0.19
Honey		1.52
Total Horticulture	1.34	17.4

ANIMAL HUSBANDRY

Livestock and Poultry Population (as per Livestock Census - 2012)	
Cow	8814042
Buffaloes	780431
Sheep	4786680
Goat	8143341
Poultry	117348894

FORESTS

Forest Area (Sq. Km.)	22877.00
Reserved Forests	19388.00
Reserved Lands	2182.94
Unclassed Forests	1306.06

LAND USE PATTERN OF TAMIL NADU

Category	Area 000' ha)	%
Forest	2125	16.3
Barren and unculturable land	488	3.7
Land put to non-agricultural uses	2190	16.8
Culturable waste land	328	2.5
Permanent pastures and other grazing lands	109	0.85
Misc. tree crops and groves not included in the net area sown	245	1.9
Current fallow	1115	8.6
Other fallow lands	1718	13.2
Net area sown	4714	36.2
Geographical area	13032	100
Area sown more than once	1184	9.1
Gross area sown	5897	45.3
Cropping intensity	113.1	0.87

Agricultural Colleges and Research Institute in Tamil Nadu

1. Agricultural Colleges and Research Institute - Coimbatore
2. Agricultural Colleges and Research Institute - Madurai
3. Agricultural Colleges and Research Institute - Killikulam
4. Anbil Dharmalingam Agricultural Colleges and Research Institute - Tiruchirappalli
5. Agricultural College & Research Institute, Eachangkottai -Thanjavur
6. Agricultural College & Research Institute, Kudumiyanmalai- Pudukkottai
7. Agricultural College & Research Institute, Vazhavachanur- Thiruvannamalai
8. Horticultural Colleges and Research Institute - Coimbatore
9. Horticultural Colleges and Research Institute - Periyakulam

10. Horticultural Colleges and Research Institute for women - Tiruchirappalli
11. Agricultural Engineering Colleges and Research Institute - Coimbatore
12. Agricultural Engineering Colleges and Research Institute -Kumulur
13. Forest Colleges and Research Institute- Mettupalayam
14. Home science Colleges and Research Institute - Madurai

Agricultural Research Station in Tamil Nadu

1. Agricultural Research Station- Aliyarnagar
2. Agricultural Research Station- Kovilpatti
3. Agricultural Research Station- Bhavanisagar
4. Agricultural Research Station- Paramakudi
5. Agricultural Research Station- Vaigaidam
6. Agricultural Research Station- Virinjipuram
7. Agricultural Research Station-Pattukkottai
8. Agricultural Research Station-Tirupathisaram
9. Coastal salinity Research Centre-Ramanathapuram
10. Cotton Research Station - Srivilliputhur
11. Cotton Research Station -Veppanthattai
12. Dryland Agricultural Research Station - Chettinad
13. Hybrid Rice Evaluation Center- Gudalore
14. Maize Research Station - Vagarai
15. National Pulse Research Station - Vamban
16. Oilseed Research Station- Tindivanam
17. Regional Research Station - Aruppukkottai
18. Regional Research Station - Paiyur
19. Regional Research Station- Vridhachalam
20. Rice Research Station - Ambasamudram
21. Rice Research Station - Tirur
22. Soil and Water Management Research Station- Thanjavur
23. Sugarcane Research Station- Cuddalore
24. Sugarcane Research Station- Melalathur
25. Sugarcane Research Station- Sirugamani
26. Tamil Nadu Rice Research Institute- Aduthurai
27. Tapioca and Castor Research Station- Yethapur
28. Centre of Excellence for Millets- Athiyandal

Horticultural Research Stations in Tamil Nadu

1. Floricultural Research Station -Thovalai
2. Horticultural Research Station - Kodaikanal
3. Horticultural Research Station - Ooty
4. Horticultural Research Station - Pechiparai
5. Horticultural Research Station - Yercaud
6. Vegetable Research Station - Palur
7. Horticultural Research Station - Thadiyankudisai
8. Coconut Research Station - Veppankulam
9. Coconut Research Station - Aliyarnagar
10. Urban Horticultural Development Centre - Chennai
11. Grapes Research Station -Theni
12. Citrus Research Station - Sankarankovil
13. Oilpalm Research Station- Pattukottai

STATE SEED FARMS IN TAMIL NADU

District	Seed farm location	District	Seed farm location
Kanchipuram	Musaravakkam(OS)	Thiruvannamalai	Athiyanthal, Vazhavachanur
Tiruvallur	Kolandalur	Thanjavur	Sakkottai
Cuddalore	Vandurayanpattu, Miralur	Nagai	Thirukadayur, Nagamangalam
Villupuram	Kakuppam, Iruvelpattu, Vanur, vadakananthal	Thiruvarur	Nedumbalam, Moongilkudi, Keeraanthi, Kanchikudikudu, Deivambalpattinam
Vellore	Navlock (SOSF)	Pudukkottai	Annapannai, vellalaviduthi (OS),Vamban (Pulses)
Karur	Inungur	Madurai	Vinayagapuram
Dharmapuri	Papparapatti	Theni	Keelangudalur
Salem	Mettur Dam, Danishpet	Virudhunagar	Devadanam
Coimbatore	Pappankulam	Tirunelveli	Karaiyiruppu, Melagaram
Tiruppur	Pongalur, Madukkulam	Kanyakumari	Tirupathisaram
Trichy	Neikuppaipudur, pudupalayam	Erode	Bhavani, Sathyamangalam

STATE HORTICULTURAL FARMS IN TAMIL NADU

District	Horticultural farm location	District	Horticultural farm location
Kanchipuram	Attur, Vichanthangal, Melkadirpur, Melottivakkam	Thanjavur	Aduthurai, Marunkulam
Tiruvallur	Madhavaram, Pichivakkam, Putlur (Mother Palm Garden)	Pudukkottai	Kudumianmalai, Nattumangalam, vallathirakottai (GO)
Cuddalore	Neyveli, Virudhachalam, Neyveli (Mother Palm Garden)	Dindigul	Kodaikanal, Bryant Park (Kodaikanal), Thandikudi, Rediyarchatram, sirumalai
Vellore	Thagarakuppam, Kudapattu, Navlock	Coimbatore	Anaikatti, Kallar, Kannampalayam
Krishnagiri	Thimmapuram, Jeenur	Sivagangai	Devakottai, Nemam
Salem	Yercaud, Karumanthurai (GO), Maniyakundram, Karumanthurai, Mulluvadi, Sirumalai (Arunuthumalai)	Nilgiris	Burliar, Katteri, Coonoor, Pomological station, Coonoor, Sims Park (Coonoor), Doddabetta, Thummanatty, Vijayanagaram Govt. Bot. Garden, Nanjanad, Devala
Namakkal	Semmedu, Padasolai	Tirunelveli	Courtallam
Theni	Periakulam	Kanyakumari	Kanyakumarai, Pechiparai
Karur	Mudalaipatti	Virudhunagar	Poovani, srivilliputhur (GO)